Energetic Materials · 2

Technology of the Inorganic Azides

ENERGETIC MATERIALS

Volume 1: Physics and Chemistry of the Inorganic Azides

Volume 2: Technology of the Inorganic Azides

Energetic Materials · 2

Technology of the Inorganic Azides

Edited by

H. D. Fair
and R. F. Walker
Energetic Materials Division
Armament Research and Development Command
Dover, New Jersey

PLENUM PRESS · NEW YORK AND LONDON

Library of Congress Cataloging in Publication Data

Main entry under title:

Technology of the inorganic azides.

(Energetic materials; v. 2)
Includes bibliographical references.
1. Azides. I. Fair, Harry David, 1936- II. Walker, Raymond F.
III. Series.
TP270.E53 vol. 2 [TP245.N8] 662'.2'08s [662'.2] 76-30808
ISBN 0-306-37077-8

A Division of Plenum Publishing Corporation
227 West 17th Street, New York, N.Y. 10011

Printed in the United States of America

Preface

This volume is about azides that detonate. As such they are unique and important substances to both civil and defense industry, but are also very real threats to life and property if handled ignorantly or irresponsibly. The authors feel strongly enough about the opportunities and the dangers that they have taken the trouble to write about the problems they encountered when the issues were of daily concern, and while the lessons learned are still fresh in mind.

During any resurgence of interest in a technological area, the mythology of the past is likely to condition the attitudes adopted in the present. This is particularly true of older, empirical technologies, and scientists and engineers approaching the subjects for the first time commonly find themselves immersed in and bewildered by a morass of conflicting opinions, scepticism, recipes, and folklore. Their caution leads them to reject nothing and to question everything. Hopefully, truths will be distilled and incremental advances in the technology will emerge that will reduce the problems for succeeding generations. It is in this spirit that the authors have made their contributions.

This volume constitutes the second of two which document research and development on the azides supported by the United States Department of the Army during the years 1966–1974 and administered through the Office of the Chief of Research and Development and the Army Material Command. In summarizing the work an attempt has been made to place the activities in perspective with those of colleagues in other countries since World War II, in commerce and other defense installations. The contents are by no means a complete description but are, nevertheless, illustrative of advances in the technology over the last 10–20 years.

A special acknowledgement must be made to Mr. G. W. C. Taylor of the Explosives Research and Development Establishment, Waltham Abbey, England, not only for his constructive comments on parts of this volume, but more particularly for the numerous insights and guidance that he provided over a period of years, and that were born of a lifetime of practical contributions to explosives research and development in the United Kingdom and the U.S.A. His professional skill and good-humored advice contributed much of incisive technical import to the U.S. Department of the Army, and made a proud contribution to a happy international relationship.

Thanks are due also to Dr. J. M. Jenkins of the same establishment and to Mr. L. E. Medlock of the Nobel's Explosive Company, Stevenston, Scotland, for the provision of photographs included in Chapter I. Among others in various countries who have provided information which has been eagerly incorporated are Drs. G. Blomqvist and S. Lamnevik, Research Institute for National Defense, Stockholm, Sweden; Messrs. P. Collvin and T. Hansson, Department of Industry, Eskilstuna, Sweden; Dr. R. Siepmann, Ministry of Defense, Bonn, West Germany; and Messrs. D'Ast, Morisson, and Cazaux, Atelier de Construction, Tarbes, France.

The editors must again give particular thanks to Dr. T. Richter for his careful reading of the manuscripts and his scrupulous attention to the reference materials, terminology, etc.; and to Dr. D. Wiegand for assuming the burdensome task of coordinating the timely preparation of manuscripts, drawings and photographs.

On behalf of all the authors, sincere appreciation is again expressed to Mrs. Beverly Malson and Mrs. Marie Smith for their patient final typing of the manuscripts, which ultimately made the volume a reality.

This volume was made possible through the grant of a Research and Study Fellowship by the Secretary of the Army to one of us (RFW). It is a pleasure to acknowledge the debt created by that opportunity.

The contributing authors are all members of the Energetic Materials Division, Armament Research and Development Command (formerly Picatinny Arsenal), Dover, New Jersey.

H. D. Fair, Jr.
Raymond F. Walker

Dover, New Jersey

Contents

Contents of Volume 1 xiii

Introduction 1
Raymond F. Walker

Short History of Azide Technology 3
Manufacture of Azides 5
Characterization 6
Handling and Storability 7
Sensitivity of the Azides 8
Explosive-Train Technology 9
References 10

Chapter 1
Processes for the Manufacture of Lead and Silver Azide 11
T. Costain and F. B. Wells

A. Introduction 11
B. Facilities and Equipment 14
1. ERDE Equipment 16
2. Typical United States Facilities 19
3. Some Tests and Equipment for Product Assessment 23
C. Processes for the Manufacture of Lead Azide 25
1. Colloidal Lead Azide 27
2. Service Lead Azide 27
3. Polyvinyl Alcohol Lead Azide 28

4. Dextrinated Lead Azide 29
5. Early CMC Lead Azide 38
6. RD1333 Lead Azide 39
7. Other CMC Modified Lead Azides 44
8. A Continuous Lead Azide Process 45
9. Gelatin Lead Azide 46
D. Silver Azide Manufacture 47
1. Early Preparations 47
2. A New Process 49
3. Semi-Pilot-Plant Production 49
4. Discussion of Process Parameters 51
References 53

Chapter 2
Analysis of Azides for Assay, and in Complex Media 55
H. Kramer

A. Introduction 55
B. Quantitative Determination of the Azide Ion 58
1. Nitrogen Evolution Method 58
2. Ceric Ammonium Nitrate Titration 60
3. Distillation Method for Azide Determination 61
4. Argentometric Determination of Azide Ion 63
5. Lead Determination 65
6. Comparison of Analyses Given by Different Methods 65
C. Determination of Microquantities of Azide Ion 66
1. Methods of Separation for the Azide Ion 66
2. Colorimetric Methods for Determination of Azide Ion 67
3. Determination of Hydrazoic Acid by Gas Chromatography and Other Methods 69
D. Trace Elements in Explosive Azides 70
E. Future Research 71
References 71

Chapter 3
Handling, Storability, and Destruction of Azides 73
B. D. Pollock, W. J. Fisco, H. Kramer, and A. C. Forsyth

A. Introduction 73
B. Safety in Handling of Azides 74
1. Philosophy and the General Problem 74
2. Explosive Azides 75

3. Toxicity ... 76
4. Reactivity ... 78
5. Indirect Hazards ... 79
6. Reference Texts ... 80
C. Destruction of Excess Azides ... 80
1. Laboratory Methods of Destruction ... 80
2. Bulk Destruction of Azides ... 82
D. Long-Term Storage of Lead Azide ... 85
1. Storage Studies ... 88
2. Current Packaging Practice in the U.S.A. ... 89
3. Assessment of Current Practice ... 90
4. Chemistry of Long-Term Storage ... 91
E. Electrostatic Charge Generation During Handling ... 96
1. General Principles ... 97
2. Charging Studies ... 99
3. Electrostatic Hazards Estimation ... 100
4. Charge Relaxation on Explosives ... 104
5. Discussion ... 106
References ... 107

Chapter 4

The Sensitivity to Impact and Friction ... 111

L. Avrami and R. Hutchinson

A. Introduction ... 111
1. Development of Sensitivity Tests ... 112
2. Sensitivity Relative to Other Explosives ... 112
B. Impact Sensitivity ... 113
1. Character and Variety of Tests ... 113
2. Design and Analysis of Impact Sensitivity Experiments ... 121
3. Impact Sensitivity of Azides ... 124
4. Effect of Impurities on the Impact Sensitivity of Azides ... 134
C. Friction Sensitivity ... 144
1. Friction Sensitivity Apparatus ... 145
2. Friction Sensitivity of Azides ... 151
D. Summary and Conclusions ... 157
References ... 159

Chapter 5

Electrostatic Sensitivity ... 163

M. S. Kirshenbaum

A. Introduction 163
B. Electrostatic Sensitivity Apparatus 166
1. General 166
2. Fixed-Gap Apparatus 166
3. Approaching-Electrode Apparatus 166
C. Current–Voltage Characteristics of Gaseous Discharge 167
D. Efficiency of Energy Delivered to Spark Gap 172
E. Contact vs. Gaseous Discharge 173
F. Optimum Gap 175
G. Effects of Cathode Surfaces 175
H. Energy Response Curves 176
I. Needle-Plane vs. Parallel-Plane Electrodes 176
J. Effect of Energy Delivery Rate on Initiation 177
1. Storage Capacitor Constant 177
2. Series Resistance Constant 178
K. Initiation Energies for Various Azides 178
1. Service Lead Azide 178
2. Polyvinyl Alcohol Lead Azide (PVA) 179
3. Dextrinated Lead Azide 179
4. RD1333 Lead Azide 181
5. Lead Azide–Aluminum–RDX Mixtures 182
L. Comparison of Electrostatic Sensitivity of Primary Explosives 183
M. Dielectric Strength of Lead Azide 184
1. Single Crystal Measurements 184
2. Pressed Pellet Measurements 185
N. Silver Azide 194
References 196

Chapter 6
Sensitivity to Heat and Nuclear Radiation 199
L. Avrami and J. Haberman

A. Introduction 199
B. Sensitivity to Heat 200
1. Explosion Temperature Tests 200
2. Thermal Sensitivities in Air 201
3. Effect of Impurities 203
4. Effect of Sample Mass 206
5. Vacuum Stability Tests 207
6. Differential Thermal Analysis (DTA) 208
7. Thermogravimetric Analysis (TGA) 210
C. Sensitivity to Nuclear Radiation 211

1. α-Particle Irradiation 211
2. Neutron Irradiation 213
3. Gamma Irradiation 223
4. X-Irradiation 226
5. Electron Irradiation 229
6. Other Types of Nuclear Irradiation 233
D. Summary and Conclusions 241
References 243

Chapter 7

The Role of Azides in Explosive Trains 249

W. Voreck, N. Slagg, and L. Avrami

A. Introduction 249
B. Classes of Igniters and Initiation Elements 251
1. Stab- or Percussion-Sensitive Elements 252
2. Heat- or Flash-Sensitive Elements 253
3. Electrically Ignited Elements 253
C. Initiation of Azides in Explosive Trains 256
1. Functioning of Primers 256
2. Growth of Reaction in the Azides 259
D. Detonation Properties of the Azides 260
1. Measurement Techniques 261
2. Dependence of Detonation Velocity on Material Properties and Confinement 264
3. Detonation Pressures of Lead Azide 268
E. Detonation Transfer from the Azides 269
1. Initiation of Secondary Explosives 269
2. Theory of Shock Transfer 272
F. Effects of Strong Shocks on Lead Azide 275
1. Equation of State of Unreacted Lead Azide via Gas-Gun Measurements 276
2. Sensitivity of Lead Azide to Short Pulses via Flyer-Plate Technique 280
3. Shock Initiation of Lead Azide with an Electron Beam 283
4. Energy Input with No Compression 285
G. Summary and Conclusions 287
1. Primer–Lead Azide Interface Mixtures 288
2. Lead Azide–Secondary Explosive Interface 288
3. Silver Azide 289
References 289

Index 293

Contents of Volume 1

Introduction
Raymond F. Walker

Chapter 1
Synthesis and the Chemical Properties
Tillman A. Richter

Chapter 2
The Growth of Crystals
Wayne L. Garrett

Chapter 3
The Crystal Structures
C. S. Choi

Chapter 4
Molecular Vibrations and Lattice Dynamics
Z. Iqbal, H. J. Prask, and S. F. Trevino

Chapter 5
Electronic Structure of the Azide Ion and Metal Azides
T. Gora, D. S. Downs, P. J. Kemmey, and J. Sharma

Chapter 6
Slow Thermal Decomposition
P. G. Fox and R. W. Hutchinson

Chapter 7
Imperfections and Radiation-Induced Decomposition
W. L. Garrett, P. L. Marinkas, F. J. Owens, and D. A. Wiegand

Chapter 8
Fast Decomposition in the Inorganic Azides
M. M. Chaudhri and J. E. Field

Chapter 9
Stability and the Initiation and Propagation of Reaction in the Azides
J. Alster, D. S. Downs, T. Gora, Z. Iqbal, P. G. Fox, and P. Mark

Introduction

Raymond F. Walker

Detonation as a phenomenon of nature is a comparatively recent scientific discovery. First observed in mercury fulminate in the late eighteenth century, it was only described in detail and defined during the late nineteenth and early twentieth centuries. It was also during the second half of the nineteenth century that most solid, high explosives were discovered; as the most powerful known sources of chemical energy, they rapidly found their place among substances of technological interest.

The rate at which the energy of high explosives can be delivered where required (generally in less than a microsecond) and the brisance or shattering power of the pressure discontinuity associated with detonation gave the materials an immediate industrial and military significance of major proportions. However, as Nobel showed in the 1860s, to a large degree their usefulness hinged on the discovery that they exhibit a spectrum of behavior, so that on the one hand they may be very powerful and energetic, but difficult to detonate, and on the other hand less powerful, but sensitive or easy to detonate.

At the former extreme are the so-called secondary explosives, such as trinitrotoluene (TNT), cyclotrimethylenetrinitramine (RDX) and its eight-membered analog (HMX), ammonium nitrate (NH_4NO_3),* and mixtures of these with each other or with other fuels, solids, or liquids. At the other extreme are the primary explosives, which include a range or organic and inorganic solids, such as fulminates, styphnates, and azides, and which are able to convert and

*Recently [1] ammonium nitrate has been cited as an example of materials that might be described as tertiary explosives, for although very energetic, it is most difficult to detonate and relatively massive quantities are required for the propagation of detonation.

magnify the energy from a gentle stimulus into an output powerful enough to detonate secondary explosives.

Between the extremes lie the more sensitive secondary explosives which are often used in an intermediate role to initiate detonation in less-sensitive substances and are then known as booster explosives. Trinitrophenylmethylnitramine (tetryl) is such a substance, but among all classes of explosives a good deal of "pharmacy" is involved in the formulation of compositions with the desired balance of properties.

The sequence of functions incorporated in such a series, as: primary explosive, booster, secondary (or "main charge") explosive, is called an explosive train. The elements or components of such a train may themselves also consist internally of a sequence of explosives, the selection of a component depending on the nature of the stimulus applied, whether from a preceding element or from a source external to the train.

Ever since their discovery in the 1890s, the azides have been prominent as primary explosives that are chemically simple and easily detonated, yet powerful enough to detonate most secondary explosives. In this capacity the azides, along with other primary explosives, are used in a relatively pure form and are the important ingredient in detonators or initiators. However, they may also be formulated with other reactive substances, principally to produce heat rapidly in the form of flame and hot particles; as priming or igniter materials, they are used to induce fast reactions in primary explosives, pyrotechnics, and propellants. In both capacities azides are critical for the successful application of explosives in mining and civil engineering enterprises, in military weapons, and in aerospace systems.

While numerous azides can be made to detonate, and others find application as gas generators, only a few are of major technological importance, and among primary explosives only α-lead and silver azides have achieved more than passing industrial significance. Owing to its unique properties, lead azide has tended increasingly to supplant its competitors, such as mercury fulminate and lead styphnate, among the primary explosives, and as will be more fully expounded in Chapters 1 and 7, it is only recently that silver azide has emerged as a material with independent virtues.

Other metal azides have industrial significance as gas generators (NaN_3), reagents (NaN_3), and potential hazards when using lead and silver azides in conjunction with metal components [e.g., $Cu(N_3)_2$, $Cd(N_3)_2$, $Hg(N_3)_2$]. However, as presented at length in Volume 1, the science of the azides has not yet progressed to where metal azides in general have found a useful industrial role.

Thus this volume is devoted principally to the technology of just two explosives, α-lead azide and silver azide. In considering the technology presented it should, however, be borne in mind that in matters of principle many of the problems and issues discussed are not just those of azides, but of all explosives and of energetic, hazardous substances in general.

SHORT HISTORY OF AZIDE TECHNOLOGY

In spite of its current unique, irreplaceable position in explosives technology, lead azide is not without its problems; in fact, it is without question the most troublesome military explosive in common use today, and were it not for the lack of a suitable alternative, it would be willingly replaced.

The problem with lead azide, as presently developed, is the unpredictable behavior manifest in its sensitivity to external stimulation and in its functional characteristics. Although not exceptional in any of these respects in comparison with mercury fulminate, the original primary explosive used by Nobel in 1867, the long-term thermal and chemical stability, and the comparative cheapness and availability of lead azide have virtually eliminated the fulminate from civilian and military use. However, the sensitivity of lead azide, which makes it so hazardous to handle, is in numerous applications also sufficient of an asset that a safety philosophy and a technology have been developed over the years which permit it to remain in use today. Even now it is beginning to have functional limitations, which account for the increasing interest in the more powerful silver azide. Although the silver compound is reported to be less capricious than lead azide, this has yet to be demonstrated during periods of high consumption.

Lead, silver, and mercury azides were all discovered in 1890–1891 by Curtius, but it was Hyronimus in 1907 who first obtained a (French) patent for the use of lead azide in the explosives industry. Following World War I interest in lead azide became more general as both the civilian and defense industries of Western Europe and the United States began to explore the use of the material [2]. It was during the 1920s and 1930s that it became more widely appreciated that lead azide is not only hazardous to handle but also hazardous to manufacture, and processes to reduce the difficulties were developed commercially and in ordnance factories to meet different specifications.

This led to the introduction in Britain, for example, of a product known as "Service lead azide," a crystalline precipitate containing some lead carbonate and designed to give the maximum of power because of a high density. In the 1930s private companies, such as the Dupont Company and Imperial Chemical Industries Ltd., introduced lead azides "phlegmatized" with dextrin to give superior manufacturing and handling properties, although with a reduction in pressed density, and these have continued to find extensive civilian and military use. The use of polyvinyl alcohol and hydrophilic colloids were other approaches introduced to improve the safety of commercial products in the U.S.A. and Germany.

Thus by the end of World War II lead azide had had a history of application in civilian and defense industry in both a pure and diluted form. Defense applications, being more demanding, tended to emphasize performance up to the presumed limits of safety; civilian industry, unable to pay the penalty of failure, made more conservative demands and emphasized phlegmatized products. The situation was not, however, entirely happy, particularly in ordnance factories. For although no explosions were reported to occur during the manufacture of phlegmatized material, Wythes [3] reported that ten explosions of Service lead azide were known to have occurred "since 1941" in process kettles in ordnance factories, and five explosions occurred during subsequent sieving of the precipitate. No clear reason for the explosions was elicited, but unstable conditions during mixing and a "spontaneous" explosion phenomenon featured prominently in attempted explanations. (See the discussion on spontaneous explosions in azide solutions in the articles by Chaudhri and Field, Volume 1, Chapter 8; by Fox in the same volume, Chapter 9; by Taylor and Thomas [3]; and by Bowden and Yoffe [4]).

Explosions also occur frequently during the compaction of lead azide in detonator cups, and Wythes [3] reported 140 such incidents in ordnance factories in Britain during the aforementioned period. A typical figure for high-volume production in an ammunition plant in the U.S.A. is 232 incidents with 1,000,000 detonators. Mechanical and electrostatic stimuli and human failure were blamed for many of these accidents, but in these cases, as with those encountered during manufacture, no loss of life or injuries resulted because of protective structures.

It was undoubtedly these experiences during and immediately after World War II that led British scientists at Woolwich Arsenal and Explosives Research and Development Establishment, Waltham Abbey, to investigate new approaches that would reduce hazards while retaining the superior functional properties of dense lead azide. They studied anew the use of phlegmatizing agents in the preparation of improved products, and in the early 1950s described the results of investigations which became the basis for manufacturing processes adopted for defense industries during the following two decades. The various chapters of this book are about the experiences of that new era and the later consequences. They emphasize investigations undertaken when the U.S.A. was engaged in the conflict in South East Asia and drew heavily on the United Kingdom experience with lead azide processes to pursue further developments in its own laboratories and to lay foundations for the future.

The immediate problems of technology transfer from the laboratory to the processing plant, and from one country to another, were accompanied by the need for a rapid buildup of production and by recurring accidents both in the laboratory and in plants. A substantial loss of life and property damage resulted. The need for explanations of the problems exposed the shallowness of knowledge both of azides and of explosives phenomena in general and led to many of the investigations described in this volume and the companion Volume 1. Numerous shortcomings in the science and technology of primary explosives and explosive trains were highlighted, and these became the basis for several further avenues of research and development.

The experiences threw into relief the need for rigorous test programs to assure both safety and reliable functioning, and at the same time provided the rationale for introducing advances in instrumentation to provide more discriminating tests and explosive train designs. In this respect the advances sought in technology paralleled the advances described in Volume 1 and the endeavor to discern directly and quantitatively the detailed behavior of explosives at all stages of their manufacture, logistics, and application.

The scope of the technology lightly touched upon and the interdependence of its various facets required broad-based technical support for the solution of problems. The competence necessary draws upon a spectrum of technical and engineering disciplines which parallel the more fundamental research described in Volume 1, and the parallelism, except for semantic and terminological peculiarities, is reflected in the plan of both volumes. In this volume, however, much of the rationale is abbreviated so as not to restate matters of principle aired more thoroughly in Volume 1, to which the reader is referred for more detail. However, it should be clear that in transferring knowledge from the laboratory to the plant and drawing board, the topic "preparation" in Volume 1 becomes manufacture and process development in Volume 2; properties, decomposition, stability, etc., become storability, sensitivity, and handling; initiation, growth of reaction, and detonation become explosives-train technology.

MANUFACTURE OF AZIDES

As may be discerned from earlier discussion, in comparison with preparation in the laboratory, the industrial manufacture of azides requires processes which are not only suitable for quantity production, but which confer superior handling characteristics to the bulk and to the milligram quantities into which the bulk is subdivided. The processes must assure a reliability of performance and margins of safety which the laboratory researcher does not require. At the same time the quantity production must be achieved within acceptable bounds of economy (in both time and money), yet with a degree of control normally obtainable only in a laboratory.

The manufacturing process is itself the key to success, but there has remained room for a difference of philosophy on both sides of the Atlantic, as discussed in Chapter 1. The difference lies in the choice between controlling processes vs. controlling performance of the products, and the issues revolve around the extent to which specific performance is uniquely defined by a process. As the scope of the book exemplifies, these are both extreme positions, for satisfaction requires control over both process and product, and over the ambient before and during utilization of azides in detonators and explosive trains (Chapter 3).

As will be seen (Chapter 1), the different philosophies have resulted in the demonstration that, even in a world of demanding specifications, a variety of approaches are possible, and the demands can be met even while seeking improved or sustained performance, with decreasing sample sizes and more restrictive geometries (Chapter 7).

Essentially, the requirements have led to reliance on batch processes, because safety is greater the smaller the quantity in process at one time, specifications can be rigorously checked, and the minute quantities used in detonators imply that batches of just a few pounds are adequate for steady production. Apart from these considerations the detonator manufacturer and the explosive designer look for a product which is dense or readily densifiable, is safe subject to prescribed precautions (Chapter 3), has flow properties suitable for pouring and pressing in detonator cups, is reasonably priced, and is adequately storable (Chapter 3).

CHARACTERIZATION

In designating specifications for products which meet the multifarious demands of the explosives industry, one of the topics that has received much emphasis in recent years has been the physical characterization (phases, particle size, shape, and form of the crystals, etc.) and the chemical species profiles of the products. By extension, and with particular reference to handling and storability, climatic conditions and the compatibility of the azides with packaging and other explosive-train components have also made demands for routine characterization of the azides as produced and utilized.

One of the more important results from early investigations of initiation was that grit and particles in comparatively minor proportions significantly sensitize azides [5,6]. Scrupulous attention to the cleanliness of manufacturing facilities and to the routine chemical analysis of products became mandatory as a consequence, although no known accidents have been unequivocally attributed to such a cause, unless the grit was deliberately added. However, the specter that minor impurities present in water or other solutions might sensitize the azides has been continually present and has resulted in several investigations to determine if the decomposition of the azides is significantly affected by impurities such as iron (see Chapters 4 and 6, Volume 1).

More dramatic in terms of suspected causes of accidents have been chemical species arising from the incompatibilities of lead azide, environmental species, and the metal components in fuze trains. Copper azide is notorious in this respect; however, it is but the most spectacular of potentially hazardous substances which have been sought or detected in the presence of azides in storage.

The foregoing are but examples of technological problems which have required continual attention to the characterization of commercial azides during the last two decades. This is one area where the application of advanced instrumental techniques has very successfully refined knowledge of the differences between samples and enabled the consequences to be investigated in a rational manner. In Volume 1, Chapter 3, a detailed summary is given of the application of X-ray and neutron spectroscopy to the physical characterization of azides in terms of crystal structures and phases. The macroscopic implications of the crystal structures in terms of particle sizes and shapes are also touched upon in Chapter 1 of this volume, where the role of the phlegmatizers and process variables in providing satisfactory shapes is discussed, and the scanning electron microscope has been introduced to acquire greater insight into the microstructure of azide particles. In Chapters 4–7 data are presented on some of the practical consequences of using azides of a different physical character.

Chapter 2 of this volume discusses the chemical characterization, basic techniques found useful, and the problems of analysis for hazardous azides in complex media. The data presented constitute a foundation upon which much of the information presented in the other chapters must stand, and the more refined techniques described are also critical to the success of much of the fundamental research described in Volume 1. The importance can perhaps best be crystallized in the statement that unless the species profiles of samples are known, data on other properties cannot be safely ascribed to any particular substance.

HANDLING AND STORABILITY

The hazards associated with handling even small quantities of azides in laboratories cannot be overemphasized and alone justify the inclusion of a chapter on appropriate procedures for handling the substances. More particularly the hazards associated with quantity production of lead azide would be unacceptable were it not for the philosophy and safety precautions which have been developed in the twentieth century for dealing with hazards in the chemical, mining, and munitions industries. The large-scale production of lead azide introduced in the U.S.A. during the 1960s and 1970s continually reiterated the necessity for safe procedures or emphasized new facets of the subject, and many of the key issues are discussed in Chapter 3.

Normally, the rate of production of lead azide is geared to the rate at which it can be incorporated in explosive trains. In this manner the total quantity on hand is minimized and is broken down into small lots and packaged so that the accidental detonation of one component will be less likely to result in the sympathetic detonation of large stores. However, during the South East Asia conflict a substantial overproduction of lead azide occurred, and bulk quantities were produced far in excess of normal rates. This situation required a special awareness of three problems which, although always of concern, do not normally assume such proportions. These were the problems associated with bulk storage and handling of the finished product; with the accumulation of hazardous waste products in sinks, sumps, ventilation hoods, etc.; and with the destruction of bulk quantities of excess materials. Each of these situations is reflected in the discussions in Chapter 3.

In civilian use the detonators produced by manufacturers are comparatively rapidly distributed to retailers and consumption is such that the azide does not normally remain in storage for periods in excess of a year or two. Military stores must, however, be maintained indefinitely, and storability is required not only for decades but also in a wide range of temperature and humidity conditions. The scope for functional deterioration of the product and for deleterious, hazardous reactions with packaging materials is, therefore, much greater in the defense industries, and the large stocks of material on hand, or called into sudden service, have also provided a basis for the discussions of Chapter 3.

SENSITIVITY OF THE AZIDES

Uncertainty with respect to the response of azides to different physical stimuli is the principal threat to safe handling and reliable performance in explosive trains. The former is of increased concern as large quantities are used, and the latter becomes more critical in trains of sophisticated design where reliable functioning depends on reduced margins of error.

As the reader of Volume 1 will have discerned, substantial efforts have been devoted to the mechanism by which different stimuli lead to detonation in azides and other explosives. The long-range objective of such studies is to learn how to control initiation, both wanted and unwanted, to quantify the energy necessary to achieve initiation by the different stimuli, and to seek new techniques for initiating explosives with safety and reliability. In the meantime, however, the researcher in the laboratory, the quality-control engineer in the plant, and the designer of explosive trains needs data with respect to common stimuli which can serve as a guide in the handling and application of the azides.

Simple tests for this purpose were introduced into explosives technology from its inception, and modifications, seemingly never-ending, were made both to reduce their shortcomings or to provide adequate discrimination between the more or less sensitive materials. The tests in common use [1,6,7] (Chapters 4–6) remain crude, however, and at best only give gross statistical pictures of the relative sensitivities of explosives to given stimuli; at worst they are misleading, for they do not give data on what they affect to measure. In most, if not all, cases the sensitivity recorded is an apparent sensitivity rather than a real sensitivity, and the data obtained apply only to the specific samples tested. There is almost never a direct observation of the material itself, and what is recorded as an event is frequently only the subjective impression of no reaction, a click, a thump, or a bang.

Nevertheless, the tests are of value, and the lack of definition inherent in many of them is consistent with the ill-defined stimuli to which explosives are commonly subjected. Unfortunately, however, they provide no insight into the mechanism of initiation, and substantial work will be necessary to yield more precise and accurate measures based on the findings of fundamental research. In the meantime refinements can be made to the existing techniques by introducing improved instrumentation to define better the mechanical, thermal, or electric stimulus applied; to study the effect of different variables, such as the confinement of the explosive; and to quantify the significance of various material parameters for the probability of initiation.

Chapters 4–6 deal explicitly with the standard "sensitivity" tests and with some of the recent attempts to improve them. Other discussions in the chapters try to understand the function of the tests, interpret their meaning, and detail the material and other factors known to affect the measures of sensitivity.

EXPLOSIVE-TRAIN TECHNOLOGY

As has been indicated earlier the function of the azide in an explosive train is to respond to an external mechanical, thermal, or electrical stimulus; release more energy; and transfer this with sufficient power and intensity to initiate the next element in the train. Each of the functions must be performed within a limited geometry (fractions of an inch), with minimal input energies, and with the maximum of assurance that the sequence will function upon demand but will remain quiescent until then.

In most respects the selection and testing of primary explosives for use in trains is qualitatively similar to sensitivity testing. Generally speaking, the standard tests [1,4,6,8,9] are attempts to simulate a function rather than to discern or quantify the mechanism of the function. Thus they are neither analytical nor useful for the development of improved trains, except on a trial-and error-basis. The data again give a statistical measure of the relative ability of an explosive to perform a limited function and apply only to the specific material under test, in the specific amount and geometry. The functioning of the azide is not directly observed, neither the profile of the stimulus nor the output is quantified, and the data obtained apply only to the explosive as specifically confined during the test.

The tests are again useful and have permitted the development of sophisticated fuzes for advanced military and aerospace systems. For limited-production items success may always be achieved by overdesign; mass production tolerates a certain dud rate, and quality assurance may often be based on no more than the subjective assessment of whether or not a component goes bang at a customary level.

However, as sophistication is increased, miniaturization, reliability, and the precision of functioning become more important. Overdesign is no longer an acceptable avenue of approach, and duds in high technology may no longer be passed off as inconsequential. Fortunately, the last decades have brought advances in mechanical, optical, and electronic instrumentation and in environmental control, and these facilitate greater insight and precision of observation into the functioning of elements. When coupled with advances in the hydrodynamic theory of the growth and propagation of detonation waves, significant improvements in the design of the components appear possible. Representative advances both in the design and development of the trains are described in Chapter 7, particularly those resulting from the use of high-speed instrumentation.

In considering the advanced diagnostics it should be remembered that the advances presented are concerned only with the azides, their properties and confinement, and the role of the substance earlier or later in the functioning sequence. Most of the work reported was done on production materials, and representative data that relate to the materials, rather than actual explosive-train components, is presented.

It should, however, be clear that we are far removed from the desirable goal of being able to select the energy of an initial stimulus and, on the basis of a required output, design the explosive elements from first principles.

REFERENCES

1. C. H. Johansson, P. A. Persson, *Detonics of High Explosives*, Academic Press, London, 1970.
2. B. T. Fedoroff, H. A. Aaronson, O. E. Sheffield, *et al.*, *Encyclopedia of Explosives*, Volume I, Picatinny Arsenal, Dover, N.J. 1960. National Technical Information Service, Washington, D.C.
3. *Proceedings of the Symposium on Lead and Copper Azides*, October 25–26, 1966, Report WAA/79/0216, Explosives Research and Development Establishment, Waltham Abbey, Essex, England.
4. F. P. Bowden, A. D. Yoffe, *Fast Reactions in Solids*, Academic Press, New York, 1958, pp. 123 *et seq*.
5. F. P. Bowden, A. D. Yoffe, *Initiation and Growth of Explosion in Liquids and Solids*, Cambridge University Press, Cambridge, 1952.
6. C. E. H. Bawn, G. Rotter, eds., *Science of Explosives*, Volumes I, II, Her Majesty's Stationery Office, London, 1956.
7. G. R. Walker, ed., *Manual of Sensitiveness Tests*, issued by Panel 0-2 (Explosives). The Technical Cooperation Program, Canadian Armament Research and Development Establishment, Valcartier, Quebec, February 1966.
8. *Blasters' Handbook*, 15th ed., Explosives Department, E. I. duPont de Nemours Co. Inc., Wilmington, Delaware, 1969.
9. *Explosive Trains*, U.S. Army Materiel Command Engineering Design Handbook AMCP706-179, March 1965, National Technical Information Service. Washington, D.C.

1

Processes for the Manufacture of Lead and Silver Azide

T. Costain and F. B. Wells

A. INTRODUCTION

There are many chemical compounds which, when used appropriately, will perform the function of a primary explosive and transfer or amplify an initial stimulus to the degree required to initiate detonation in secondary explosives. In fact, a secondary explosive itself can be used, and the need for hazardous and expensive materials like the azides (in particular, lead azide) may be questioned.

The answer lies in a consideration of not one property, but of all the properties of explosives. Some explosives are safer to handle than others but require complicated mechanisms or very high input energies, such as burning-to-detonation devices or exploding bridge wires, to initiate detonations. In some such devices secondary explosives have been found to be useful for particular applications. Primary explosives are easy to initiate but are not as stable (e.g., mercury fulminate) or are not as powerful (e.g., diazodinitrophenol) as secondary explosives. By thus trading off a variety of technologically important properties, the azides have achieved a unique position in the explosives industry and have, in particular, displaced other initiating explosives in the overwhelming majority of applications.

A similar selection process occurred within the family of inorganic azides soon after their discovery at the end of the nineteenth century. Silver azide was recognized as at least the equal of lead azide in performance, but lead has always been cheaper than silver. At that time also most fuze designers were satisfied with mercury fulminate, which was cheaper than either azide. Comparisons with other metal azides gradually brought out the fact that lead azide is more stable than some, less sensitive than others, and more powerful than most. It began to replace other primary explosives during the 1920s, first in civilian applications and then more conservatively in military applications; however, it remained very deficient in handling and safety characteristics.

A breakthrough came with the development about 1930 of dextrinated lead azide, which was considerably safer to handle than the unmodified material but still powerful enough to function satisfactorily in contemporary detonators and related devices. Dextrinated lead azide was the first initiating explosive to be manufactured* under carefully controlled conditions so as to produce an explosive with a more desirable range of properties. Subsequent developments followed the same principles and employed crystal-modification agents, or "phlegmatizers" (e.g., dextrin, polyvinyl alcohol, etc.), to produce a balance between reliable initiation, output energy, and safety in handling. Products were successfully developed to be free flowing for ease of introduction into small-diameter detonators and to have a high bulk density to provide the maximum energy in pressed compacts.

The trend in detonator and explosive-train design, which has continued into the 1970s, has been to smaller components, requiring decreased amounts and diameters of more efficient explosives. This trend itself has tended to emphasize the technological importance first of lead and then silver azide and to assure the continued modification of their properties by process development and control.

In the United Kingdom the prewar development of Service lead azide represented a trade-off of some handling convenience and safety for increased output. During World War II, polyvinyl alcohol (PVA) lead azide was developed in the United States and resembled Service lead azide in appearance, output, and other properties. Then at the end of World War II, work began on lead azide modified by precipitation in the presence of sodium carboxymethylcellulose (CMC) [1]. Originally, CMC-type lead azide was developed to have the self-binding properties of dextrinated lead azide, so that it could be pressed into sleeves, yet have the initiation and output properties of the Service material. Later development in the United Kingdom led to a variety of new explosive products given "RD" serial numbers and in particular to RD1333 and RD1343 CMC-type azides. American versions of RD1333 were produced in both government plants and private industry and were ultimately produced in large batches as Special Purpose

*The numerous laboratory methods for preparing inorganic azides on the small scale (a few grams) are presented in Volume 1, Chapter 1. With the advent of differing preparative techniques, it also became increasingly clear that lead azide can exist in more than one crystallographic form (see Volume 1, Chapter 2). Unless otherwise stated, throughout Volume 2 the alpha form is implied by the term lead azide.

lead azide. CMC-type lead azide had supplanted to a high degree but not completely replaced Service lead azide, PVA lead azide, or dextrinated lead azide for military use in the United States, the United Kingdom, and most countries of the North Atlantic Treaty and South East Asia Treaty Organizations. Dextrinated lead azide remains the principal product for civilian use, for mining and demolition applications in general, and some military items.

Other azides have been considered for use as detonants, but besides lead azide only silver azide possesses a combination of properties which has found favor in industry. However, in spite of its long-recognized virtues [2], no process was developed that would make silver azide in a form suitable for pouring and pressing into detonators. During the early 1950s, Taylor [3] and Williams and Peyton [4] developed processes for making granular silver azide. Taylor's process [3] was adopted for use in British ordnance factories, but it yielded only small (1.5 kg) batches in the standard British production vessels, and four hours were required to complete a batch.

Silver azide has not been used in production detonators in the United States and has only limited use in other countries (e.g., Netherlands, Sweden, Germany), where its superior chemical stability and detonation velocity have justified its use despite a high cost. Recently (see Chapter 5), it has been shown that in important detonator applications the performance of lead azide is already marginal and that silver azide will function satisfactorily within the critical weight-diameter-volume relations of miniature detonators. An improvement to the Taylor process decreased the processing time with consequent savings. The efficiency had been increased by a factor of nearly two when an entirely new approach was discovered. Using the new approach, a method for making granular silver azide suitable for small detonators was devised and is described later in this chapter.

While the finished products sought in different countries are usually quite similar, if not identical, in performance characteristics, somewhat different philosophies exist with respect to control of the manufacturing processes. Thus, in the United States the manufacture of lead azide for military purposes is not so rigorously specified as it is in Britain, and more emphasis is given to satisfactory performance in routine tests as the criterion for an acceptable product. This difference of emphasis, together with the smaller scale of production, have led to a greater standardization in Britain, not only of the processes but also of equipment, and the processes have been "exported" or licensed for use in several countries of Europe and in Asia and Australia.

On the other hand a more thorough knowledge of the effect of process variables has been required in the United States both to contend with large fluctuations in the demand for materials and to understand the consequences of looser manufacturing specifications. Thus, although the processes introduced into the U.S.A. are often based on the Taylor processes, these have also evolved with distinguishing features.

Notwithstanding this tendency, in all countries, notably the U.S.A., U.K., Sweden, Germany, etc., the manufacture of lead azide both for civil and military use is usually strictly controlled within individual private companies. Safety alone predicates control over all operations, but process parameters may vary significantly from one company to another, and the physical appearance of the product may vary accordingly. It is the recognition of these variations and the competitive bidding used in countries such as the U.S.A. and Germany for the purchase of the materials which is the basis for a looser approach to manufacturing specifications.

This chapter presents in detail the more recent advances in azide manufacturing technology. More emphasis is given to advances made since World War II and particularly to those of the 1950-1974 period. In endeavoring to understand the rationale for, or relevance of, some of the more specific studies, it is helpful to appreciate the practical significance of the differing philosophies expressed above. However, it is also pertinent to ask the question, what does one look for, both chemically and physically, as criteria for judging the acceptability of a product? And in what way is the basic chemistry and physical appearance of the products related to their performance, safety, and handling properties? These questions are difficult to answer, as is implied in the preceding paragraph, and are the subject of controversy and a certain degree of "mystique." The lack of fundamental understanding is perhaps the best argument for process control, so that reproducibility will be within narrow limits. It also accounts for the considerable activity devoted to arriving at a better understanding of azide behavior and a better discrimination in measurements, both of which constitute much of the content of these two volumes on azide research and development.

B. FACILITIES AND EQUIPMENT

The hazardous nature of azide manufacture necessitates simple facilities, which are easy to keep clean and free from grit* and residues, provide protection to the operator through the use of barriers and remote control, yield only limited-size batches, and permit unimpeded access to reaction vessels. In general, because of the similarity between the processes utilized by different manufacturers, there is little difference between the facilities and equipment used at the various locations.

Because the chemical reaction used is a simple metathesis or double decomposition of two salts, the reaction vessels can also be simple. An open kettle† with a jacket for temperature control, and a simple stirrer rotating with plenty

*See Chapters 3 and 4 for the critical role grit and copper may play in increasing hazards.
†A "pan" in British usage.

of clearance, have been used for azide manufacture almost universally. Highly polished stainless steel provides a noncorroding and easy to clean surface. The use of copper or copper alloys near the reaction kettle or reactant supply is strictly forbidden. All filling, stirring, tilting, and draining operations are arranged to be operated remotely to the maximum possible degree.

Typically, stock solutions of the lead salt (acetate or nitrate) are prepared and filtered to remove grit and lead carbonate. The required quantity of stock solution is then transferred to a calibrated hold tank from whence it can be fed to the reaction kettle in a separate room. Similarly, the other reactant, sodium azide, is prepared as a stock solution, filtered, and apportioned to another hold tank. It is good practice to segregate the lead salt and the sodium azide areas from each other as well as from the reaction-kettle room.

The solutions are transferred to the reaction kettle through metered tubes, and the precipitated lead azide is washed and either filtered and dried or packed wet for shipment. In some countries the practice is to introduce the azide into detonators or explosive trains only at the place of manufacture; in other countries, notably the U.S.A., transportation is permitted, and the packed azide may be stored and used in detonators or fuze trains at distant locations. However, while such peripheral practices and associated equipment are important, it is the control exercised during the metathesis in the kettle that affects the product and, more importantly, introduces subtle differences that make one form of lead azide different from another in performance characteristics, if not in gross physical appearance. Therefore, the emphasis of the present discussion will be on the reaction kettles and their associated equipment, and the following section will emphasize the control on the metathesis.

At the Explosives Research and Development Establishment (ERDE) facilities at Waltham Abbey and Woolwich Arsenal, England, a small but prolific band of investigators, headed by Taylor, developed during the past two or three decades not only a great variety of initiating and igniferous compounds but also a systematic method for developing manufacturing processes. The approach is based on the use of standardized equipment and the systematic progression of developments from "bench-scale" experiments through a pilot scale to full-scale production. This approach has been utilized not only within the United Kingdom's ordance factory system but has been exported to several other European and British Commonwealth nations. The approach is particularly valuable within the Commonwealth because both the development laboratories and ordnance factories are under common government control. However, it is not so readily introduced in private industry or in countries such as the United States where proprietary processes and competitive bidding are liable to dominate manufacturing and procurement practices. There is thus a greater variety of equipment in the latter case, although similar principles are used in its design, and even in the U.S.A. the ERDE approach has profoundly affected local practice.

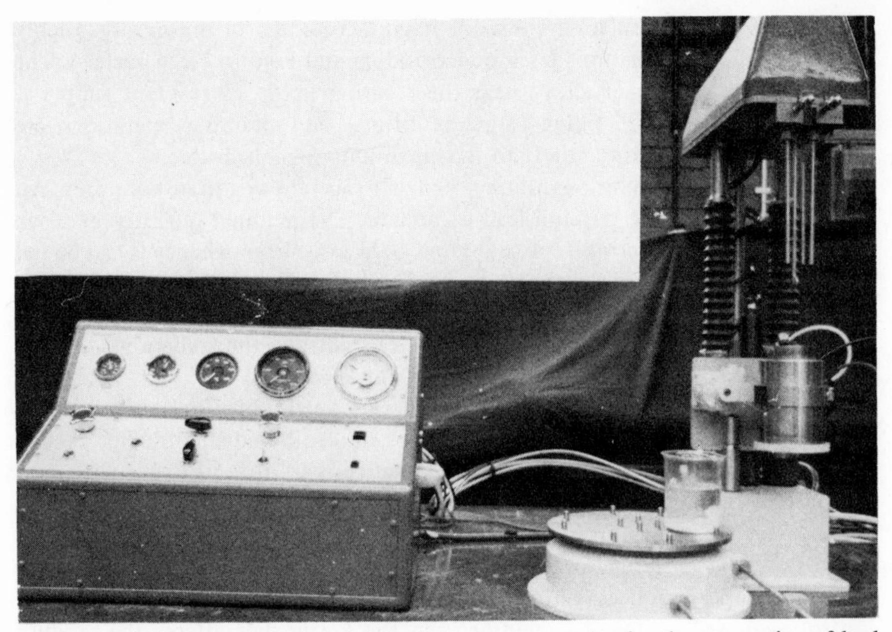

Figure 1. Bench-scale apparatus of 600-ml maximum capacity for the preparation of lead azide and other sensitive explosives. The apparatus on the right is enclosed with a barricade during use and remotely operated through the control box shown in the left foreground. The turntable in the right foreground, when properly positioned beside the reaction apparatus at the rear, permits the selection of such operations as decant, wash, and filter. The stainless-steel reaction vessel is pivoted to pour into stations on the turntable. (Photo courtesy of ERDE, Waltham Abbey, England.)

1. ERDE Equipment

The original standard bench-scale kettle was a 600-ml* stainless-steel beaker, temperature was controlled by immersion in a water bath, and agitation was effected with a 2-in.-wide, oval-shaped, flat-bladed, air-driven stirrer. Operations were performed semiremotely behind a shield in a fume hood. However, considerable exposure of the operator's hands and forearms took place even when tongs and other tools were used.

Subsequently, a completely remote, bench-scale apparatus was developed in the late 1960s; Figure 1 shows the apparatus. The control box shown to the left can be mounted outside a fume hood and used to control and monitor a reaction remotely. Next to the control box is a turntable with four indexing stations (i.e., decant, wash, filter, etc.), while on the far right is the 600-ml reaction kettle

*The working capacity of any of the kettles is about two-thirds the capacity given.

Figure 2. Pilot-plant, or 10-liter, kettle of Explosives Research and Development Establishment (U.K.) design as installed at Picatinny Arsenal. As shown the original flat oval agitator has been replaced with a propeller and a swirl stopping baffle has been added to increase agitation needed for silver azide manufacture [19].

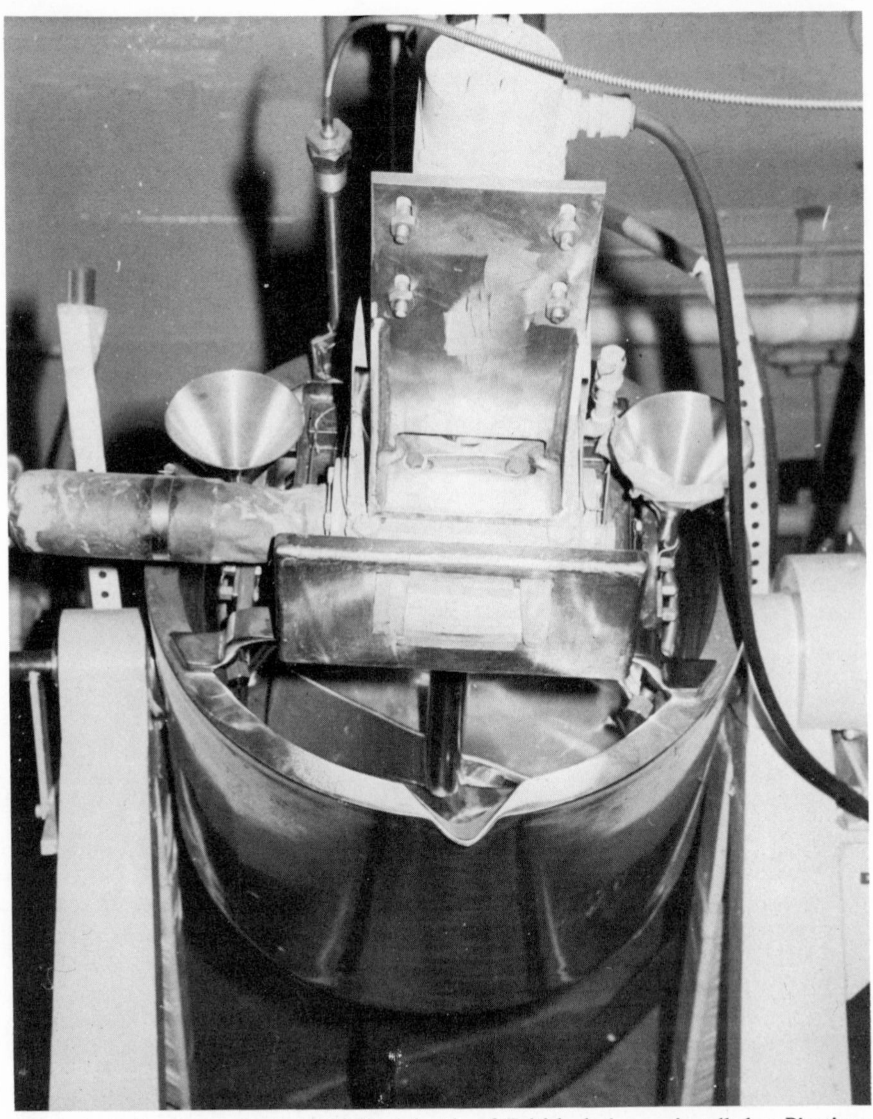

Figure 3. Full-scale production (60-liter) kettle of British design as installed at Picatinny Arsenal. Used at Royal Ordnance Factories for primary explosives manufacture in 2½–5-kg batches.

agitator and feed lines are raised or lowered pneumatically on the pistons covered by rubber accordion shields, which can be seen behind the vessel. The flat-bladed, oval agitator (poorly visible in the illustration) is 5 cm wide and 2 cm high. Like the interior of the vessel, it is highly polished stainless steel. Batches of 10–30 g can be made in the 600-ml kettle or beaker.

The next step in scaling up is a pilot-plant, or 10-liter, kettle, the latest model of which is shown in Figure 2. Like the 600-ml kettle, the height–diameter ratio is near unity, and a larger, 10-cm version of the flat stirrer is normally used. Shown in Figure 2 is a propeller-type stirrer which may be used in conjunction with a baffle (also shown) to achieve vigorous agitation, as needed for making silver azide by the new process described later in the chapter. The kettle is capable of producing batches of about 500 g (1 lb) lead azide and is used to refine the processing parameters, such as temperature, before the final scale-up in the production-plant facilities. Operation is semiremote and can be set to carry the process through a washing-by-decantation stage before it is necessary to enter the reaction room to change the collection equipment.

The largest kettle produces 3.5-kg (7.5-lb) batches of lead azide and has a 60-liter capacity; the geometry is again similar but not identical to the smaller vessels. The agitator blade, as can be seen in Figure 3, is no longer oval but is propeller-shaped with the blades tilted 30 degrees. With the agitator operating at approximately 90 rpm and the kettle tilted slightly off the vertical, the same degree of agitation can be obtained as from 300-rpm operation of the flat stirrer in either the 600-ml or 10-liter kettles. As with the 10-liter kettle, operation is semiremote. With this kettle funnels are mounted on top to facilitate introduction of the reactants without splashing and spillage.

2. Typical United States Facilities

As indicated, a wider range of kettle designs and ancilliary equipment is in use in the United States than in British defense establishments, particularly in private industry. A brief description of equipment installed at Picatinny Arsenal, Dover, New Jersey, is typical of American practice and will further serve to illustrate the structural and other facilities desirable for pilot-plant and production operations. The fully remote 600-ml kettle is not available in the United States, but in addition to the British-design 10- and 60-liter kettles, a kettle of the E.I. duPont type with a capacity of 160 liters is used, producing batches of up to 7 kg (15 lb) azide at a time. The duPont kettle (Figures 4 and 5) was used to manufacture the overwhelming majority of lead azide produced in the United States over the last 30 years, as it is the standard kettle used in all government-owned plants and at duPont's own considerable manufacturing facilities. The three sizes of kettle are maintained in a pilot facility at Picatinny Arsenal and used in the development and manufacture of primary explosives. The reaction room (Figure 5) is humidity controlled and separated by a 30-cm concrete wall from the control room (Figure 6).

Also at Picatinny is another pilot-plant 10-liter kettle of a somewhat unique design. Unlike its British counterpart which seeks to maintain similar agitation

Figure 4. The standard 160-liter vessel used in the U.S. for manufacture of up to 8 kg of lead azide. It is often referred to as "the duPont design" but originated in Germany. The large four-bladed agitator is raised by a wire rope and pulley to permit the tilting of the vessel for discharge of the product.

Figure 5. The pilot plant at Picatinny Arsenal showing from left to right: British 60-liter, duPont 160-liter, and British 10-liter kettles.

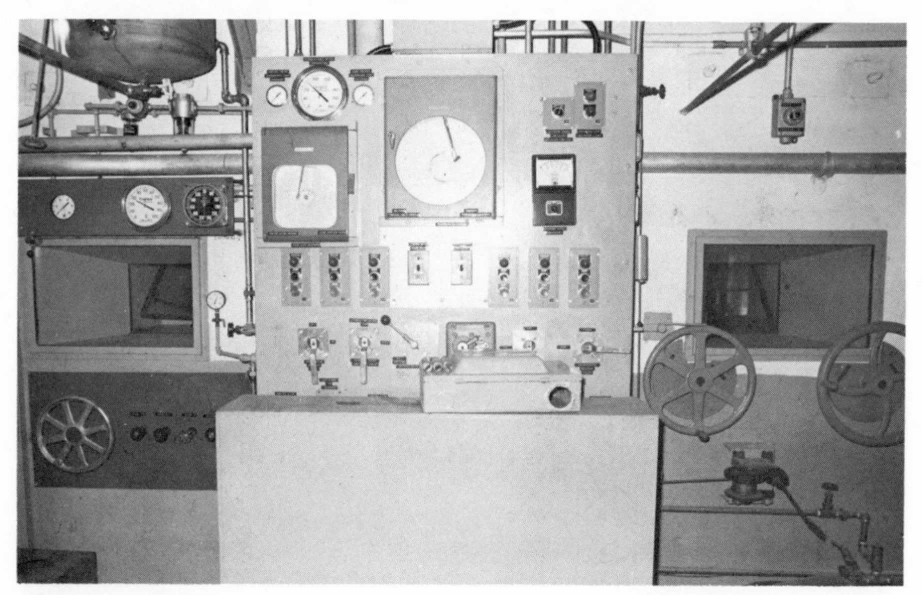

Figure 6. Control room for remotely operating the vessels shown in Figure 5. All process parameters (i.e., temperature, rpm, etc.) can be monitored and/or recorded. The two periscope windows allow observation during reaction.

in the different kettle sizes rather than reproduce geometry, it is an exact miniaturization of the standard British 60-liter size. It was made specifically for a study of the reaction parameters of the RD1333 lead azide process. In addition the facilities were developed to permit fully automated remote handling from the raw-material stage to the packaging of the dry product. As can be seen in Figure 7, the product can be washed in the kettle, air-dried on the filter, and the dry product distributed to 10 small conductive containers by use of a sample

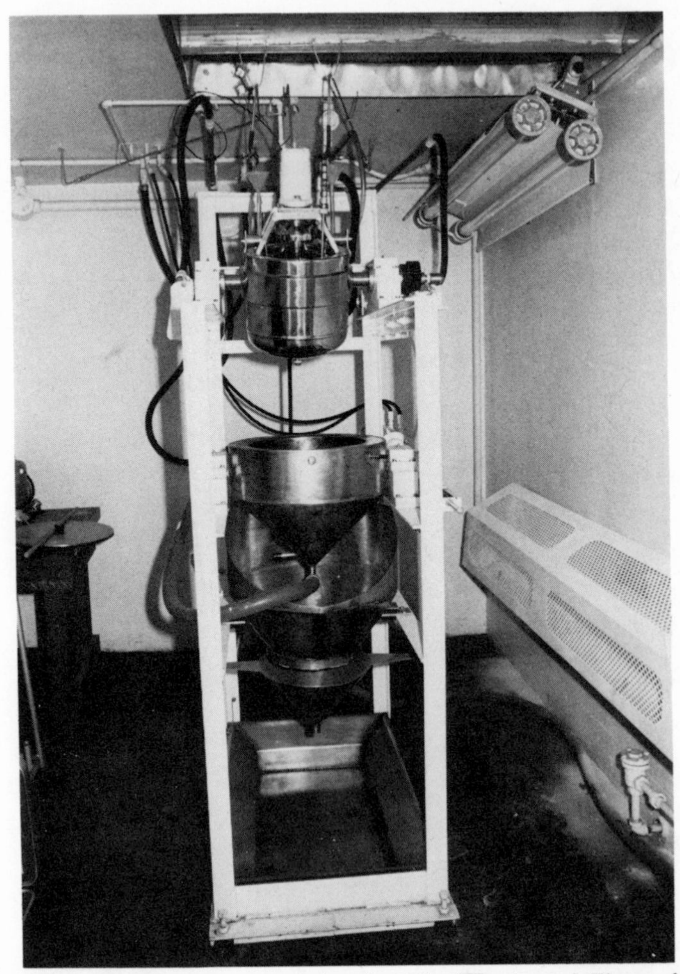

Figure 7. Primary explosive apparatus designed at Picatinny Arsenal. The product can be remotely made, filtered, washed, air dried, sieved, and poured into containers if it is made in a free-flowing form which easily drops by gravity from one stage to another.

Figure 8. Control room for the Picatinny Arsenal primary explosive apparatus shown in Figure 7.

selector (not shown). However, unless the product is free flowing, it may fail to transfer from one stage to the next. The atmosphere of the reaction room can be kept at 50% relative humidity by humidification, but it is not completely air-conditioned. Operations are conducted from the control room (Figure 8), where the reactant lead acetate and sodium azide solutions may be fed to the kettle through precision proportioning pumps and all other process parameters may be monitored and controlled.

3. Some Tests and Equipment for Product Assessment

Before going on to the description of the manufacturing processes for lead and other azides, it is necessary to describe briefly some of the test procedures used for quality control or acceptance criteria. Test procedures for the safety, handling, and functional properties are described in much greater detail later in this volume (Chapters 2-5), but it will be necessary to refer to the results of some tests when pointing out differences between products from different manufacturing processes or variations of manufacturing parameters. In general the tests given here are the minimum necessary for even a superficial assessment of product differences.

a. Physical and Chemical Characteristics of the Products

Moisture Content. Moisture content is determined by a standard procedure of drying to constant weight over P_2O_5 and reporting the result as the percent loss in weight.

Hygroscopicity. In this test a preweighed dry sample is held for 48 hr in 90% relative humidity at 30°C and the result reported as percent gain in weight of the dry sample.

Bulk Density. Bulk density is determined by a method in which 1 g of lead azide is added incrementally to a vertical, graduated tube about 0.5 cm diam and partially filled with normal butyl alcohol. Particles adhering to the sides are washed down with butyl alcohol, addition of which is continued until the volume of material in the tube is exactly 5 ml. After standing 15 min without jarring or vibrating, the level of the lead azide column is read, calibration corrections applied, and the bulk density, calculated from the volume occupied, reported in g/ml.

Purity. In this determination [7] the volume of gaseous nitrogen liberated by treatment of the lead azide with ceric ammonium nitrate is used to calculate the weight, reported as percentage, of lead azide in the test sample.

b. Sensitivity of the Products to Impact, Heat, and Electrostatic Discharge

Ball Drop Test. A $\frac{1}{2}$-in.-diam hard steel ball weighing 8.33 g is dropped onto a 0.014-in.-thick layer of sample on a hardened steel block. The ball is initially retained by a supporting rod whose vertical position is graduated at every inch from 1 to 45 in. and a remotely operated gate is used to release the ball. The block bearing the sample is moved slightly after each drop to assure that the ball will not make impact at the same point twice on an undetonated sample. The results are given as the minimum distance (in inches) through which the ball must fall freely to cause at least one firing of the test material in 10 trials.

Heavy Impact Sensitivity Test. The test results obtained are the minimum distances (in inches) through which a 2-kg hammer must fall to cause at least one sample to detonate in a series of trials with 10 test samples of approximately equal weight.

Explosion Temperature Test. Results of this test are given as the temperature which will cause a 10-mg sample, held in a copper detonator shell, to flash or explode in 5 sec. The test is conducted by successively immersing samples in a molten-metal bath held at temperatures at which explosions are obtained in both more and less than 5 sec. Averages of the times required to achieve explosion of 10 samples at the various temperatures are plotted against the temperatures and a curve drawn through the points obtained. The 5-sec temperature value is read from the curve.

Electrostatic Sensitivity Test. Electrostatic sensitivity tests are conducted at 25-30°C and a relative humidity of 20% or less. The azide is commonly exposed to a discharge between metal electrodes having a fixed gap of, typically, 0.019 in. Results are reported as the lowest electrical discharge energy capable of initiating the sample in at least one of 10 trials.

c. *Efficiency of the Product as an Initiator of Detonation*

The test is often known as a "functioning" test and determines the relative efficiency with which primary explosives, such as lead azide, initiate secondary explosives, such as RDX. Typically, a detonator cup is loaded with an upper charge of 15 mg priming mixture, an intermediate charge ranging from 25 to 155 mg (using 15-mg steps) of the azide being tested, and a lower charge of sufficient RDX to provide total height of materials loaded of 0.27 in. and a finished detonator length of 0.29 ± 0.005 in. In loading, each material is pressed as a single increment at 15,000 psi with a 3-sec dwell. Groups of at least 50 detonators containing the same loading are fired in a test apparatus to determine the minimum quantity of azide under test which gives 50 consecutive firings that result in perforations of a lead disk having a diameter of at least 0.156 in. Test results are reported as the minimum weight of lead azide used, the weight of RDX required to fill the detonator, the number of detonators fired, and the number firing high order, together with the maximum, minimum, and average diameters of the perforations obtained in the firing of each group.

C. PROCESSES FOR THE MANUFACTURE OF LEAD AZIDE*

If one simply mixes a solution of lead nitrate (or acetate) with a solution of sodium azide, a precipitate of nearly pure lead azide forms which can be washed, filtered and dried, and utilized in a detonator or other explosive device. However, this lead azide is at best a fluffy fine powder (not unlike that produced by the reaction of hydrazoic acid with lead nitrite (Figure 9), very sensitive to electrostatic discharge, difficult to pour into small detonator cups, and difficult to compact by pressing. It is, however, a stable and powerful initiator.

The requirements for a product which is both safer to handle, yet retains the power of the pure, dense material, led to the development of the various phlegmatized materials. Thus most of this section deals with the development and improvement of processes, based on such crystal-modifying agents as dextrin and carboxymethylcellulose. Much of the technical detail presented has not been

*The reader is referred to Volume 1, Chapter 1 for details of other laboratory methods for the preparation of lead azide on a small scale, and to references [7] and [8] for the origins and details of early manufacturing processes.

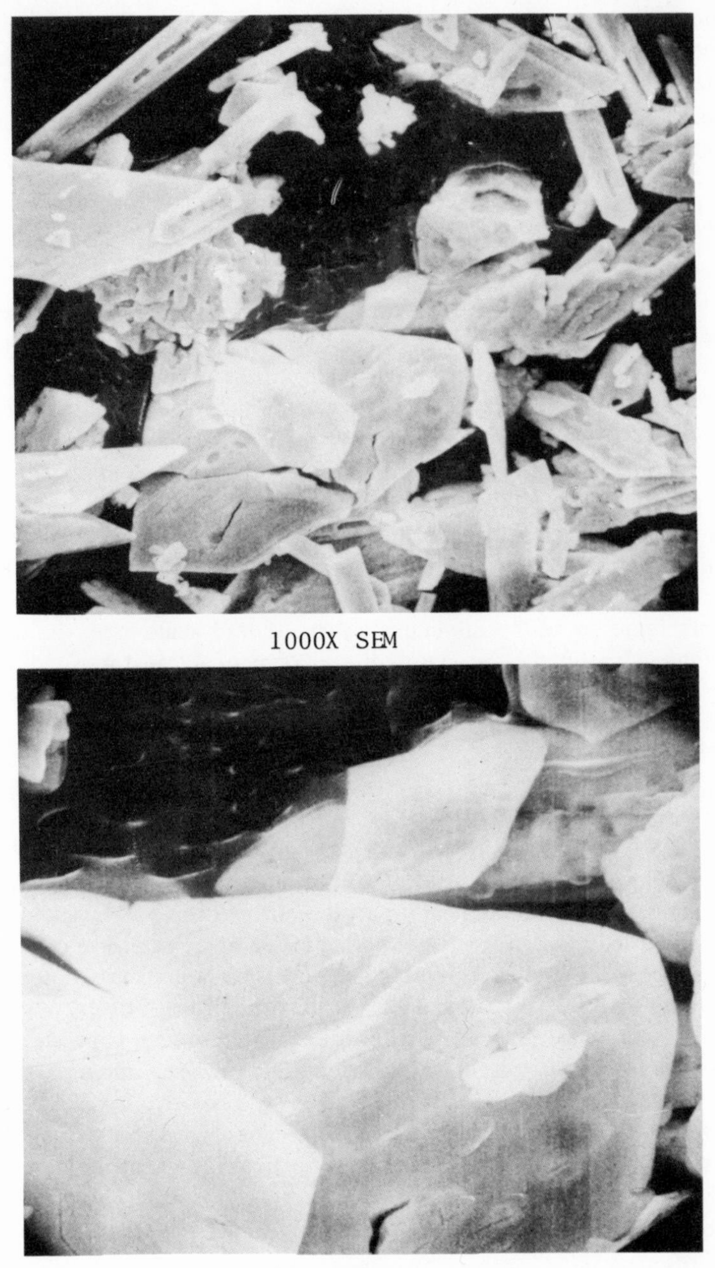

Figure 9. Pure lead azide made from hydrazoic acid and lead nitrite.

published previously or has been published only in internal government reports. Other varieties of azide are also still encountered, and for completeness brief reference will, first, be made to these and to the processes for manufacturing them. The phlegmatizing agents are capable of producing azide particles in numerous shapes, sizes, and states of aggregation, which are believed to play a role both in the density and sensitivity of the powder, and in the flow and storage properties. Microscopic and X-ray examinations of the powders are, therefore, often used to distinguish between the different products.

Among the early consequences of such examinations was the discovery that variations in the shape of azide particles were sometimes due to differing crystallographic phases (Volume 1, Chapter 3). It was early suspected that differing sensitivities of the products was attributable to significant differences in the intrinsic sensitivities of the phases. Although this suspicion has not been substantiated by the most recent measurements, it is widely believed that some phases appear in large flat or needle-shaped crystals whose reduced handling properties contribute to an increased effective sensitivity of such products. In the CMC processes developed by Taylor, as described below, the use of 25°C as a process temperature is one instance of an attempt to minimize the precipitation of the higher temperature β form of lead azide. However, it is pertinent that modifications of the basic process have raised the precipitation temperature significantly, and also as described below, the one continuous method known to have been developed for lead azide manufacture deliberately produces β-lead azide as an intermediate product.

1. Colloidal Lead Azide

The only nearly pure (99+%) form of lead azide used today in the United States is the so-called colloidal lead azide made by quickly mixing dilute solutions of lead and azide salts. Despite the information in reference [5], colloidal lead azide is neither colloidal in particle size (the specification [6] calls for an average of particle size of 5 μm) nor safer to handle, as this form of lead azide is particularly sensitive to electrostatic discharge. It is used in contact with bridge wires in electric detonators (sometimes applied by mixing with a lacquer binder). The process as developed commercially is a proprietary product, and only small amounts are made as needed.

2. Service Lead Azide

There are phlegmatized lead azides which in the physical appearance is entirely different from the aggregates which are typical of most processes. Instead of being opaque granules, they are equant transparent crystals (Figure 10). The first of these, called Service lead azide, is made by the near-simultaneous

Figure 10. British Service lead azide, 170×. (Photo courtesy of ERDE, Waltham Abbey, England.)

addition of a 1 M lead acetate solution and a 2 M sodium azide solution to a kettle which contains a quantity of sodium carbonate solution. The lead acetate addition is started just ahead of the sodium azide so that a quantity of lead carbonate is produced which serves as a nucleating or seeding agent for controlling crystalline growth. Service lead azide is 96% pure and considerably more energetic than dextrinated material, but also more sensitive. The process conditions represent a considerable advance over previous processes in that quite concentrated solutions (up to four times the solid content) of reactants are used, permitting much larger batches from the same size kettle. For all practical purposes this product has today been supplanted in Britain and the U.S.A. by RD1333 and RD1343 described below. However, the process is still used for production purposes in Sweden. No accidents have been encountered there during its use, and azide contents up to 98% are claimed.

3. Polyvinyl Alcohol Lead Azide

The first cousin of British Service lead azide is the crystal form precipitated in the presence of polyvinyl alcohol: PVA lead azide. At the present time, little (if any) PVA lead azide is produced in the United States, although it is believed that it is manufactured in Germany by the Dynamit-Nobel. In the United States it is manufactured solely by the Olin Matheson Corporation using patented procedures [9] which give little insight into the actual manufacturing techniques used or the problems encountered. PVA lead azide is 96% pure and may contain some polyvinyl alcohol combined with lead, but not in the same manner as the lead dextrinate in dextrinated lead azide, for the product consists of transparent, well-defined crystals. It has an initiating efficiency equal to Service lead azide and, according to data published in the patent, about the same handling and sensitivity properties.

4. Dextrinated Lead Azide

As indicated in the introduction to this chapter, one of the major advances which affected the civil and military technology of lead azide occurred with the development of the dextrinated products. Today these continue to play the most significant role in civilian industry and are still used for some military applications.

The basic process has been documented in readily available sources [7,8], thus it will suffice to mention only the critical parameters for reference in comparing the more recent modifications to the processes and products. Dextrinated lead azide is made by adding over a period of 30 min a dilute (3%) solution of sodium azide to an equal volume of a 7% lead acetate solution, maintained at 60°C. The sodium azide solution, which contains an amount of potato dextrin roughly equal to 6% of the expected yield of lead azide, is made slightly alkaline with sodium hydroxide. The lead azide is thus formed mainly in the presence of excess dissolved lead salt and with a reasonably constant concentration of dextrin. The individual granules of dextrinated azide are formed by accretion in concentric layers of microcrystals bound with lead dextrinate (Figure 11), giving a product which is about 92% pure.

In Sweden the Förenade Fabriksverken of the Ministry of Industry utilizes more dilute solutions in making dextrinated azide. Although normally precipating the azide at 60°C, this company has tried precipations from 75°C in (unsuccessful) attempts to reduce the HN_3 content of the product and thus avoid the formation of copper azide in detonators fabricated from copper-containing metals. Azide contents up to 94% are obtained by their process.

Over the years numerous proprietary modifications have been made by private industry in order to improve the flow, pressing, and general handling and safety characteristics of dextrinated material. Figure 11d, for example, is a micrograph of a product known as Elcoat, manufactured by the duPont Company, which has 0.25% calcium stearate added as a coating to the dextrinated particles to decrease the sensitivity and improve the compactability of the powders.

During the 1950s a spheroidal form of dextrinated lead azide was developed

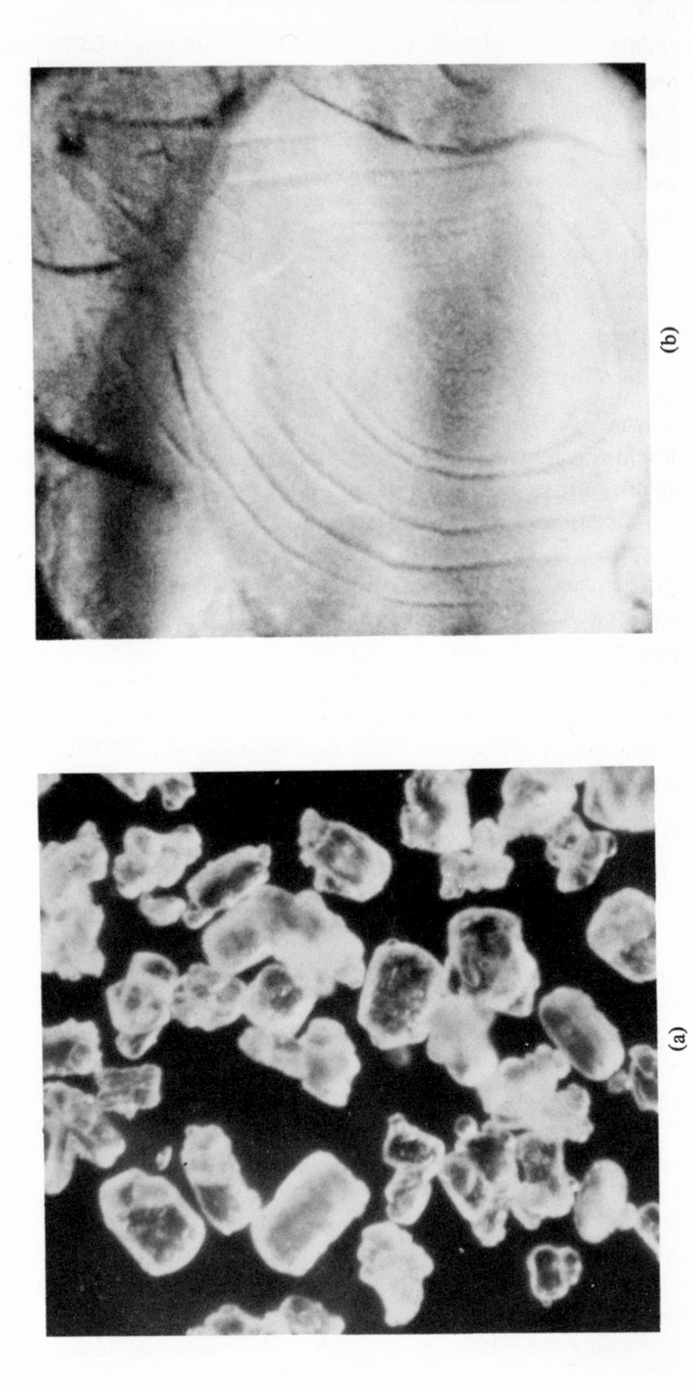

(b)

(a)

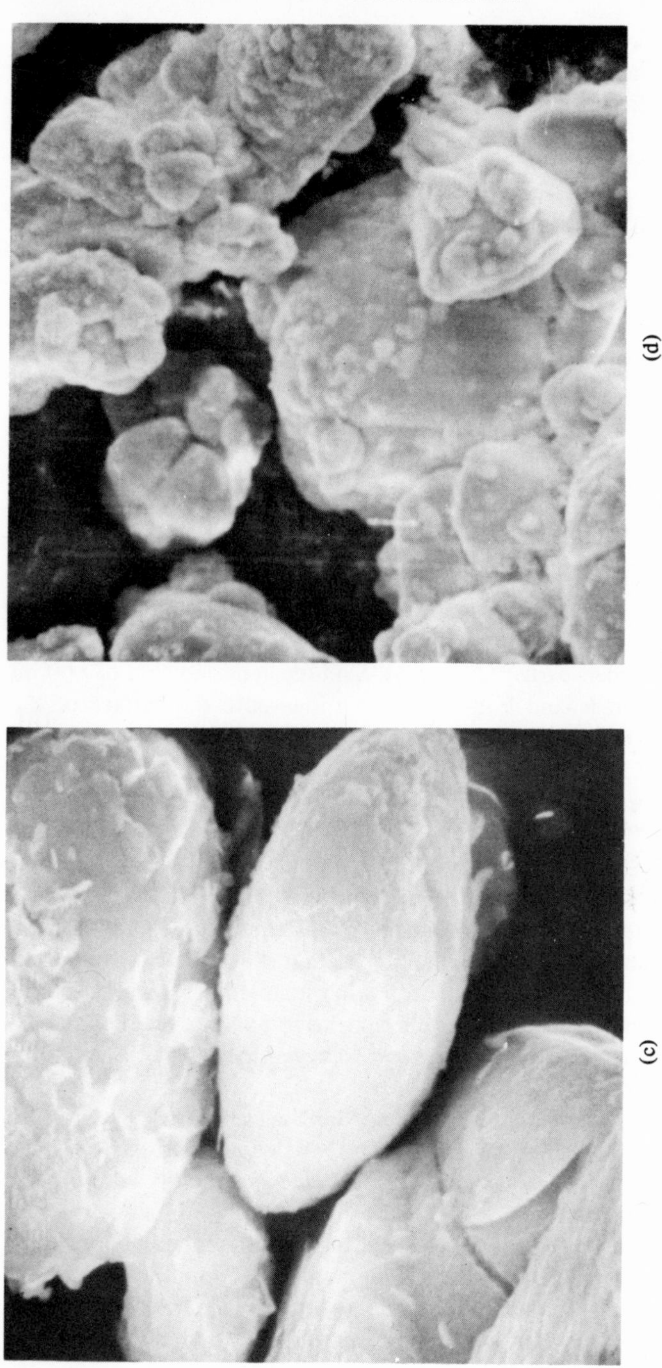

Figure 11. Dextrinated lead azide of duPont manufacture: (a) 450×; (b) cross-section of one particle showing growth rings, 10,000× (SEM); (c) surface details of particles, 10,000× (SEM); (d) dextrinated lead azide mixed with calcium stearate (small particles are calcium stearate) 1000× (SEM). (From [20].)

by Taylor and coworkers and was serialized as composition RD1352 [10]. The low dextrin content and the physical shape of the particles of RD1352 indicated that it should be an efficient initiator and more easily handled than the irregularly shaped dextrinated azides available commercially.

An investigation was, therefore, undertaken by Wells to determine if United States domestic dextrin can replace British-made material in the preparation of RD1352 and to establish its usefulness in military detonators. The products obtained were in the form of dense spheroidal particles having bulk densities of about 2.2 g/ml, closely resembling in both size and appearance the particles of RD1352. The spheroidal form rendered it easier to use than the irregularly shaped particles of domestic dextrinated azide, with which it was compared, and it was found to be a more efficient initiator of secondary explosives such as RDX. Because details of this investigation have not been published elsewhere, they are presented in the following discussion.

a. Preparation of Spheroidal Dextrinated Lead Azide

The five azide batches were prepared in accordance with the procedure [10], except that the solutions were added at a constant rate with metering pumps. Preparation was carried out by remote control using a stainless-steel kettle of 10-liter capacity described earlier (Figure 7). Simultaneous addition of 800 ml each of the sodium azide and lead nitrate solutions onto the surface of the reaction mixture was controlled by adjustment of the metering pumps to deliver 800 ml of solution in 20 min for batch 1, 25 min for batch 2, and 30 min for each of the other three batches. All preparations were carried out at 60°C using an agitator speed of 275 rpm.

i. Materials and Solutions. Stock solutions for the preparations were prepared as follows, two forms of dextrin being compared as described.

Dextrins. Dextrine, White Farina 30, the dextrin manufactured by the British firm of Elliot and Crabtree and used in the preparation of RD1352, was supplied by Liang National Ltd. as Dextrine, White Farina required for manufacture of lead azide. There is no specification for this dextrin, and as received it contained 10.73% moisture. A portion was air-dried and was used to prepare the solutions for the three batches of dextrinated azide.

The Dextrine, White Farina, is similar to the Staley Company's No. 600 Potato Dextrin, labeled Morningstar 600 Dextrin, which contained 10.84% moisture. A portion of the latter was air-dried and used to prepare the solutions for two batches of dextrinated azide.

The air-dried dextrins were analyzed by established methods [11], except for absolute dextrin content, which was determined by a modification of the Munson-Walker method for determining reducing sugars [12]. In the determination of absolute dextrin, the described method was applied to the hydrolysate to determine the reducing sugars present. The difference in quantities of reducing sugars found was equated to dextrin and recorded as percent absolute dextrin. The analyses of dextrins are given in Table I, where particle size refers to the percentage of material which passes a No. 80 sieve.

In the preparation of dextrin solutions, the dextrin was stirred into about 1500 ml of boiling, deionized water, allowed to cool overnight, filtered, and the filtrate made up to 2000 ml at 20°C with water. The three solutions for batches 1, 2, and 3 were each prepared from 100.24 g of the Dextrine, White Farina, to yield solutions of 45 g dextrin per 1000 ml. The solution for batch 4 had a concentration of 45 g/1000 ml, prepared from 100.45 g of Morningstar dextrin. For batch 5 the solution concentration was 18 g/1000 ml and was prepared from 41.80 g of the Morningstar dextrin.

Tamol SN. The surface-active agent was certified by Rohm and Haas to contain 93% active ingredients and was similar chemically to the active ingredients of Belloid TD used in the preparation of RD1352. A solution containing 50 g Tamol SN per 1000 ml was used in making the azides.

Sodium Azide Solution. Sodium azide (601.2 g of 99.80% purity), 1.52 g Rochelle salt, and 6 g sodium hydroxide pellets were dissolved in 3000 ml water, filtered, and the filtrate made up to 4000 ml at 20°C with water. The concentration of this solution was 150 g sodium azide per 1000 ml.

Lead Nitrate Solution. Lead nitrate, 1532 g, was dissolved in 3000 ml water, filtered, and the filtrate made up to 4000 ml at 20°C to give a solution containing 383 g lead nitrate per 1000 ml.

ii. Procedure. The following procedures were used for all batches: A heel of 4100 ml water, 40 ml Tamol SN solution, and 37 ml water rinsings from the graduate used to measure the Tamol SN solution were placed in the reactor, the agitator run slowly, and circulation of water at about 63°C started through the reactor jacket. When the temperature of the heel reached 50°C, 1283 ml of dextrin solution and 300 ml of water rinsings from the graduate used to measure the dextrin solution were added. The agitator speed was increased to 275 rpm.

Table I. Analyses of Air-Dried Dextrins

Test designation	Dextrine White Farina	Morningstar 600 Dextrin
Particle size	2.5	1.5
Moisture (%)	10.21	10.40
Ash (%)	0.23	0.37
H_2O-insoluble matter (%)	0.5	0.6
Acidity as acetic acid (%)	0.12	0.18
Reducing sugars as dextrose (%)	4.7	3.6
Absolute dextrin (%)	80.3	77.0

When the reactor contents reached 60°C, 16 ml of the sodium azide solution was added. Within 5 min the metering pumps were started and operated for the preselected period. After cooling with circulating water, the mother liquor was removed by decantation and the product washed with three 7000-ml portions of water. The product was then transferred to a vacuum filter, sucked to dryness, and washed with two 500-ml portions of methylated spirits. It was then dried for 12 hr by drawing 6000 liters of air over solid KOH or NaOH and through the filter. The product was placed onto a vibrated U.S.S. No. 40 screen from which it passed into conductive rubber containers.

During the preparative runs, four to seven small samples were dipped from the reactor at equally spaced intervals to follow the course of particle formation and growth microscopically. These mother-liquor-wet samples and portions of the final product were microphotographed to record the changes which took place during the preparation and the final size and shape of the particles produced.

b. *Microscopic Examination of Dextrinated Lead Azide*

Figures 12 and 13 show the progress of particle growth and the final product resulting from the preparations. Figure 12, which relates to batch 3, shows particles closely resembling those of composition RD1352, while Figure 13 shows particle growth during batch 5 to give grains which largely resemble those of batch 3. No large agglomerates were found in the latter case.

Examination of the other batches showed similar progress in the formation of particles, although in all products of the Dextrine, White Farina, some agglomerates were formed, and in batches 1 and 2 some more complex forms and hollows between particles were evident. Some of the properties and functional behavior of the preparations were compared with two commercial dextrinated materials manufactured for ordnance applications. Under the microscope the particles of these samples were more irregular in shape than shown in Figure 11, and in addition to possessing numerous sharp corners and protuberances, they consisted of agglomerations of different-sized fragments in many instances.

c. *Some Comparative Properties of Dextrinated Azides*

Functional tests and property determinations were made for each of the batches prepared, and a summary of results is given in Tables II and III. Where available, data are compared with typical results in performing the same tests on the duPont dextrinated lead azide (DUP) and the product of the Kankakee Ordnance Works in the United States (KNK).

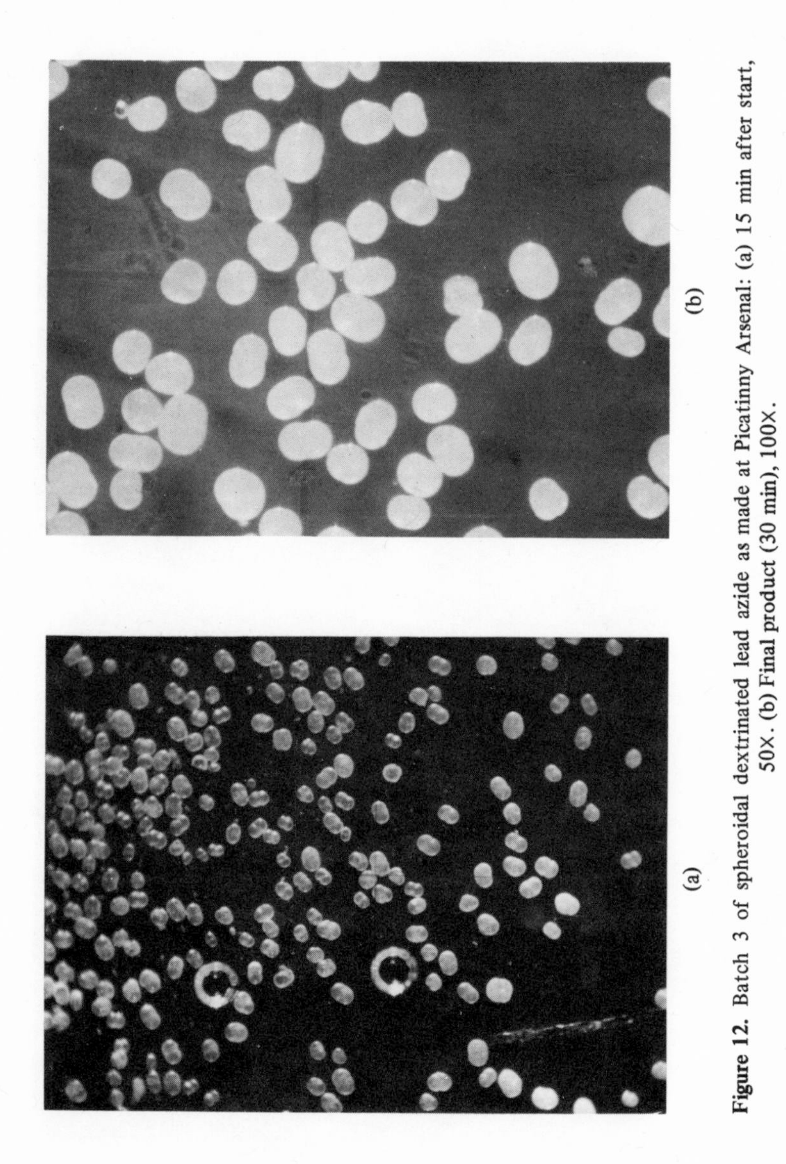

Figure 12. Batch 3 of spheroidal dextrinated lead azide as made at Picatinny Arsenal: (a) 15 min after start, 50×. (b) Final product (30 min), 100×.

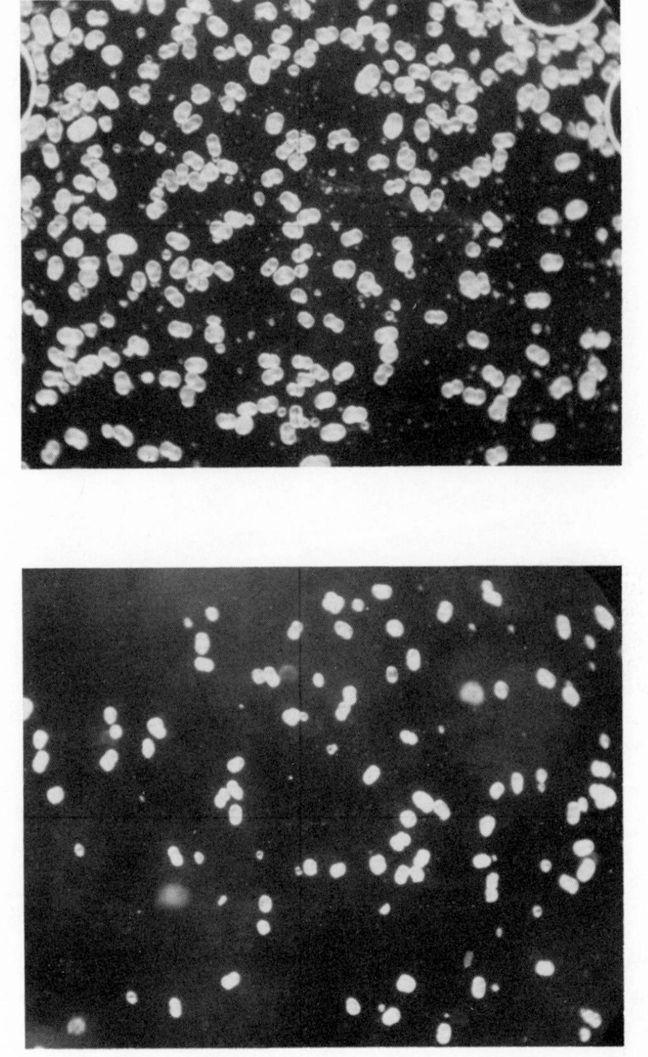

(b)

(a)

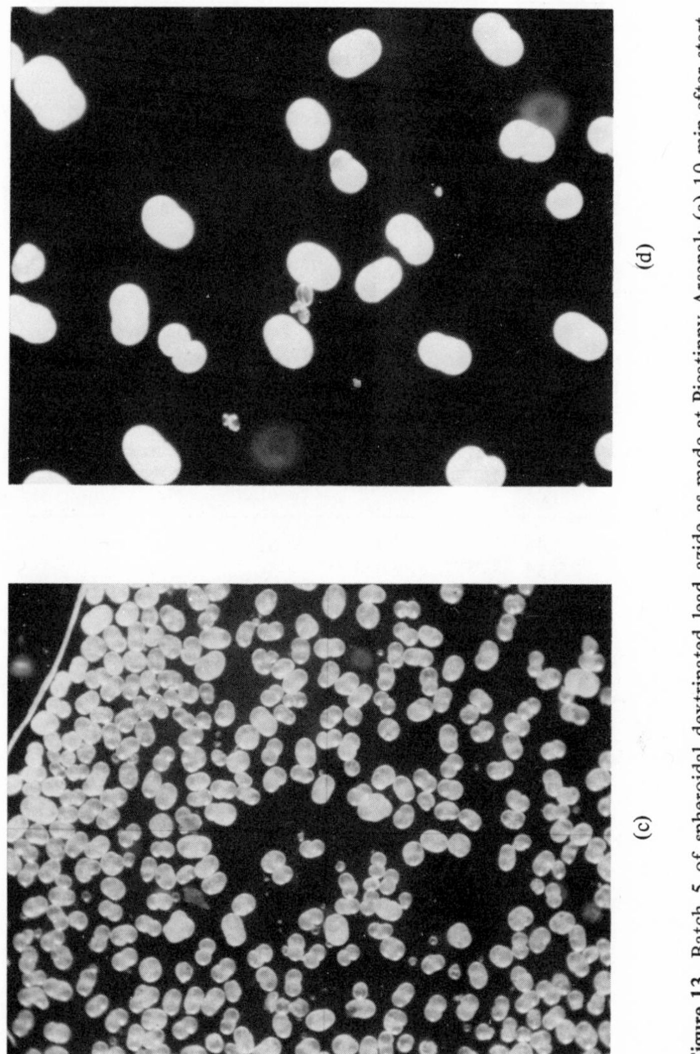

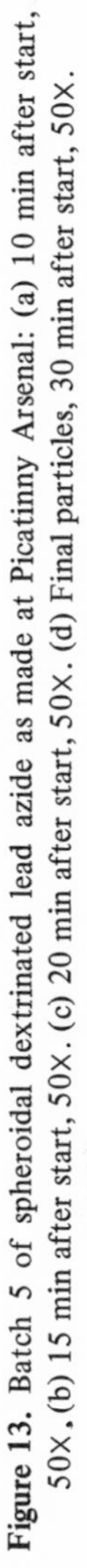

Figure 13. Batch 5 of spheroidal dextrinated lead azide as made at Picatinny Arsenal: (a) 10 min after start, 50×, (b) 15 min after start, 50×. (c) 20 min after start, 50×. (d) Final particles, 30 min after start, 50×.

Table II. Properties of Spheroidal and Conventional Dextrinated Lead Azide

Property	Batch number or product						
	1	2	3	4	5	DUP	KNK
Yield (%)	85.85	96.55	94.43	94.17	95.20		
Purity (%)	92.90	93.16	93.41	91.98	93.44	91.50	
Moisture (%)	0.46	0.61	0.57	0.66	0.74		
Hygroscopicity (%)	1.00	0.98	0.63	1.19	0.99	0.87	
Bulk density (g/ml)	2.15	2.17	2.21	2.12	2.27	1.86	
Sensitivity to impact							
Ball-drop (8.3 g/in.)	37	39	32	34	29	32	40
Impact (2 kg/in.)	7	5	5	4	5	7	
Explosion temp., 5 sec (°C)	301	302	301	299	308	294	301
Electrostatic sensitivity (J)	0.0032	0.0058	0.0053	0.0044	0.009	0.0096	

Table III. Detonator Functioning Tests of Spheroidal and Conventional Dextrinated Lead Azide

Batch number	Weight of detonator charges (mg)		Number tested	Number of high-order functions	Perforations (mils)		
	Lead azide	RDX			Max.	Min.	Avg.
1	80	48	53	53	274	215	247
3	65	55	50	50	278	244	255
4	65	55	50	50	268	212	246
5	50	62	50	50	271	188	246
DUP	95	40	52	52	253	204	235
KNK	140	20	56	56	272	170	206

5. Early CMC Lead Azide

The most important forms of lead azide are the products modified by sodium carboxymethylcellulose (CMC). The first of these was described in 1946 [1], and although the exact method of manufacture is no longer important, what is of interest is its stepwise development. Starting with the original Service lead azide process, CMC and other hydrophylic colloids were added in turn with generally satisfactory results. When the sodium carbonate was omitted, the product was greatly improved, particularly if CMC was used. Finally a usable product was made through adjustment of parameters such as the position of the feedlines, the time and rate of addition of reactants, etc. Also, a disparity between bench-scale experience (30 g) and plant-scale products (3 kg), emphasized the need for an intermediate step in the scale-up process. In fact, this experience seemed to be the beginning of the British development philosophy, mentioned earlier, with primary emphasis on well-defined process parameters.

6. RD1333 Lead Azide

Much has been said concerning the differences between the British and United States methods of specifying explosives such as RD1333 lead azide, but little said about the similarities. The British military specification CS2637 contains a statement, "The composition RD1333 shall be manufactured by a process which will receive authoritative approval." The United States specification MIL L-46225C requires only that "a performance test detonator loaded with lead azide RD1333 shall form a hole in the lead disk 0.156 inch diameter." In essence the two specifications are the same.

In practice in Britain the CMC lead azide is manufactured in a government ordnance factory, so process control is relatively easy. Even though no mention is made of a required output from the azide in the British specification, it is understood that it should perform to a certain standard. In the United States no process is specified for the lead azide, except that it be precipitated in the presence of CMC; the batch size is also specified. The process used by independent manufacturers closely resembles the British RD1333 process [21], and its significance is perhaps best illustrated by the fact that up to 1970 155,000 lb of azide was produced and utilized in the production of nearly one billion detonators and igniters. However, at the time (about 1964) when large quantities of RD1333 lead azide were being manufactured for the first time in the United States, there remained the question whether small process differences affect the safety and quality of the product.

In contrast to British experience with RD1333, its use in the United States was accompanied by numerous accidents that raised questions concerning differences between British and American practice and doubt concerning the flexibility available for the adjustment of process parameters. The situation was further complicated by the introduction by the British of a new military-grade material RD1343 [21] (discussed in the following subsection).

As a consequence, Hopper, Wells, and associates, during the years 1967-1969, conducted an investigation in which 34 one-pound batches of CMC-type azides were made to determine the effect (Tables IV and V) of varying key process parameters. The results have not been published previously, and the following summarizes the scope of the investigation:

Parameter varied	Range of variation[a]
Sodium carboxymethylcellulose	Five different types
Nonisol 100 wetting agent	0, 2, 21, 42 ml/batch
Agitator speed	100, 200, 300, 350 rpm
Time of addition of reactants	30, 40, 60 min
Temperature of reaction	20, 25, 30, 35, 40°C

[a]Normal level underlined.

Table IV. Characteristics of Carboxymethylcellulose (CMC)

Designation of the CMC tested	Moisture content (%)	Sulfated ash (%)	Water insoluble (%)	Lead equivalent (%)	Viscosity, 2% sol (cp)	Sodium chloride (%)	Degree of ether	Purity	Reaction of aqueous solution	
									pH of water	pH of solution
Brit. B50 powder low degree of sub.	11.1	26.5	0.26	24.4	420	3.13	0.55	93.2	7.0	6.8
Herc. 4MP "Bag 6"	10.8	17.2	0.0	20.3	730	0.27	0.56	99.3	6.3	7.0
Herc. 4M6F Lot 60095	7.2	15.7	0.02	13.7	1120	0.29	0.45	99.4	6.9	6.9
Herc. 4M6 Lot 63018	4.9	16.9	0.56	12.7	600	0.36	0.45	99.7	6.3	7.0
Herc. 4M6F Lot 63058	5.8	17.2	0.20	19.4	740	0.33	0.49	98.6	7.0	6.9
Herc. 4H1F Lot 63776	4.9	16.0	0.05	18.1	22,500	0.22	0.46	98.8	6.7	6.9
Herc. 4M71 Lot 63215	10.2	17.8	0.00	17.1	19,875	0.49	0.49	99.1	6.8	6.9
Herc. CMC3 Lot 1152-84	4.2	9.8	46.7	18.5	490	0.12	0.28	99.6	6.9	6.8
MIL-S-511132A required	10 max.	27 max.	1 max.	N.A.	240–325	2 Max.	0.5–0.8	97 min.	6–7	

The original method for making RD1333 lead azide [21] is given in the footnote below.*

The process used by Hopper was similar except in the scale of operation and the mechanical conditions of agitation. The addition of the reactant solutions was also automated by the use of precision metering pumps.

The first batches were concerned with various commercial varieties of sodium carboxymethylcellulose, as it became evident that the CMC used by Taylor *et al.* (CMC-NM83) could no longer be procured. After testing several candidates, Hercules-type 4M6F was used throughout the study as a standard of comparison. The results of tests performed on 4M6F-CMC and other CMCs are listed in Table IV, along with the latest specification covering CMC for use in making RD1333 for U.S. military purposes [13]. None of the CMCs used met the current U.S. specification, which was recently tightened to exclude CMCs with a viscosity outside the range 240–325 centipoise.

Of the CMCs tested, types B50, 4MP, and 4M6F produced a lead azide which met the requirements of the detonator functioning test at the 35-mg loading level. Curiously, the failing lead azides exhibited "dead pressing" properties not usually associated with lead azide (Chapter 5).

The wetting agent, Nonisol 100, used by Hopper was chemically equivalent to the Empilan AQ100 wetting agent used by Taylor *et al.* Through an error, three batches (19,20, and 21) were made with 1/10 the recommended concentration of Nonisol. Later, batches 22 and 23 were made with and without twice the concentration of Nonisol, respectively. Disregarding their bulk densities, the

*Process instructions for RD1333

Product Discrete granules of alpha-lead azide. Bulk density = 1.6 g/ml.

Plant Standard British Initiator equipment.

Ingredients
1. Sodium azide solution: 108 g NaN_3/liter.
2. Lead acetate solution: 315 g $Pb(CH_3COO)_2 \cdot 3H_2O$/liter.
3. Sodium CMC Solution: 10 g NaCMC/liter.
4. Empilan AQ100 solution: 100 g of 100% ethylene derivative of lauric acid/liter.

Process
1. Add to pan 1.2 liters of sodium CMC solution and 0.145 liter of Empilan AQ100 solution making up to a volume of 15 liters with water with stirring.
2. Maintain temperature of all solutions at 25°C.
3. Add simultaneously during a period of 60–70 minutes 14 liters each of the sodium azide and lead acetate solutions maintaining the stirring.
4. The product is washed with water by decantation, filtered and dried.

Yield 3.2–3.3 Kg lead azide; azide value 96.3.

Table V. Effect of Process Parameters on Lead Azide

Batch	CMC	Agitator (rpm)	Temp (°C)	Bulk density (g/ml)	Electrostatic sensitivity (J)	Functioning test[a] 15 K psi 35 mg	Functioning test[a] 15 K psi 25 mg	Functioning test[a] 1 K psi 35 mg
RD1333								
1[b]	MN83	200	25	1.33	0.0102	F	–	P
2[b]		300		1.49	0.0090		–	
3[b]				1.49	0.0078		–	
4[b]		350		1.46	0.0096		–	
5[b]		125		1.01	0.0360		–	
6[b]	7MSP	300		1.14	0.0048		–	F
7[b]	B50			1.48	0.0040	–	–	P
8[b]	4MP			2.11	0.0040	P	–	–
9[b]	B50			1.53	0.0029	–	–	
10[b]	4M6F			2.02	0.0048		–	
11[b]	4M6			2.04	0.0044		–	
12[b]	B50			1.89	0.0044		P	–
13[b]	4NO		37/38	1.40	0.0053	F	–	
14[b]	4M6F		25	2.05	0.0023	P	F	–
15[b]			20	2.00	0.0026			–
16[b]			30	2.13	0.0026		P	–
17[b]			40	1.88	0.0016		F	–
18[b]			35	1.90	0.0013		P	–
19[c]		200		1.88	0.0036		–	–
20[c]		100		1.11	0.0282		–	–
21[c]		350		2.13	0.0014		–	–
22[d]		300	25	1.95	0.0036		–	–
23[e]				2.09	0.0058		–	–
24[b]		100		0.85	0.0346	F	F	–
25[b]		200		1.77	0.0063	P		–
26[b]		350		2.00				–
27	CMC-3	300				F	–	–
28[f]	4M6F			2.09	0.0032	P	P	–
29[g]				1.81	0.0032	F	P	–
30	4H1F			1.05	0.0063		F	–
RD1343								
1	MN83	250		1.46	0.0038		–	F
2				1.85	0.0040		–	
3	7MSP			1.11	0.0048		–	
4	4MP			1.27	0.0032		–	
duPont RD1333				1.49	0.0040	P	F	–
Special purpose				1.46	0.0063		P	–

[a]F represents failure, P represents passing functioning test. Dash (–) indicates test not conducted.
[b]21 ml Nonisol 100 used.
[c]2 ml Nonisol 100 used.
[d]No Nonisol 100 used.
[e]42 ml Nonisol 100 used.
[f]Reactants were added over a 60-min period.
[g]Reactants were added over a 30-min period.

batches all had nearly identical properties. The bulk density of batches 1-5, 19-21, and 24-26 varied systematically, the obvious cause being the speed of agitation. From 200 rpm to 350 rpm the bulk density increased from about 1.9 to 2.1 g/ml. At 100-125 rpm the bulk density was only 0.85-1.1 g/ml. This effect was independent of the CMC used or the concentration of Nonisol. The change in agitator rpm may have affected the functioning characteristics of batch 24, because it did not function satisfactorily at the 35-mg level. However, it is evident from examining the electrostatic sensitivity data from batches 5, 20, and 24 that low rpm can be associated with less electrostatic sensitivity.

Comparing the products from batches 28 and 29 with batch 14, it can be deduced that varying the time of addition of the reactant solutions from 30 to 60 min does not have a significant effect. There is one anomalous result in that batch 29 did not pass the detonator functioning test at the 35-mg level but did pass at the 25-mg level.

The most significant result of the investigation was the effect of increasing temperature. Batch 16 conducted at 30°C and batch 18 conducted at 35°C passed the 25-mg detonator test. Only two other batches (28 and 29) passed this specification requirement. Batch 17 (at 40°C) did not pass at the 25-mg level.

In summary, the parametric study resulted in the following conclusions:

1. The sodium CMC is a critical parameter. Unless the product has the right viscosity and degree of etherification, the resultant lead azide is not of the highest quality.
2. The wetting agent, Nonisol 100, is not critical and can be omitted without affecting product quality.
3. Below a certain degree of agitation (i.e., 200 rpm) product quality deteriorated drastically, but increased agitation beyond the minimum brought only marginal changes in product properties.
4. Changing the time of addition or reactants from 40 min to 30 min or 60 min does not affect product quality.
5. It could be concluded that the reaction at 30°C or 35°C produced a better product than that from reaction at 25°C. (See also Special Purpose lead azide, Section C.7.b, below).

The investigation confirmed, therefore, that certain changes in the original process were advantageous. The vast majority of RD1333 lead azide procured in the United States in recent years was made in duPont-type reactors at 37°C using 300 centipoise viscosity CMC (see Section C.7, below). It should be noted that the tightening of the specification requiring 25 mg to pass the detonator functioning test, rather than 35 mg, was introduced to induce procurement of superior material made under the above conditions.

7. Other CMC Modified Lead Azides

While a number of different CMC lead azides may be encountered in industry, they are all microcrystalline agglomerates containing about 98% lead azide and less than 1% lead carboxymethylcellulose. As Hopper's study has shown, useful products can be obtained over a wide latitude of process parameters, but subtle differences can make one product considerably better than another.

a. RD1343 Lead Azide

In British ordnance applications RD1343 has been substituted for RD1333 and is made in an almost identical manner, except the wetting agent is omitted and the sodium azide solution is made more alkaline. The principal reason for the change was to obtain a process which did not cause a deposit to build up at the water line in the reaction kettle, leading to reduced cleaning requirements.

The Hopper study (see Table V) included making and testing four batches of azide designated RD1343. None of these products functioned as well as the corresponding RD1333 batches in the detonator test. Because of this result, the use of RD1343 process modifications has not been adopted in the United States. However, U.K. tests have been said to show RD1333 and RD1343 are equal in performance.

b. Special Purpose Lead Azide

The most extensive parameter modifications applied to the original process were those used in the United States to make a product known as Special Purpose lead azide (SPLA), which was made and used in very large quantities (over a million pounds) for antipersonnel mine applications. For this purpose an azide of greater sensitivity and lower hygroscopicity than dextrinated lead azide was needed, and U.S.-type RD1333 was found to be suitable. In order to produce the very large quantities mentioned, the individual batch size was increased from 3.5 to 7.7 kg (17 lb) and the time for addition of the much larger volume of reactant solution reduced from 4 min to 3 min. This resulted in a fourfold increase in output from existing equipment and also considerably reduced the cost of manufacture.

In the 100-liter working capacity duPont kettle the following procedure was used for making a 7.7-kg batch of Special Purpose lead azide:

Ingredients

1. Sodium azide solution: 108 g NaN_3/liter.
2. Lead acetate solution: 315 g $Pb(CH_3COO)_2 \cdot 3H_2O$/liter.
3. Sodium CMC solution: 10 g NaCMC/liter.
4. Nonisol* 100 solution: 100 g Nonisol/liter.

*Trademark Geigy Chem. Co. for an ethylene derivative of lauric acid.

Process instructions
1. Add to kettle
   25.6 liters water
   6.25 liters NaCMC solution
   0.34 liter Nonisol solution
   0.576 liter NaN_3 solution
2. Adjust temperature to 37 ± 2°C.
3. Add simultaneously 32 liters each of sodium azide and lead acetate solution during 30 min.
4. Wash twice by decantation using 50 liters of distilled water.
5. Transfer batch to cloth filter bag to remove excess water.
6. Pack in 50% ethanol/water solution for shipment to loading plant.

The specifications for Special Purpose lead azide given in U.S. military specification MIL-L-14758 are the same as those for RD1333 lead azide except that the detonator test and other minor requirements are dropped.

Of over one million pounds of Special Purpose lead azide produced, approximately half was not consumed but stockpiled. Later this stockpile material was tested for compliance with the RD1333 specification and was found to be suitable. After other tests, not covered in the specification and described elsewhere in this volume, the use of Special Purpose lead azide was authorized in place of RD1333 lead azide in the United States. Since there is such a large stockpile of Special Purpose lead azide, all U.S. CMC-type lead azide needs for several decades could be supplied by this material if it could be preserved for this length of time. The most interesting point brought forward here is that certain process parameters can be varied seemingly without affecting product quality. Others, such as CMC type, have to be held constant to maintain quality. Unfortunately, without empirical studies as conducted by Taylor *et al.* and Hopper, the knowledge to predict which parameters are critical is not available.

8. A Continuous Lead Azide Process

The overwhelming majority of lead azide manufactured for civil and defense applications is the product of batch processes. Continuous processes are not, in general, required because the small quantities used in detonators are easily furnished from the batches described in the foregoing, provided that the processing time is accommodated to daily plant schedules. Although lead azide is an expensive explosive, costing in the region of $20–25/kg, the milligram quantities used in detonators add little to the cost of a single detonator, and there is thus again little reason to achieve the ultimate in production rates. It is pertinent to mention, however, that at least two continuous processes have been devised, although little is known about the details of the processes.

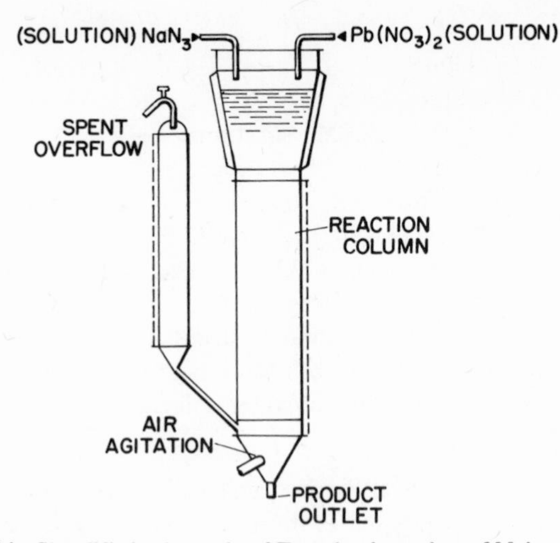

Figure 14. Simplified schematic of French adaptation of Meissner process.

One process, developed by the Bofors-Liab Company in Sweden leads to the precipitation of β-lead azide in a pure form. The precipitate is washed and allowed to stand in water overnight, where it recrystallizes to the α form. The product is claimed to be very pure but with a density less than that of Service lead azide.

The other process is the Meissner process, described in detail in Urbanski's Volume III [8]. The process as shown or as applied by the French [Atelier de Construction, Tarbes (ATS)] is continuous only during methathesis, the subsequent filtering, washing, and drying steps are batch operations.

In the manufacture of lead azide, ATS uses the Meissner process in which lead acetate is added to sodium azide and precipitated at a temperature of 30-40°C. This is described as a continuous process, and in this respect it contrasts markedly with the processes which are used in the United States and Great Britain. A further contrast is that they manufacture only a little CMC-stabilized material, concentrating their efforts on the production of pure or dextrinated products. In fact, in the last three years there had been no production of CMC material, as they are quite satisfied with the dextrinated product. A sketch of the process which they use for manufacture by the Meissner process is shown in Figure 14.

9. Gelatin Lead Azide

Animal gelatin can be used for crystal modification and desensitization of lead azide in a manner similar to dextrin. A product recently described by

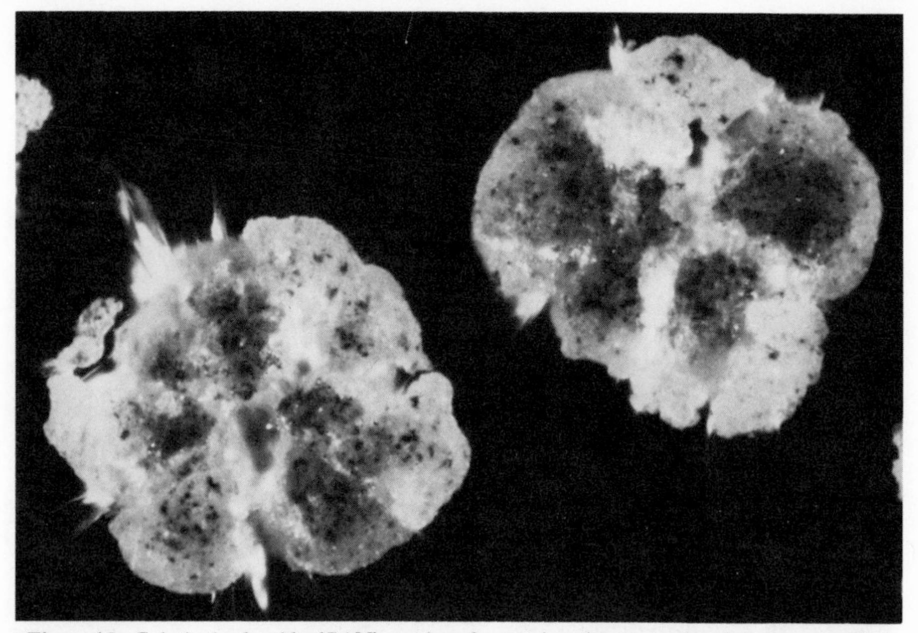

Figure 15. Gelatin lead azide (GAM) sectioned crystal agglomerate showing concentration of molybdenum disulfide in inner part of lobes, 308×. (Photo courtesy of ICI, Nobel's Explosives Company Ltd., Stevenston, U.K.)

Medlock [22] (Figure 15) contains, in addition, a small amount of molybdenum disulfide encapsulated and is called GAM for the initials of the principal constituents. No details of the process as used by Nobel's Explosives G. Ltd. (U.K.) were given, but since the product is somewhat weaker in output than dextrinated lead azide, it can be assumed that GAM is less pure than the dextrinate. It is mentioned that precipitation is carried out at elevated temperatures. Since the major claimed advantages for this product relate to safety in handling, it is described in more detail in Chapters 4 and 5.

D. SILVER AZIDE MANUFACTURE

1. Early Preparations

Silver azide was first prepared in 1890 by Curtius [14] by passing HN_3 into silver nitrate solutions. Various other methods (discussed in Volume 1, Chapter 1) were developed, but the most feasible is the reaction of the readily available sodium azide with silver nitrate.

Silver azide has a low solubility in water (variously reported as 0.001-

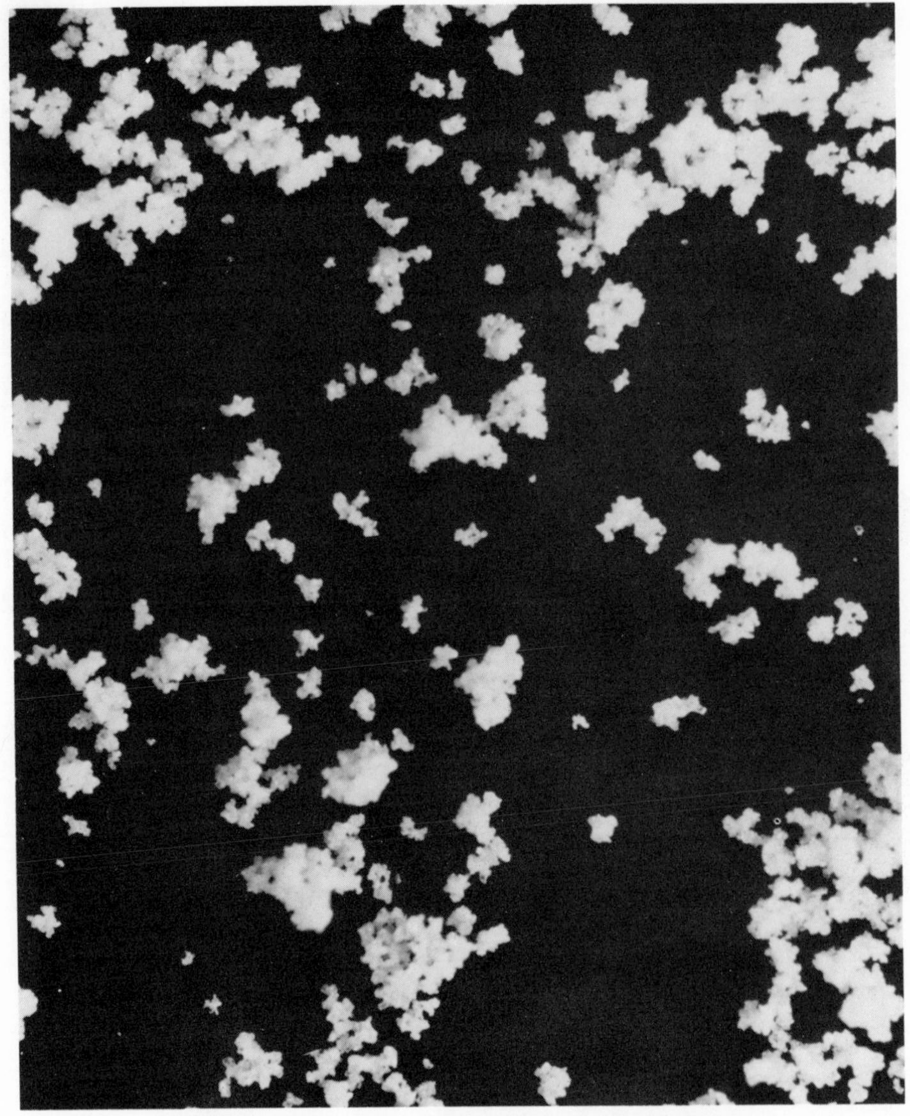

Figure 16. Silver azide made by British process [3], 60×.

0.008 g/liter [15-18]) and a tendency to nucleate profusely, so it normally precipitates as a fine powder. Taylor and Rinkenbach [2] reported a reduced sensitivity for finely divided silver azide, prepared by adding a 5% sodium azide solution to a 25% silver nitrate solution in stoichiometric proportions. However, later efforts were directed at producing silver azide in a coarsely granulated, free-flowing form which would be amenable to uniform compaction. In the early 1950s Williams and Peyton [4] developed a process for making silver azide by neutralizing ammoniacal solutions of silver azide with gaseous or solid carbon dioxide. Sodium thiosulfate and carboxymethylcellulose were used as crystal-modifying agents to induce formation of cubic crystals. Undocumented difficulties perhaps arose, because shortly afterwards another process [3] was developed in which silver azide was produced by slow, simultaneous addition of dilute sodium azide and silver nitrate solutions to a dilute ammonium hydroxide base, followed by a slow neutralization of excess ammonia with dilute nitric acid. The process, which produced free-flowing crystalline aggregates in the 100–200-mesh range (Figure 16), was subsequently adopted as the standard method of preparation in the Royal Ordnance Factories in the United Kingdom and in Swedish defense industries.

2. A New Process

Silver azide has not yet been used in the United States and has had only limited use in other countries where its superior chemical stability and detonation properties have mandated its use despite its cost. However, of late there developed a need for miniature explosive trains where it was likely that the more efficient silver azide would function where lead azide could not. Tests conducted by Costain (see Chapter 5) showed that silver azide was needed for proper functioning below certain critical weight–diameter–volume relations in small detonators. Therefore, an effort was undertaken to improve the existing British process by reducing processing time with consequent cost savings. The improvement program had progressed to the point where the productivity had been increased by a factor of nearly two when an entirely new approach was discovered. Using the new approach, a new method for making silver azide in a granular form suitable for detonator loading was devised [19].

The new process (Figure 17) consists of mixing concentrated solutions of silver nitrate, ammonium hydroxide, and sodium azide, and heating the resulting solution to drive off the ammonia and precipitate silver azide.

3. Semi-Pilot-Plant Production

The laboratory method was scaled up to the minimum amount that could be successfully processed in a 10-liter stainless-steel kettle (Figure 2), fitted with a 12-cm propeller-type agitator and a 2.5-cm-wide baffle. The reactants were: 2.00 liters of filtered 2 N (340 g/liters) silver nitrate solution, 2.00 liters of filtered 2 N (130 g/liters) sodium azide solution, 1.07 liters of filtered reagent-grade (28%) ammonium hydroxide.

The silver nitrate solution was transferred to the kettle, followed immedi-

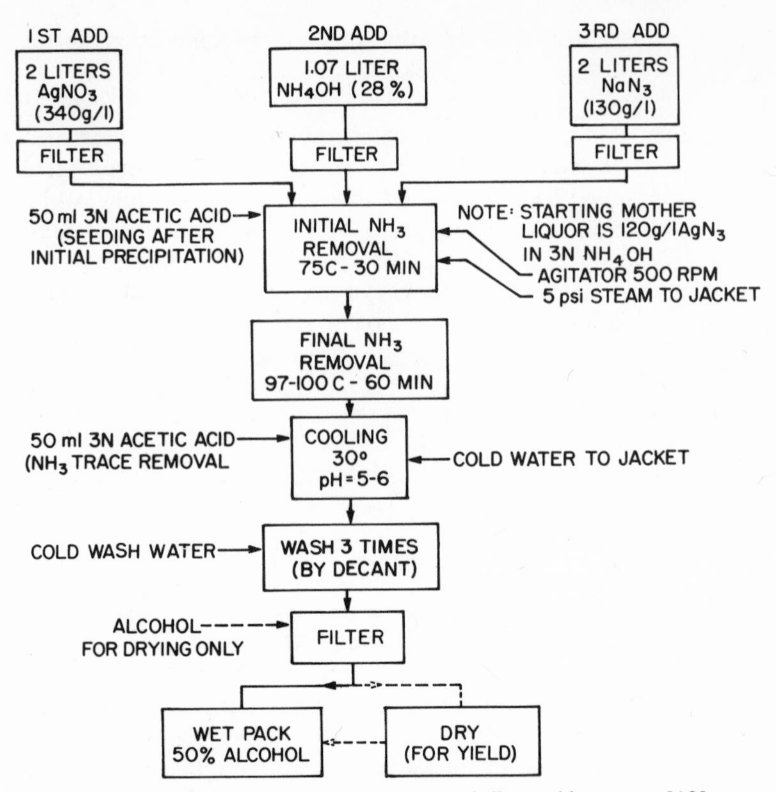

Figure 17. Flow chart for Picatinny Arsenal silver azide process [19].

ately by the ammonium hydroxide. The sodium azide solution was then added from a remote location while the contents of the kettle were agitated at 300 rpm. The agitation speed was increased to 500 rpm and 5 psi steam was introduced into the jacket to raise the temperature to 75°C within 5 min. Approximately 5 min later the clear solution in the kettle turned milky, indicating that silver azide had started to precipitate. At this point 50 ml of 3 N acetic acid was introduced over a 5-min period to induce seed-crystal formation. Distilled water was added at a rate of 40–60 ml/min to maintain the original liquid level and counteract evaporation.

When the total hold time at 75°C reached 30 min, the steam flow was inincreased raising the kettle temperature to 97°C within 10 min. Over the next 60 min the temperature in the kettle was raised to 99°C (BP water at 1000 ft above sea level), indicating that virtually all the ammonia had boiled off. The makeup water was stopped (3.5 liters had been added), and immediately thereafter cold water was circulated through the jacket. During the cooling period, 50 ml of the 3 N acetic acid was introduced into the kettle. After 10 min the

Table VI. Properties of Silver Azide

Property	Costain [19]	Taylor [2]	Lead azide [5][a]
Density (g/ml)	1.6	1.1	1.3
Granulation (%)		S[b]	
On 100	18–17		1
140	27–32		5
200	33–24		14
325	18–22		42
Thru 325	4–5		38
Purity (%)	99	99+	97–98
Hygroscopicity	Nil	Nil	Slight
Vacuum stability,			
1 g/40 hr at 150°C	0.49–0.34	0.40	0.40
Impact sensitivity			
Picatinny Arsenal 10%	11–7		7
50%	17–10		8
Ball-drop 10%	11–10		10
Electrostatic sensitivity (J)	0.0094–0.018	0.0094	0.0005

[a]RD1333.
[b]S, same as RD1333.

contents of the kettle were at 30°C. At this point agitation was stopped, the silver azide allowed to settle (very rapid), and the mother liquor was decanted. The product was washed twice by decantation using 3 liters of distilled water for each wash. Finally, the product was washed onto a cloth-bag-covered filter. A small amount of product remained crusted on the kettle walls, agitator shaft, and baffle at the water line.

After washing with 95% ethanol and drying for 40 hr at 50°C, the yield was 90% (540 g out of a possible 600 g). A flow chart of the procedure is given in Figure 17.

Two such batches were produced, giving essentially identical materials, as shown in Table VI and Figure 18.

4. Discussion of Process Parameters

The concentration and qualities of the reactants (i.e., silver nitrate, ammonia, sodium azide), as given, are not critical and could probably be varied by a factor of two. The concentration of the ammonia (presently 3 N) would have to be adjusted to maintain the solubility of the silver azide. For instance, a 200 g/liter concentration of silver azide in a 5 N ammonium hydroxide could easily be achieved as a starting mother liquor. In the present process the 120 g/liter concentration of silver azide represents a 10-fold increase in productivity over the Taylor process [3].

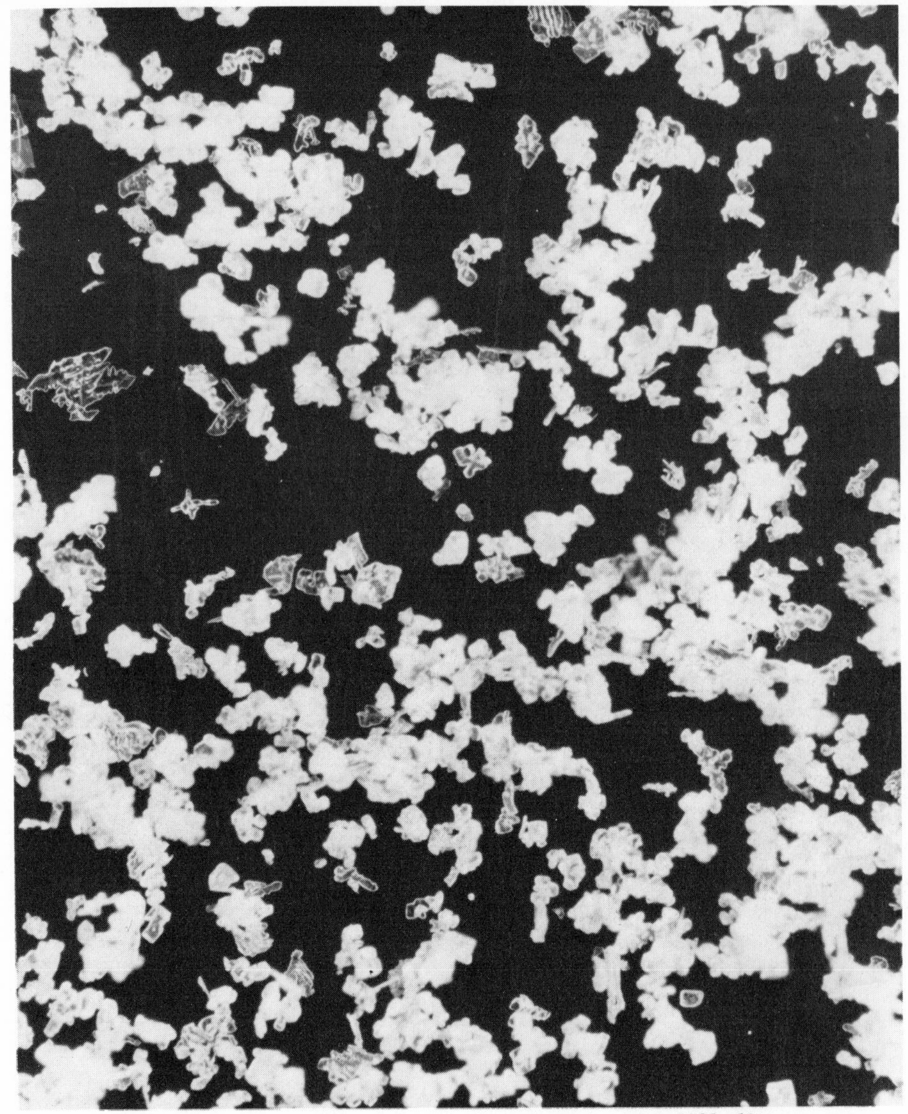

Figure 18. Silver azide made by Picatinny Arsenal process [19], 60×.

Vigorous agitation during evaporation is very critical. Preliminary tests showed that a reduction of bulk density and an increase in large flat crystals occurs unless the agitation produces turbulence, particularly top to bottom turnover.

The selected time–temperature program evolved directly from the avoidance of too rapid boiling, which could lead to frothing and loss of material. If the kettle were instrumented to indicate ammonia concentration, the temperature could be adjusted to follow the boiling point of the mother liquor more closely.

The quantity of acetic acid added for seeding was suitable for achieving uniform bulk density and desirable granulation range of the product. Silver azide is quite resistant to acetic acid. Its addition results in profuse nucleation.

The second addition of acetic acid serves the purpose of neutralizing the last traces of ammonia in the mother liquor and reduces losses of valuable silver in the mother liquor.

During the course of the many laboratory-scale experiments conducted, it was found that the addition of surface agents (i.e., anionic, cationic and non-ionic) did not improve the crystalline habit of the resultant product. An attempt to use an air sparger to simultaneously agitate the mother liquor and speed evaporation of the ammonia resulted in a rapid clogging of the sparger by precipitated silver azide.

Silver azides made by both the Taylor and the new process were compared and found to be chemically and explosively (see Chapter 5 and Table VI) the same. The product of the Costain process is coarser in granulation and of a different appearance, as can be seen in the microphotographs (Figures 16 and 18).

REFERENCES

1. G. W. C. Taylor, S. E. Napier, Phlegmatised Lead Azide, Armament Research Dept. Explosives Report 607/46, Woolwich Arsenal, London, England, 1946.
2. C. A. Taylor, W. H. Rinkenbach, *U.S. Army Ordnance*, *5*, 824 (1925).
3. G. W. C. Taylor, Br. Pat. 781,440 (applied for 1949, published 1957).
4. E. Williams, S. Peyton, Br. Pat. 887,141, Jan. 17, 1962.
5. W. H. Rinkenbach, U.S. Pat. 1,914,530, June 20, 1933.
6. U.S. Army Military Specification, MIL-L-3055A, Lead Azide, September 28, 1962.
7. B. T. Fedoroff, O. E. Sheffield, *Encyclopedia of Explosives and Related Items*, Vol. V, pp. A545 *et seq*. Picatinny Arsenal, Dover, N.J., 1972.
8. T. Urbanski, *Chemistry and Technology of Explosives*, Pergamon Press, Oxford, 1964.
9. J. Fleischer, J. G. Burtle, U.S. Pat. 2,421,778, June 10, 1947.
10. K. J. Holloway, G. W. C. Taylor, A. T. Thomas, Br. Pat. 1,045,271 (applied for 1962, published 1966); U.S. Pat. 3,173,818 (1965).
11. Military Specification MIL-D-003994A(MU), Dextrin Technical, August 12, 1968.
12. Official Methods of Analysis of the Association of Official Agricultural Chemists, 9th ed., Washington, D.C., 1960.
13. U.S. Army Military Specification MIL-L-51132A, Sodium Carboxymethylcellulose, Sept., 1964.
14. T. Curtius, *J. Chem. Soc.*, *60*(1), 57, 112 (1891).
15. J. Eggert, *Chem. Abstr.*, *48*, 1864 (1954); *Z. Naturforsch. 8b*, 389–95 (1953).
16. A. C. Taylor, *J. Am. Chem. Soc.*, *60*, 262 (1938); *Chem. Abstr.*, *32*, 3703 (1938).
17. E. H. Riesenfeld, *Chem. Abstr.*, *29*, 2897 (1935), *Z. Elektrochem*. *41*, 87–92 (1935).
18. R. E. J. Williams, ERDE Tech. Memo. 20/M/56 (1956).

19. T. Costain, A New Method for Making Silver Azide, Technical Report 4595, Picatinny Arsenal, Dover, N.J., February 1974; U.S. Pat. 3,943,235 (1976).
20. W. E. Voreck, Photomicrographic Examination of Explosives, Technical Report 4093, Picatinny Arsenal, Dover, N.J., August 1970.
21. G. W. C. Taylor, S. E. Napier, Br. Pat. 849,101 (1960); U.S. Pat. 3,291,664 (1966).
22. L. E. Medlock, J. P. Leslie, Some Aspects of the Preparation and Characteristics of Lead Azide Precipitated in the Presence of Gelatin. Paper No. 10, Int. Conf. Res. Primary Explos., Explosives Research & Development Establishment, Waltham Abbey, Essex, England, 17-19 March (1975).

2

Analysis of Azides for Assay, and in Complex Media

H. Kramer

A. INTRODUCTION

The chemical analysis of explosive azides is principally concerned with their azide content; however, there is occasional need for azide analysis in complex mixtures or in substances with which the explosive comes in contact during storage and application. The analyses are required because methods of manufacture are not rigorously defined, and products such as lead azide are contaminated with basic lead salts or other metals which hydrolyze during precipitation. The impurities may be incorporated in the lattice or deposited on the crystal surface. Although assay of relatively pure materials is of paramount importance, there is also a need for azide analysis in the stored elements of explosive trains and in bulk-stored lead azide to determine if decomposition or hydrolysis is taking place. In the United States and in countries that permit the transportation and storage of wet lead azide, surveillance of the stored materials is also necessary because it hydrolyzes in the water or alcohol–water desensitizing solutions, as discussed more extensively in Chapter 3. Single crystals and research samples require semimicro- or microanalytical techniques. Consequently, it is desirable to be able to analyze for azides at the macro and micro levels in solids, liquids, and gases.

Azides must also be analyzed for trace-metal content, such as copper, mercury, or thallium, because of the possibility of enhanced sensitivity. Other tests include analysis for grit, average particle size, and crystal morphology in conformance with stringent specifications. Table I gives some characteristics of commercial explosive azides prepared by processes described in Chapter 1 and of laboratory-prepared thallous azide and barium azide.

This chapter emphasizes the methods of analysis for azides in current use and the applicability, simplicity, accuracy, and precision of the methods. The information is practical in nature, with the physical and inorganic chemistry held to a minimum.

The analysis of azides was reviewed by Blay [1, 2] and Fedoroff *et al.* [3]. Useful analytical guides include the specifications issued by governments for the purchase of military-grade azides: for example in the United States, U.S. Military Specification, PA-PD 2825, February 5, 1966, or in Great Britain, Admiralty Specifications, A-163-J, Ministry of Supply, Chemical Inspectorate, Method E.C.3, Examination of RD1333, January 7, 1958.

In general, solutions of the azides are not very reactive to oxidizing and reducing agents at room temperature. Oxidation with iodine [4] proceeds only slowly in the presence of a catalyst; oxidation with permanganate [5] results in complex and variable products; hypochlorous and nitrous acids [6] bring about complete and rapid oxidation, but standard solutions of these reagents are unstable. The reaction of azides with ceric salts [7] in acid solution is quantitative, stoichiometric, and rapid. Homogeneous reduction of the azide ion in strongly basic media with titanous ion [8] is complex, with the formation of approximately one third of a mole of ammonia, two thirds of a mole of nitrogen, and traces of hydrazine per mole of azide. Heterogeneous reduction in alkaline solution with aluminum gives similar stoichiometric problems and also in the liberation of hydrogen. The argentometric determination of azides [9–11] in neutral media has been investigated because the azide group displays many reactions of the halides and haloids. The reaction with silver is quantitative, stoichiometric, and rapid.

The analyst has little choice in the selection of a redox analytical method. In acid solution ceric ion is the most effective oxidant; however, in acid solution there is possibility of analytical error because of the volatility of hydrazoic acid. In alkaline solution, reduction of the azide ion by homogeneous or heterogeneous reduction methods results in variable stoichiometry and cannot be used. The argentometric determination of azide ion in neutral solution is satisfactory but has one serious drawback: the manipulation of silver azide, a high explosive. Two basic approaches have evolved for the assay of azide ion:

1. Precipitation of silver azide from acetate solution and gravimetric determination of the silver as silver azide or silver chloride, or volumetrically by addition of an excess of standard silver nitrate solution and back-titration with standard ammonium thiocyanate using ferric alum as internal indicator.

2. Oxidimetrically with ceric ammonium nitrate using one of the three pos-

Table I. Properties of Explosive Azides

Properties	Lead azide				Experimental preparation	
	Colloidal	Special purpose (RD1333)	Dextrinated	Silver Azide	Barium azide	Thallous azide
Assay based on % azide	99.0+	98.5	91.5	99.9	99.9	99.9
Color, visual	White to buff	White to buff	White to buff	White	White to yellow	White to pale yellow
Form under microscope	Average particle 5 μ with largest 10 μ	Opaque, irregular, no translucent crystals	Average particle 70 μ, no needle-shaped crystals	Average size 105 μ opaque and translucent crystals	Rectangular platelike crystals	Maltese crosses and rectangular plates
Solubility in water (g/100g)	0.026 (20°C)	1.0	1.0	0.006 (20°C)	Slightly soluble in cold water; soluble in hot water	Slightly soluble in cold water; soluble in hot water
Solubility in other solvents	Ammonium acetate, dimethyl sulfoxide, triethanolamine hot 20% NaOH			Soluble in NH_4OH; KCN	Slightly soluble in alcohol/water	Slightly soluble in water
Concentration of deleterious contaminants allowed (%)	<0.002 Cu and Fe	<0.002 Cu and Fe	<0.002 Cu and Fe	None established	<0.00001 Cu and Fe found	<0.0001 Cu and Fe found
Phlegmatizing agent (%)	None	PbCMC, 0.6–1.2	Dextrins, 0.34–0.41	None	None	None

sible alternatives: (a) acidification of the azide sample, distillation of the hydrazoic acid into a known excess of standard ceric ammonium nitrate and a back-titration using standard ferrous ammonium sulfate solution with ferroin as internal indicator; (b) addition of a known excess of standard ceric ammonium nitrate to the azide sample directly and back titrating the excess ceric ammonium nitrate with standard ferrous ammonium sulfate solution as noted in (a); (c) addition of excess ceric ammonium nitrate and quantitatively measuring the nitrogen liberated.

B. QUANTITATIVE DETERMINATION OF THE AZIDE ION

1. Nitrogen Evolution Method

The most common assay method for lead azide used in the United States [12, 13] involves measurement of the nitrogen evolved during reaction with ceric ammonium nitrate according to the following equation:

$$Pb(N_3)_2 + 2(NH_4)_2Ce(NO_3)_6 \longrightarrow Pb(NO_3)_2 + 4NH_4NO_3 + 2Ce(NO_3)_3 + 4N_2$$

The nitrogen is measured in a water-jacketed gas buret as shown in Figure 1. The procedure works well with all inorganic azides and makes provision for scrubbing

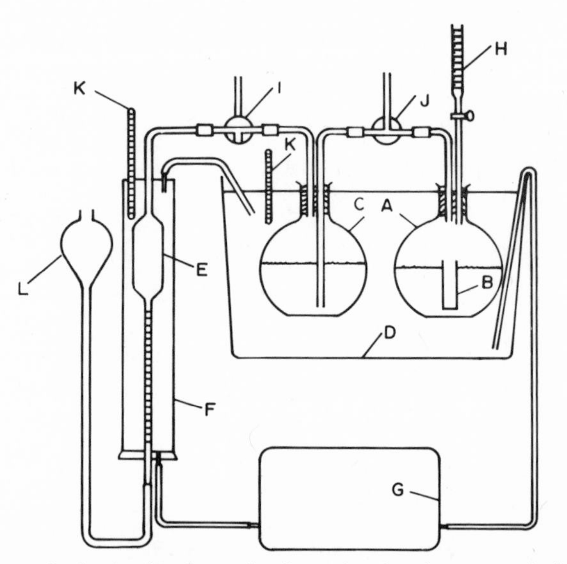

Figure 1. Apparatus for lead azide determination using the nitrogen evolution method [3]. A, 125-ml flat-bottom reaction flask; B, 1.4 × 4.5-cm vial; C, 125-ml flat-bottom absorption flask; D, water reservoir; E, 500-ml gas buret; F, water jacket; H, 50-ml buret; I, 3-way T-stopcock, position B; J, 3-way T-stopcock, position A; K, thermometers; L, leveling bulb.

carbon dioxide from the evolved gas. The carbon dioxide results from oxidation of the phlegmatizing agent or release of carbon dioxide from basic lead carbonate salts. The disadvantages of the method are the hazards of manipulating large quantities of lead azide and the complexity of the apparatus. The method can easily be modified for use with smaller-sized samples and simpler apparatus, as used for the determination of diazo nitrogen [14]. Details of the method follow.

a. Apparatus

The apparatus is shown in Figure 1. The buret, contained water saturated with nitrogen, is inserted into one hole in the rubber stopper of the reaction flask. Sodium hydroxide, 90 ml of a 10% solution, saturated with nitrogen is added to the 125-ml carbon dioxide absorption flask, and the gas buret and leveling bulb filled with a 0.1% solution of Nacconol (linear alkylate sulfonate), or equivalent, saturated with nitrogen. The temperature of the system is controlled by a circulating water pump between the water reservoir, which serves as a jacket for the reaction flask, and the glass jacket of the gas buret.

b. Determination

A weighed portion of approximately 1.7 g of the dried sample is transferred to the glass vial (Figure 1), 3 ml of water is added to the sample, and the vial is held erect in the reaction flask containing 75 ml of 15% ceric ammonium nitrate solution saturated with nitrogen. The reaction flask is connected to the apparatus together wth an absorption flask, using a coat of molten paraffin wax for all rubber-to-glass joints to ensure the absence of air leaks. The three-way stopcocks are opened to the atmosphere by adjusting them to position A (Figure 1, legend), and the water level in the gas buret is set to zero with the leveling bulb. After the system has come to a steady temperature (i.e., after about 10 min), the thermometer in the water jacket is read to 0.1°C. The stopcocks are turned to position B, and the reaction flask is shaken so that the vial is upset and lies horizontally on the bottom of the reaction flask. As gas is evolved from the reaction mixture, the leveling bulb is lowered so that the liquid level in the bulb is slightly below that in the gas buret. The reaction flask is gently agitated occasionally to complete the decomposition of the lead azide. When all the lead azide has decomposed, as indicated by the fact that gas bubbles are no longer forming in the mixture, the flask is filled with a measured volume of water (if necessary) from the water buret until the water level in the gas buret is between 45 and 50 ml. The temperature of the system is allowed to adjust itself to within 0.1°C of its temperature at the beginning of the determination, and the volume of gas in the gas buret is then measured at the atmospheric pressure, determined to the nearest 0.1 mm Hg with a mercury barometer and corrected to 0°C.

$$\text{Percent lead azide} = \frac{0.1558\,(A - B)(C - D)}{(273 + T)W}$$

where: A = ml of gas measured in gas buret; B = ml of water added to reaction flask, C = atmospheric pressure in mm of mercury; D = vapor pressure of water in mm of mercury at temperature T; W = weight of dry sample in grams; and T = temperature of water in the jacket surrounding gas buret in °C.

2. Ceric Ammonium Nitrate Titration

In this procedure [3, 12, 13] a known excess of standard ceric ammonium nitrate solution is added to an azide solution or slurry. The excess ceric ammonium nitrate is titrated with standard ferrous ammonium sulfate or sodium oxalate, using ferroin as indicator. The method is extremely simple and free from hazard once the reagents have been mixed. A serious drawback is that dextrin and polyvinyl alcohol are oxidized by ceric ion. Blay [1] reports gelatin and carboxymethyl cellulose are not oxidized. The method is as follows.

a. Reagents

An 0.025 M solution of *ortho*-phenanthroline-ferrous ammonium sulfate indicator is prepared by dissolving 1.487 g of *ortho*-phenanthroline monohydrate in 100 ml of an aqueous ferrous ammonium sulfate solution, containing 0.980 g of ferrous ammonium sulfate hexahydrate per 100 ml.

Dilute perchloric acid solution prepared from 60 ml of 70–72% perchloric acid solution is diluted to 1 liter with distilled water.

The standard ceric ammonium nitrate solution is prepared by diluting 60 ml 70–72% perchloric acid solution to approximately 700 ml with distilled water and adding this solution to 60 g reagent-grade ceric ammonium nitrate. The mixture is heated on a steam bath for 2 hr and stirred occasionally, then cooled to room temperature and diluted to a volume of 1 liter. After thorough mixing, it is allowed to stand for approximately 48 hr, filtered through a Gooch crucible, which has been ignited to remove any organic matter, and cooled.

The ceric ammonium nitrate solution is standardized against reference sodium oxalate, available from the National Bureau of Standards, in the following manner: Transfer a weighed portion of 0.25–0.30 g of the sodium oxalate to a 250-ml beaker, add 100 ml dilute perchloric acid solution, stirring the mixture until the salt is completely dissolved. The solution is cooled to room temperature, 2 drops of the *ortho*-phenanthroline-ferrous ammonium sulfate indicator is added, and the solution is titrated with the ceric ammonium nitrate solution until its color changes from faint red to faint blue.

The normality of the ceric ammonium nitrate solution is given by:

$$\frac{AW}{6.701B}$$

where A = purity in percent of the sodium oxalate; W = weight of sodium oxalate taken in milligrams; and B = ml of ceric ammonium nitrate solution used.

To prepare the ferrous ammonium sulfate solution 75 ml of 70–72% perchloric acid solution is diluted to 750 ml with distilled water; 39.21 g of $FeSO_4(NH_4)_2SO_4 \cdot 6H_2O$ is added, mixed, and made up to 1 liter.

The volume of standard ceric ammonium nitrate solution equivalent to 1 ml of ferrous ammonium sulfate solution is determined as follows. A measured portion (approximately 40 ml) of the standard ceric ammonium nitrate is placed in a 250-ml beaker and diluted to 100 ml with the dilute perchloric acid solution. Two drops of the *ortho*-phenanthroline-ferrous ammonium sulfate indicator is added, and the solution is titrated with the ferrous ammonium sulfate solution until a color change from light blue to red is produced. The volume of standard ceric ammonium nitrate solution equivalent to 1 ml of ferrous ammonium sulfate solution is calculated by dividing the volume of the former solution, which was taken for test, by the volume of the latter solution used for titration.

b. Procedure

A measured portion of about 50 ml of the standard ceric ammonium nitrate solution is mixed with 25 ml of dilute perchloric acid in a 250-ml beaker and cooled in an ice bath to about 5°C. A known weight (about 0.65 g) of dry lead azide in a 1-ml weighing bottle or porcelain crucible is dropped into the cooled solution and covered immediately with a watch glass in order to prevent loss of solution by effervescence. The mixture is stirred occasionally to dissolve the lead azide, after which the solution is removed from the ice bath. After the addition of two drops of the *ortho*-phenanthroline-ferrous ammonium sulfate indicator, the solution is titrated with the ferrous ammonium sulfate solution until a color change from light blue to red is produced. The percent lead azide in the sample, on a moisture-free basis is given by

$$\frac{14.56\,(A - BC)N}{W}$$

where A = ml of ceric ammonium nitrate solution taken; B = ml of ferrous ammonium sulfate solution required; C = ml of ceric ammonium nitrate solution equivalent to 1 ml of ferrous ammonium sulfate solution; N = normality of the ceric ammonium nitrate solution; and W = weight of dry sample taken.

3. Distillation Method for Azide Determination

To eliminate the problem caused by oxidation of the phlegmatizing agents, it is common practice to distill hydrazoic acid from the azide solution or slurry using perchloric acid. The hydrazoic acid (boiling point 35.7°C) is distilled into a known volume of standard ceric ammonium nitrate. The excess ceric ammonium nitrate is titrated with either standard ferrous ammonium sulfate or sodium oxalate. The distillation method is somewhat more lengthy and complicated than others; however, it gives reliable results and is not limited in its applicability. The method of analysis with distillation follows.

a. Apparatus

The apparatus shown in Figure 2 utilizes standard-taper ground-glass joints to connect the distilling flask, side-arm adapter, condenser, and adapter. The three-way side-arm adapter has a female joint at its upper end to which the buret is fitted with a No. 5 rubber stopper. A thin film of silicone grease is applied to all of the ground-glass joints. A portion of the lead azide sample is air-dried on a Buchner funnel and heated in an oven at 65-70°C until constant weight is obtained. A weight of 2-3 meq (0.2911-0.4366 g) of the dried sample is covered with water in a small porcelain glass crucible or boat. To prevent foaming, a small amount of Dow Corning Anti-Foam AF Emulsion, or equivalent, is added to the distilling flask before the introduction of the sample. The sample is transferred to the distilling flask, using rubber-tipped forceps, and the ground-glass joints of the apparatus are connected. A measured volume of (40-50 ml) of 0.1 N ceric ammonium nitrate, which is 2 N with respect to perchloric acid, is added to the Erlenmeyer flask. The flask is positioned so that the adapter from

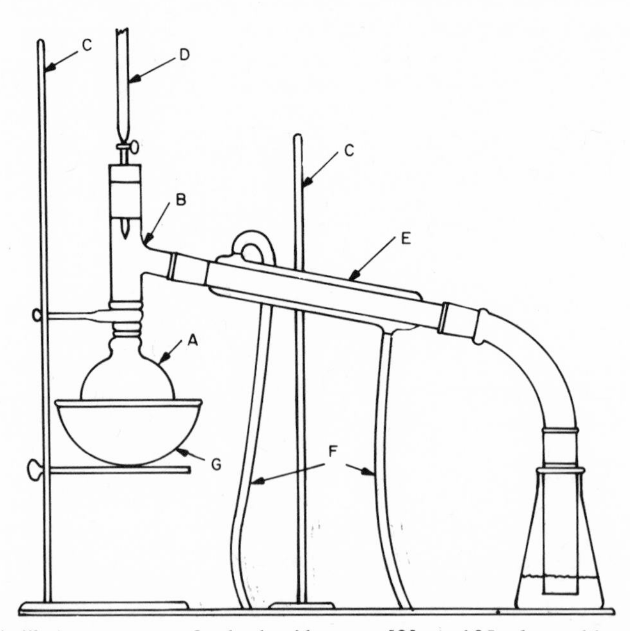

Figure 2. Distillation apparatus for lead azide assay [3]. A, 125-ml round-bottom distilling flask; B, side-armed adapter; C, clamp stands; D, buret; E, condenser; F, rubber tubing; G, heating mantle.

the condenser extends below the surface of the ceric ammonium nitrate and the buret filled with 3 N perchloric acid. The distillation flask is heated before adding perchloric acid. After checking that all glass joints are sealed, the stopcock of the buret is opened and 50 ml of 3 N perchloric acid is cautiously added. Distillation is allowed to proceed for approximately 12 min. The adapter is disconnected and rinsed thoroughly with water which is collected in the receiver flask. Two drops of a 0.025 M solution of *ortho*-phenanthroline–ferrous ammonium sulfate indicator is added to the receiver flask, and the excess cerate is titrated with 0.1 N sodium oxalate to the endpoint, denoted by a sharp color change from red to pale green-blue. A blank determination is made on the reagents and a correction applied if necessary. The ceric ammonium nitrate solution is standardized by titrating with the standard oxalate solution, and the percent lead azide is given by

$$\frac{(B - AN)\,14.56}{W}$$

where B = volume of sodium oxalate solution used for blank in ml; A = volume of sodium oxalate solution used for sample in ml; N = normality of sodium oxalate solution; and W = sample weight in g.

4. Argentometric Determination of Azide Ion

The procedure is essentially the classical Volhard titration, where azide ion is precipitated in neutral solution by excess silver nitrate solution, silver azide is filtered off, and the excess silver ion is determined in strong acid solution with standard ammonium thiocyanate, using ferric alum as an internal indicator. The method is favored by the British Chemical Inspectorate [15] and is required in the British military specification for lead azide. In the procedure, the precipitation, filtration, and destruction of silver azide are detailed. Blay [1, 15] reports the method rapid and reasonably safe for routine control analyses of lead azide. The method gives slightly low values for azide content compared to the gasometric procedure, but this may be due to silver azide peptizing and running through the filter paper. Addition of nitrobenzene to form a protective colloid around the coagulated silver azide is not effective in retaining the precipitate on the filter paper. The endpoint with ferric alum is best seen under fluorescent light.

a. *Standardization of Ammonium Thiocyanate*

Solution Required. A 0.1 N solution is prepared by dissolving 17.1 g of silver nitrate in chloride-free distilled water and diluting to 1 liter in a volumetric flask. The silver nitrate solution is standardized using accurately weighed 0.1-g portions of dried reagent-grade sodium chloride. The factor [F] is calculated from the chloride consumption. The standardized solution is stored in a dark bottle.

Preparation. The silver nitrate is ground in a clean mortar, dried at 140–150°C for 2 hr, cooled in a desiccator, and immediately weighed. It is advisable to dry only sufficient material for the preparation of one lot of solution. The factor for the silver nitrate will not alter significantly over a period of one month, provided it is stored in well-stoppered, dark-glass storage bottles.

Standardization. A 50-ml aliquot of the silver nitrate solution is added to 5 ml of concentrated nitric acid (free from nitrous acid) and 2 ml of a saturated solution of ferric alum. The silver nitrate solution is titrated with the ammonium thiocyanate solution until a fleeting reddish endpoint is reached. The strength of the ammonium thiocyanate solution is calculated from the titer.

b. Determination of Azide

A 0.5-g sample is weighed using a varnished cardboard scoop and transferred to a 500-ml conical flask containing 100 ml of 20% (w/v) ammonium acetate (pH 6.5–8.0) at 40°C. The mixture is swirled gently until all the lead azide has dissolved, and 50 ml 0.1 N silver nitrate is added. The flask is swirled to coagulate the silver azide precipitate. The precipitate of silver azide is cooled to 25°C and filtered on a Buchner funnel through two thicknesses of No. 42 Whatman paper, repeating if required, until the filtrate comes through clear. The precipitate on the filter paper is washed off with approximately 50 ml distilled water. The combined filtrate and washings are added to 20 ml of concentrated nitric acid (free from nitrous acid) and 2 ml of saturated ferric alum and titrated with standardized 0.1 N ammonium thiocyanate solution (T_1). A blank using the same volume of the same reagents in the same order (T_2) is carried out. The percent lead azide is determined by

$$\frac{[(T_2 - T_1)N]\,14.56}{W}$$

where N is the normality of the ammonium thiocyanate solution, and W is the weight of dried sample. Ammonium azide may be lost if the ammonium acetate solution used to dissolve the lead azide is heated above 50°C.

c. Destruction of the Silver Azide

On completion of the determination, the silver azide on the filter paper must be destroyed as follows:

The silver azide is transferred by a jet of water from the filter to the destroying vessel containing 30 ml sodium nitrite solution. The filter paper is removed using a rubber-tipped wooden rod and placed in the sodium nitrite solution. The funnel is washed with distilled water and the washings collected in the destroying vessel. Twenty ml of nitric acid solution are added, and the solution is stirred gently. After a few minutes the solution is washed down the sink.

The destruction solution is made up as follows: (a) nitric acid solution: 200 ml of conc. nitric acid (SG 1.42) per liter, and (b) sodium nitrite solution: 100 g sodium nitrite per liter.

5. Lead Determination

The total lead content of the sample may be determined by taking a known weight (about 1 g) of the dried sample in a 400-ml beaker, adding 50 ml of a saturated solution of ammonium acetate, and dissolving the azide by stirring and warming. The solution is diluted with 200 ml of distilled water and heated to boiling, after which 10 ml of a 10% solution of potassium dichromate is added slowly and with rapid agitation. The mixture is allowed to digest on a hot plate or steam bath for 1 hr, with frequent stirring. The precipitate of lead chromate is filtered through a tared crucible, washed with hot distilled water, dried for 2 hr at 100°C, then cooled in a desiccator and weighed. The weight of the precipitate as percent lead in the sample on a moisture-free basis is as follows:

$$\frac{64.108A}{W}$$

where A = weight of lead chromate, and B = weight of dry sample.

6. Comparison of Analyses Given by Different Methods

Comparative analyses of the various methods are presented in Table II. From the data available the gas-evolution method is generally the most precise and least subject to error once the operator has developed the manipulative skill and experience. It tends to yield the highest azide contents, particularly for lead azide. Loss of hydrazoic acid is minimal, and provisions are made for collecting any carbon dioxide generated by oxidation of the phlegmatizing agent. The 2-g

Table II. Percent Purity of Commercial Azides

Material		Ceric ammonium nitrate		
	Nitrogen evolution	No distillation	Distillation	Argentometric
Service lead azide [2]	96.87 ± 0.18	96.58 ± 0.25	ND	96.65 ± 0.32
RD1333 lead azide [2]	98.06 ± 0.12	98.63 ± 0.40	ND	97.75 ± 0.53
RD1333 lead azide [16]	98.7[a]	ND	ND	97.8[a]
RD1343 lead azide [16]	98.7[a]	ND	ND	97.0[a]
Dextrinated lead azide [12]	93.81 ± 0.12	ND	93.62 ± 0.15	93.52 ± 0.14
PVA lead azide [12]	96.66 ± 0.10	ND	96.55 ± 0.18	96.44 ± 0.18
Sodium azide [2]	99.42 ± 0.25	99.46 ± 0.35	ND	99.62 ± 0.18
Sodium azide [16]	99.89[a]			99.65[a]

[a]Data represents average of at least 6 determinations but no standard deviation reported.

sample of lead azide utilized is a serious hazard, but as soon as the sample is in the apparatus, operator risk becomes a minor problem.

The titration of excess standard ceric ammonium nitrate solution after addition of azide either as a slurry, solution, or as distilled hydrazoic acid gives precise and accurate results. The ferroin endpoint is sharp and virtually instantaneous. Loss of hydrazoic acid in the distillation is apparently nil, and the reaction of hydrazoic acid with ceric ion is extremely rapid.

In Table II the results are presented of investigations conducted in the United Kingdom [2] and the United States [15, 21] to compare the data for azide ion content of various commercial azides as determined by the above four methods. Different samples were used for each of the investigations, and thus it is not possible to make direct comparison with respect to the reproducibility of data from investigation to investigation. Despite the precise results, manipulation of azides in acid solution is always subject to error through possible loss of hydrazoic acid; it is preferable to work in neutral solution.

The argentometric determination requires a minimum of equipment, but time must be spent on standardization of solutions. The silver azide-precipitation procedure tends to give low results for commercial lead azides. Possible sources of this problem are coprecipitation of lead or carboxymethylcellulose. Surface charge on the silver azide may cause peptization, allowing the precipitate to run through the paper and giving low results. Although the endpoint is a deep-red-colored complex, vigorous shaking is required between additions of thiocyanate. This is because silver ions are adsorbed on the precipitate and are desorbed only slowly.

C. DETERMINATION OF MICROQUANTITIES OF AZIDE ION

The determination of small quantities of azide and the analysis of single crystals require sensitive but nonsubjective methods of analysis. This section discusses relevant methods of sample preparation and analysis as they relate to small samples; the next section discusses techniques suitable for trace-element analysis.

1. Methods of Separation for the Azide Ion

Techniques are required for the concentration of small quantities of azide ion in solution or for separating the azide ion from interferences. There are several approaches:

1. Silver nitrate-carrier precipitations can be used advantageously to collect microgram to milligram quantities of azide ion by coprecipitation, using 10 mg chloride ion as a carrier in dilute acid solutions. Although no data have been reported using the technique, it cannot be excluded on a theoretical basis.

2. Distillation of hydrazoic acid from strong acid solutions is the most common method of separation [12, 17]. For small quantities of azide ion in relatively large volumes of solvent, evaporation in alkaline media or carrier precipitation is necessary for the preliminary concentration of azide ions prior to distillation. The distillations are usually made from perchloric acid solutions. A 50-ml round-bottom flask, with a side-arm attached (so that a stream of inert gas passes through the solution) and an air condenser is preferable for distilling hydrazoic acid. The addition of a diluent carrier gas provides added safety. Absorbing solutions for the hydrazoic acid include known, excess quantities of ceric ion, standard base, or a known quantity of ferric perchlorate for colorimetric determination of small quantities of azide ion. The distillation separation is complex and lengthy, but the method is reliable and has universal applicability.

3. A useful method for the separation of hydrazoic acid is by a column extraction technique using a mixed-bed ion-exchange resin, a strongly acidic resin in the [H^+ form], and a weakly basic resin in the (OH^- form). All cations and most anions are held on the column while hydrazoic acid runs through the column. Other cations and anions elute as water. Weak acids, e.g., boric, silicic, and carbonic will also run through the column. The technique has not been applied to the analysis of explosive azides; however, it has been used for the analysis of alkali azides and for the preparation of standard solutions of hydrazoic acid [18].

4. Solvent extraction of the slightly ionized mercury (II) azide complex, HgN_3^+, or neutral $Hg(N_3)_2$ in the pH range 4-6 has been investigated. With *n*-butanol the distribution coefficient is 10.6 at 25°C. The complex obeys Beer's law at 248 nm in both aqueous and butanol solution [23].

5. The dissolution of lead azide in excess hot sodium hydroxide, forming lead plumbate ion and sodium azide, is an excellent and safe method for preparing lead azide for analysis by removing many inorganic interferences and destroying some organic material. For example, it is used in the analysis of a primer composition (NOL-130), containing lead styphnate, lead azide, tetracene, antimony sulfide, and barium nitrate; heating this mixture in excess 20% sodium hydroxide destroys the lead styphnate and the tetracene. Barium hydroxide and antimony sulfide precipate, and lead plumbate ion, sodium nitrate, and sodium azide remain in solution [23].

2. Colorimetric Methods for Determination of Azide Ion

There are two general colorimetric methods of determining small quantities of azide ion, either directly as the ferric azide complex or indirectly by addition of a known excess of oxidant such as ceric ammonium nitrate or nitrous acid. The excess oxidant is determined colorimetrically.

The determination of azide ion as the ferric azide complex is well documented [17,19]. The two ions form a 1:1 complex with a molar absorptivity of 3.68 × 10^3 at 460 nm. The net reaction is

$$Fe^{3+} + HN_3 \rightleftharpoons FeN_3^{2+} + H^+$$

indicating the reaction proceeds only to a small extent at high hydrogen concentrations. The optimum pH lies between 1.7 and 2.4, and most investigators adjust the pH of the iron solution to a value in this range. The detailed procedure follows.

a. Apparatus

A simple distillation apparatus consisting of a 10-ml round-bottom microflask with a 4-cm neck fitted with a one-hole rubber stopper is fabricated. The latter supports a glass tube drawn out at one end to approximately 0.4 mm inside diameter on the receiver end. The tube is bent so that it will extend to the bottom of a 50-ml volumetric flask. Absorption measurements are made with matched 1-cm Corex cells, or equivalent, using a prism spectrophotometer. Any final pH adjustments on the test solution are made with a pH meter.

b. Reagent Solutions

A reagent solution 0.011 M in iron(III) is prepared by dissolving 0.6142 g of electrolytically pure iron wire in 75 ml of 6.0 M perchloric acid. The solution is heated below boiling until the volume is reduced to 25 ml, transferred to a 1-liter volumetric flask, and diluted to the mark with distilled water, giving a solution 0.01 M in iron(III). Test for iron(II) and for chloride ions with potassium ferricyanide and silver nitrate, respectively, should be negative.

A solution 0.011 M in azide ion is prepared by dissolving 0.7150 g of commercial purified sodium azide in 1 liter distilled water.

c. Procedure

A 25-ml aliquot of 0.11 M ferric perchlorate solution is pipetted into a 250-ml beaker marked to indicate the 100-ml level, and 25 ml of distilled water is added. The pH is adjusted to 2.2 with 0.1 N perchloric acid or 0.1 N sodium hydroxide (carbonate-free), using a pH meter. A sample containing no more than 10 mg of azide is distilled into the ferric perchlorate. The solution is mixed and diluted to 90 ml with distilled water, the pH is checked, and the solution is transferred to a 100 ml volumetric flask and diluted to volume. The absorbance is measured at 460 nm against a ferric perchlorate blank. The amount of azide in the sample is calculated from a calibration curve made from standard sodium azide solutions ranging from 0.5 to 10.0 mg azide per 100 ml.

The method described was tested on primer composition, NOL-130. The composition was decomposed in hot excess 20% sodium hydroxide and analyzed without a distillation (23). Seventeen 1-g samples were decomposed, and the lead azide averaged 17.7% with a standard deviation of ±0.50%. The volumetric gas procedure gave a lead azide concentration of 18.2 ± 0.20%.

d. *Indirect Colorimetric Procedures*

Indirect colorimetric methods for the determination of azide involve the decrease in absorption at 390 nm of known excess volumes of ceric ammonium nitrate added to azide solutions [20]. A calibration curve is prepared by adding known quantities of azide ion to the same volume of ceric ammonium nitrate. Similarly, azide ion may be determined at concentrations of 10^{-7} M by the reaction of azide ion with a known excess of nitrous acid and then colorimetrically determining the excess nitrite ion [21]. However, these methods are not recommended because of the possibility of interference from any other reducing agent present. The methods are, nonetheless, reliable, if the azide ion can be separated from all possible interferences.

3. Determination of Hydrazoic Acid by Gas Chromatography and Other Methods

Trace quantities of hydrazoic acid at levels of 10^{-7} M can be determined by gas chromatography [22]. The sample is freeze-dried in alkaline media on a vacuum line, sulfuric acid is added to the azide sample without breaking vacuum, and helium is then used to sweep the hydrazoic acid through a Linde No. 5 molecular sieve powder, binder-free, packed in a 2-mm glass capillary column. The retention time for a 6-ft column and a 50 ml/min helium flow rate at ambient room temperature is about 15 min. A thermal conductivity cell with gold-plated filaments is used as a detector.

As little as 1 mm of hydrazoic azid [23] in a volume of 250 ml has been determined on a vacuum line by passing the hydrazoic acid through a packed tube of ceric ammonium nitrate, which had been evaporated on an asbestos fiber substrate, dried, and ground with mortar and pestle. The packing effectively reduced hydrazoic acid to nitrogen, which can be measured on a calibrated Toepler–McLeod gauge.

Both the above techniques require standardization using a known quantity of sodium azide carried through the same procedure. A semimicro gas-evolution method has been described by Blais [24] which utilizes only a few milligrams of lead azide. Nitrogen is determined by reduction of hydrazoic acid with heated cupric oxide.

a. Colorimetric Method for Azide Ion by Reduction to Ammonia

A Kjeldahl procedure which reduces the azide ion to ammonia followed by colorimetric determination of the ammonia has been described [25].

The method has the advantage of using small quantities of material, but has the drawback in that one third of a mole of ammonia is produced for each mole of hydrazoic acid.

D. TRACE ELEMENTS IN EXPLOSIVE AZIDES

Impurities may have a profound effect on the chemical and physical properties of primary explosives. Of particular importance are metals, such as copper, that sensitize lead azide. An increased knowledge of the effect of impurities on explosive properties emphasizes the need for accurate determinations of these impurities.

For instrumental analysis explosive azides must be converted to the chlorides or nitrates using semiconductor-grade acids. The sample is dissolved in semiconductor-grade nitric acid and ashed. The lead nitrate is then analyzed using flame, emission, X-ray, or mass spectroscopy; neutron activation analysis; or X-ray

Table III. Analysis of Lead Azide

Element	Single crystal[a]	Commercial-grade CMC
B	2.6	2.9
F	2.1	ND
Mg	3.5	58
Al	11	190
S	45	26
K	7.6	17
Ca	22	45
Ti	2.9	ND
Cr	0.85	2.5
Fe	7.8	62
Co	0.71	ND
Ni	1.1	3.5
Cu	0.30	7.0
Zn	4.3	13
Br	1.4	14
Ce	2.4[b]	ND
Si	ND	134
Bi	ND	6.4
Mn	ND	11

[a] As prepared by the method described in Chapter 2, Vol. 1.
[b] Room contaminant.

fluorescence, to mention only a few instrumental approaches. For accurate work suitable standards must be prepared, e.g., lead nitrate with known concentrations of elemental impurities of interest. As an example analyses of lead azide by spark-source mass spectrometry [23] are presented in Table III.

E. FUTURE RESEARCH

In the science and technology of explosives the lack of definitive chemical characterization including elemental impurities and molecular impurities has in some cases proved to be a stumbling block in the improvement of explosives behavior. Indeed, it should be axiomatic that an investigator should completely characterize an explosive if an in-depth understanding of explosive behavior is to be gained. Only under the urgency of World War II were silicon and germanium of sufficiently high purity obtained to demonstrate clearly the vital importance of trace impurities in semiconductor technology. The extrapolation to explosive semiconductors may not be entirely valid; however, the oftimes erratic behavior of lead azide make it imperative to study all facets of its chemistry.

There is also a definite need for standard explosive reference materials, such as lead azide and sodium azide. Standards of this nature would more easily permit the interchange of data, and round-robin analysis of samples among international laboratories would serve a useful purpose.

REFERENCES

1. N. J. Blay, Proceedings of the Symposium on Lead and Copper Azides, ERDE, Waltham Abbey, Essex, England, October 25–26, 1966, Ministry of Technology Report No. WAA/79/0216 Paper B-4.
2. N. J. Blay, Methods for the Assay of Lead Azide for Specification Purposes, ERDE, Waltham Abbey, Essex, England, Ministry of Technology, Report No. ERDE 11/M/67, December 28, 1967.
3. B. T. Fedoroff, H. A. Aaronson, E. F. Reese, O. E. Sheffield, G. D. Cliff, *Encyclopedia of Explosives and Related Items*, Vol. I, Picatinny Arsenal, Dover, NJ., 1960.
4. F. Feigl, E. Z. Chargaff, *Z. Anal. Chem.*, *74*, 376 (1928).
5. J. H. Van Der Meulen, *Rec. Trav. Chim.*, *67*, 600 (1948).
6. J. F. Reith, J. H. A. Bowman, *Pharm. Weekbl.*, *67*, 475 (1930).
7. J. W. Arnold, *Anal. Chem.*, *17*, 215 (1945).
8. T. A. Richter, W. Fisco, private communication (1974).
9. M. Marcquerol, P. Loriette, *Bull. Soc. Chim. France*, *23*, 401 (1918).
10. R. Haul, G. Uhlen, *Z. Anal. Chem.*, *74*, 376 (1918).
11. S. M. Moskovich, *Ber. Inst. Phys. Chem. Akad. Wiss. U.S.S.R.*, *6*, 179 (1936).
12. R. Croom, F. Pristera, Investigation of Methods for the Analysis of Lead Azide, Technical Rep. No. 2486, Picatinny Arsenal, Dover, N.J., 1958.

13. A. R. Lusardi, A Manual of Laboratory Procedures for the Analysis and Testing of Explosives and Pyrotechnics, Picatinny Arsenal, Dover, N.J., 1955.
14. S. Siggia, *Quantitative Organic Analysis*, 3rd ed, Wiley, New York, 1963, p. 545.
15. Ministry of Supply, Chemical Inspectorate, CIP/353, Method EG3, 1.7.58 Test 5.8, Woolwich Arsenal, England.
16. F. Pristera, Unpublished Report No. 66-VG1-177 December 29, 1966, Picatinny Arsenal, Dover, N.J.
17. C. E. Roberson, C. M. Austin, *Anal. Chem.*, *29* 854 (1957).
18. D. Bunn, F. S. Dainton, S. Duckworth, *Trans. Faraday Soc.*, *57*, 1131 (1961).
19. R. J. Rapley, Ministry of Supply, ERDE TN 29, Waltham Abbey, Essex, England, January 1971.
20. H. G. Higgs, H. Batten, unpublished, ERDE, Waltham Abbey, England, 1971.
21. J. Sutherland, H. Kramer, *J. Phys. Chem.*, *71*, 4161 (1967).
22. D. Christman, unpublished, Brookhaven National Laboratory, Brookhaven, N.Y.
23. H. Kramer, unpublished, Picatinny Arsenal, Dover, N.J.
24. M. Blais, *Microchem. J.*, *7*, 464 (1963).
25. L. P. Pepkowitz, *Anal. Chem.*, *24*, 400 (1952).

3

Handling, Storability, and Destruction of Azides

B. D. Pollock, W. J. Fisco, H. Kramer, and A. C. Forsyth[†]

A. INTRODUCTION

This chapter is concerned with the handling and storage of azides, particularly lead azide, in the laboratory and in bulk. The two principal requirements are that the material reach the end of this part of the "trail" in serviceable form, and above all, that it do so without harm to personnel. The sensitivity and explosivity of these substances render the subject of safety of major concern, not only in the plant but also in the laboratory. The latter is also true because a continuing program of research and development constantly presents new candidate materials and processes of unpredictable behavior. Also a lively interest in industrial hygiene and pollution abatement combine with the need for a well-considered and comprehensive safety philosophy. For these reasons Section B below is devoted to a description of safety and health hazards and recommended means of dealing with them, especially in the laboratory. In view of the authors' personal experience and the sometimes tragic experience of friends and colleagues, this emphasis may be well understood.

In normal practice it is a matter of principle to manufacture only limited-size batches, sufficient for immediate needs. Ideally, the material should be used at the point of processing, thus avoiding problems (and costs) of transportation and storage. This practice is followed more consistently in Europe, especially in the U.K., and the limited amounts stored are often packaged dry and as free as possible from hydrolysis agents. However, in the United States the production of larger quantities and logistic practices result in the need to transport and store large amounts of azides, usually under water–alcohol.

A particularly troublesome situation arose in the United States in the 1960s when large quantities of lead azide were manufactured under great time pressure. As a result of the urgency a number of waivers were issued for the commercial, nonstandard packaging of products. Deviations included non-plastic-lined steel drums, substitution of rubberized bags for plastic bags, use of substitute alcohols, relaxation of controls on alcohol–water ratios, and use of sawdust from different woods. When the demand was reduced, military depots were left with stores far in excess of needs for the foreseeable future. After one or two years of storage some of the lots showed deterioration, liquid levels in storage drums were low, the odor of alcohol was detected in the storage igloos, discoloration of the liquid in the drums was evident, and rusting of the drums had occurred. Moreover, analysis of the liquids from the drums revealed the presence of as much as 10^{-3} M hydrazoic acid and ammonium and ferric ions, as well as wood hydrolysis products. Under the stimulus of these findings, investigations were carried out on the chemical equilibria involved in storage, on methods of safe disposal of large amounts of excess or questionable supplies, and on long-term surveillance of azides to assess effects of storage on sensitivity and functionality. These topics are the subject of Sections C and D.

Among the current trends in azide technology, which is justified both on economic grounds and by the desirability of reducing the exposure of operators to large quantities of azide, is the introduction of automated handling facilities. The generation of electrostatic charge during the handling of explosives has been of long-standing concern, and several accidents have been attributed to this cause. Concern has also been expressed that the introduction of autoremote handling techniques will further contribute to the problem, increasing property damage, if not personal injury. The paucity of data on charging and charge relaxation made investigations of the phenomenon of critical interest, and the result of recent progress is presented in Section E.

B. SAFETY IN HANDLING OF AZIDES

1. Philosophy and the General Problem

An approach applicable to any scale of operation is to recognize not only direct hazards due to the explosivity of explosive azides, but also the indirect hazards resulting mainly from hydrolysis. Hydrolysis gives hydrazoic acid, HN_3, a toxic substance which is physiologically very active and is also the medium by which very sensitive heavy-metal azides, most notably copper azide, can be formed.

It is prudent to assume that sensitive material will explode at some time, but that one cannot predict when, and to plan experiments accordingly. It is thus necessary to adopt the practice of minimizing quantities, use distance or space between personnel and samples as well as protective barriers, avoid working with primary and high explosives in the same place, and in general practice good housekeeping.

2. Explosive Azides

a. Safe Practice and the Assessment of Hazards

It is essential to minimize the quantity of material and number of personnel at any location and to utilize all appropriate protective equipment (such as barriers, safety glasses, and flak jackets) to minimize personnel injury in the event of an unexpected detonation. Cleanliness, the avoidance of gritty or other explosive substances, the wearing of appropriate (antistatic) clothing, and the electrical grounding of personnel are standard precautions taken to avoid accidental initiations.

A primary requisite for working with explosive azides is to have a clear understanding of their sensitivity to mechanical shock, friction, heat, and electrostatic discharge. This knowledge is essential in developing the methods whereby the sample is to be handled, confined, and processed and knowing the quantities with which it is safe to work. Knowledge of the maximum temperature a sample can withstand without detonation can be obtained by means of differential thermal analysis (DTA), thermogravimetric analysis (TGA), and explosion temperature measurements using milligram quantities (Chapter 6). Impact, friction, and electrostatic sensitivity tests should be conducted as described in Chapters 4 and 5. This information is essential to assess properly the hazards involved in a particular azide and the manner in which it is to be handled. For information on specific azides consult the literature and Chapter 1, Volume 1. In most countries, government laboratories should be utilized if the requisite knowledge and facilities are not otherwise available.

Before beginning work with a compound for the first time, consult the references given at the end of this section (p. 80) that describe the properties and potentially hazardous situations.

b. Hydrazoic Acid, HN_3

Hydrazoic acid is unique in that both the aqueous and gaseous phase are explosive hazards. Detonations of hydrazoic acid solutions are capable of causing serious injury. It is possible to work with gaseous HN_3, if an inert diluent gas is used.

Lead azide, which is normally stored in alcohol-water, reacts with the liberation of hydrazoic acid. It is possible to detect hydrazoic acid in the vapors over lead azide immersed in alcohol-water. With other azides, such as that of barium, water can generate hydrazoic acid readily. The normal laboratory preparation of hydrazoic acid is to acidify sodium azide and distill the hydrazoic acid with an inert gas carrier. The explosion limits of this substance have not been established with any degree of assurance; however, it is known that hot spots and sparks will initiate gaseous mixtures of hydrazoic acid in air [1,2].

c. Laboratory Precipitation and Short-Term Storage of Lead Azide

Recrystallization of lead azide from an ammonium acetate solution, the classical way of preparing lead azide crystals, can be an extremely hazardous operation which, depending upon the recrystallizing conditions, can give rise to spontaneous detonations in the crystallization apparatus, extensive damage to equipment, and injury to personnel. For details see Chapter 2, Volume 1. Recrystallization should always be carried out with apparatus which is barricaded and from which the operators are protected at all times. It is suspected that the crystallization rate is one of the primary factors which influence the detonation probability of solution, but whether the crystallization rate is the cause *per se* or whether factors such as crystal strain, concentration gradients, etc., within the liquid are the causes of detonation is as yet unknown. It has been observed that a rapid cooling rate gives rise to a rapid rate of precipitation, which gives rise to a maximum possibility of detonation.

In addition to minimizing the amount, lead azide should not be stored with booster or main-charge explosives; it should be kept in a separate magazine. Normally, any quantity of explosive azide greater than 10 g should be stored in 50% alcohol-50% water in order to minimize its sensitivity. It is sound practice to sample azides while they are wet and to dry only small portions as required.

The azide should be stored in conductive, nonglass containers. If the use of glass is unavoidable, it should be shielded with an inert container, so that the glass fragments will be contained in the event of an explosion.

Care must also be exercised in the storage of azides to see that no material can react to form hydrazoic acid. In long-term storage of explosive azides in the presence of atomospheric or absorbed moisture, explosive azides may form on brass or copper heating pipes, radiators, refrigerator coils, etc., and for this reason compatibility of metals with HN_3 must be kept in mind.

3. Toxicity

a. Symptoms

Of concern to the industrial worker, as well as to laboratory personnel, are both the long- and short-term toxicological effects of azides. One of the earliest articles dealing with azide toxicology is that of Loew [3], who studied the effect of "azoimides" on living organisms. He found that injecting NaN_3 into animals causes either paralysis or death, depending on the concentration. Smith and Wolf [4] made the first systematic study of the physiological action of HN_3. They state HN_3 is a protoplasmic poison similar to HCN. Inhaling HN_3 vapors causes respiratory excitation with a lowering of blood pressure and muscle and nerve paralysis. Stern [5] gave the first detailed account of an individual response to inhaling a "significant" amount of HN_3. The individual concerned was in apparent good health prior to an accidental inhalation of HN_3. He was admitted to the hospital with the following symptoms: painful conjunctivitis, tormenting cough, tremors, knee joints swollen with large blue spots on legs, fever, bronchial sounds in lungs, swollen liver, swollen and pressure-sensitive spleen, and albumin in urine. The patient was hospitalized for 12 days with symptomatic treatment and released with no apparent permanent damage. The authors of this chapter have encountered brief exposures to HN_3 leaking from fume hoods and were subject to fainting and headaches. There were no lasting effects, however. Although the symptoms were not as severe as those described above, the experiences lend credence to the findings.

The effects of azides on specific organs has been examined. Using frogs as test animals, Biehler [6] demonstrated the effect of HN_3 and NaN_3 on the colon, heart, central nervous system, and certain muscles. In the same year Hildebrand and Schmidt [7] found that lethal doses of HN_3 are toxic to blood, causing bleeding of the inner organs.

In a series of articles Graham [8–10] stated that HN_3 caused a lowering of blood pressure, accompanied by headache below the lethal amount. His studies showed no pathological change in workers exposed to HN_3 regularly for up to 15 years.

Werle and Fried [11] studied the biological effect of several organic azides on blood pressure and certain bacteria. In similar studies Roth *et al.* [12] found certain organic azides were as effective as NaN_3 in reducing blood pressure, with less toxicity.

Werle and Stucker [13] found that NaN_3 undergoes rapid decomposition in the liver and, using rats, found the liver's capacity does not increase with chronic feeding of NaN_3.

b. Limits

Toxicological limits of azide poisoning have been established for industrial workers. The years of World War II generated studies on the toxicological properties of explosive azides. Schwartz [14] conducted a dermatological study of industrial workers engaged in production of war materials. He found no evidence of PbN_6 causing dermatitis. Also, during the war years, Fairhall *et al.* [15] determined the minimum lethal dosage of NaN_3 was 35–38 mg/kg body weight. Using rats, they determined that HN_3 above 1160 ppm is fatal when breathed for 1 hr. As a comparison, H_2S has a reported lethal toxicity of 400–700 ppm in man when inhaled for 1/2–1 hr [16].

More recent studies have been directed at the maximum allowable concentration of azide in an industrial environment. Bassendowska and coworkers [17], using rats and guinea pigs as test animals, studied chronic poisoning by sodium azide. Their results showed the maximum allowable concentration of NaN_3 in air is 2 mg/m^3, provided the skin of the worker is entirely protected.

After many years of laboratory experience with various azides, the authors can report no occurrence of poisoning symptoms attributable directly to solid azides. However, with substances such as thallous azide, the physiological effects of thallium, in addition to the azide toxicity, are of equal concern.

c. *Detection and Treatment*

Since exposure of vapors of HN_3 has an effect which is immediately apparent to the individual affected and to coworkers, it is easy to avoid serious HN_3 poisoning. It has been observed on repeated occasions that the blood vessels of the eyes turn a very apparent red before swelling of the mucosae occurs. Advantage of this fact may be taken to provide early warning of HN_3 poisoning by providing a mirror for frequent eye examinations. A physician should be consulted promptly when such symptoms are observed. First aid consists of prolonged exposure to fresh air.

4. Reactivity

The reactions of azides are described in detail in Chapter 1, Volume 1; however, those germane to industrial and laboratory safety are discussed here. Inorganic azides are subject to solvolysis in acid media with resulting liberation of hydrazoic acid. When working with azides hydrolysis is a distinct possibility, whether working with inorganic or organic systems, and proper precautions should be taken to handle the hydrolysis products.

The azide ion is a moderate reducing agent, and this fact is normally exploited in its destruction. The azide group can be destroyed by mixtures of nitrous and nitric acids, and by cerium(IV) in the form of ammonium hexanitratocerate(IV).

Azides do not react with metals unless hydrazoic acid is formed as an intermediate. For example, freshly prepared surfaces of lead and copper are readily attacked by HN_3 to form their respective azides. Lamnevik [18] interpreted the reactions as follows:

$$2Cu(s) + 3HN_3(g) \longrightarrow 2CuN_3(s) + NH_3(g) + N_2(g)$$

In a second stage, copper (I) is oxidized by air:

$$2CuN_3(s) + \frac{1}{2}O_2(g) + H_2O(l) \longrightarrow Cu(N_3)_2 \cdot Cu(OH)_2(s)$$

5. Indirect Hazards

a. Solutions

During crystal growth, a metastable condition exists such that spontaneous detonation can occur [19]. One should refer to the discussion in Chapter 2, Volume 1.

b. Melts

It is possible to maintain potassium azide in the molten state; however, extreme care must be taken to keep the melt free from contamination. For example, a speck of rust will cause a potassium azide melt to decompose with a spontaneous flash. Although there is normally no detonation *per se*, extremely rapid deflagration occurs. The behavior is indicative of the hazards involved in dealing with azides under such conditions.

c. Copper

Copper is particularly incompatible with lead azide because the very sensitive copper azide forms in the presence of even small percentages of adsorbed moisture [20,21]. The copper azide appears as a black tarnish on the copper or copper-bearing surface. The film is very sensitive to electric, impact, and frictional stimulus.

d. Gritty and High-Melting Inclusions

All azides can be sensitized by the addition of hard gritty materials, such as ground glass, rust, or fine silica. Instances have been reported in which moist lead azide, mixed with fine silica and allowed to dry, detonated when a wooden spatula was inserted to collect a small subsample. The sensitivity is particularly enhanced if the inclusion, in addition to being gritty, has a melting point higher than the ignition temperature of the azide. For additional information see Chapter 4.

e. Spills

Spills of azide samples should be confined and cleaned up as soon as possible. Accumulations of solid azides in waste lines and receptacles are a definite hazard in that inadvertent addition of acidic materials or oxidizers may cause an exothermic reaction which will detonate the explosive or cause formation of toxic HN_3. Alternately, later additions of a waste to the accumulation may cause formation of an explosive azide. In no case should azides be washed into drain or liquid-waste systems because accumulation of such substances are a hazard to maintenance personnel who conduct repair or construction operations.

Spilled solid azides can normally be cleaned up by saturating them with an appropriate killing solution (described in the next section) or by flushing them into a wet sawdust trap, which is then destroyed by chemical action or heat. Dry azide samples can sometimes be conveniently cleaned up by use of a simple suction flask containing kill solution, and the "house vacuum."

6. Reference Texts

The following publications have been found useful for guidance with respect to the safe handling of explosives, including azides:

N. Irving Sax, *Dangerous Properties of Industrial Materials*, 3rd ed., Van Nostrand-Reinhold Co., New York, 1968.

Volume VII, Propellants, Explosives, Chemical Warfare; Recommended Methods of Reduction, Neutralization, Recovery of Disposal of Hazardous Waste, Report EPA-670/2-73-053-g, Office of Research and Development, U.S. Environmental Protection Agency, Washington, D.C., 1973, pp. 125–135.

Dangerous Articles Emergency Guide, Bureau of Explosives Pamphlet No. 7A, Bureau of Explosives, Association of American Railroads, Washington, D.C., 1970.

Manual of Hazardous Chemical Reactions, NFPA Report No. 491M, National Fire Protection Association, Boston, Mass., 1971.

General Information Relating to Explosives and Other Dangerous Articles, Bureau of Explosives Pamphlet No. 7, Bureau of Explosives, Association of American Railroads, Washington, D.C., 1972.

Laboratory Waste Disposal Manual, 2nd ed., Manufacturing Chemists Association, Washington, D.C., 1969.

Recommended Electrostatic Practices, NFPA Pamphlet No. 77, National Fire Protection Association, Boston, Mass., 1972.

C. DESTRUCTION OF EXCESS AZIDES

1. Laboratory Methods of Destruction

a. Ammonium Hexanitratocerate(IV) Method

The most frequently used method to destroy unused or unwanted azides requires ammonium hexanitratocerate(IV), $(NH_4)_2Ce(NO_3)_6$ (common name, ceric ammonium nitrate). The advantage of the method is that acid is not required, thereby minimizing the formation of HN_3. The procedure commonly used in the laboratory is to add the solid azide or azide solution to a saturated solution of ammonium hexanitratocerate(IV). The reaction proceeds smoothly without the evolution of large quantities of heat according to the equation

$$Pb(N_3)_2(s) + 2(NH_4)_2Ce(NO_3)_6(aq) \longrightarrow 2Ce(NO_3)_3(aq) + Pb(NO_3)_2(aq) + 4NH_4NO_3(aq) + 3N_2(g)$$

While the ceric ammonium nitrate method is well suited for small-scale use in the laboratory, it is not economically feasible for the destruction of commercial quantities of azide.

b. Nitrite Method

The nitrite method involves wetting or dissolving the azide with excess water. A freshly prepared 25% solution of $NaNO_2$ is added and slowly acidified with a 36% solution of nitric acid. The equation describing the decomposition is

$$4HN_3(aq) + 2HNO_2(aq) \longrightarrow 6N_2(g) + N_2O(g) + 3H_2O(l)$$

The undesirable characteristics of this reaction are that HN_3 may escape and other toxic vapors may be generated [22]. However, the method destroys the azide ion effectively, as no color reaction can be observed with the Griess reagent for nitrite [23]. The method may be economical for large-scale destruction of azides, but the presence of HN_3 may preclude its use.

c. Electrolysis

The destruction of lead azide by electrolysis in hot 20% NaOH has also been studied [24]. The advantage of this method is that the lead metal is deposited on the cathode and may be recovered directly. The disadvantage is that hydrogen and oxygen are generated.

d. Reduction

Azides may be decomposed in alkali media to form N_2 and NH_3 [25]. This method is desirable in that HN_3 is not a precursor for reaction, thus eliminating potent hazard. Magnesium, aluminum, zinc, Devarda's alloy, and Raney nickel are used.

e. Heating

Certain azides may be conveniently destroyed by heating [26]. The alkali and alkaline-metal azides decompose smoothly to the respective metal and nitrogen. This procedure is not recommended for the explosive heavy-metal azides.

f. Destruction by Explosion

The normal and preferred means for destroying an explosive azide is to explode it. In a laboratory where small quantities are used, it is strongly recommended that samples, as they outlive their usefulness, be exploded in small quantities in an appropriate laboratory area. Nonexplosive azides, of course, may be decomposed thermally, if they do not present any flame or flash hazard. However, for such materials, the preferred method is chemical destruction.

g. Detection of Residual Azides

The chemical test most frequently used for determination of azides is the ferric chloride test. The azide ion forms a deep red complex with ferric ion, which is readily detected by the unaided eye.

2. Bulk Destruction of Azides

When large industrial stores of a material such as lead azide must be destroyed, considerations of safety; transportation and handling costs; conformation with atmospheric, soil, and water pollution regulations; and public reaction preclude the straightforward methods that are feasible for *in situ* destruction of laboratory samples.

Several approaches have been investigated for bulk destruction of lead azide in the last decade. The detonation of material in bulk was described by Williams [27]. Valentine *et al.* [26] outlined several unique thermal decomposition reactions. Kramer and Warman [28] investigated the oxidation of lead azide by chemical means in acid solution, and Richter and Fisco [25] investigated the reduction of lead azide in alkaline solution. The electrolytic destruction of lead azide in alkaline solution was investigated by Stull and Stouder [24]. Drugmand [29] proposed conversion of the lead azide to the starting materials, lead nitrate and sodium azide, by preliminary addition of sodium hydroxide or adding a large excess of sodium hydroxide solution and converting the lead azide to lead plumbate ion and sodium azide solution. Other methods are feasible, e.g., the reaction of lead azide with sodium chloride and sodium carbonate to form lead azidochloride, basic lead salts, and sodium carbonate. The following discussion presents the approaches in more detail.

a. Detonation of Lead Azide in Bulk

Perhaps the largest quantities of lead azide ever destroyed were disposed of by detonation. This approach is also often used as an alternative to burning for the destruction of military and civilian detonators and fuzes containing sensitive substances.

Large-scale detonation of lead azide was conducted at a United States Army Depot in Colorado. The lead azide packed in drums, as described in the next section, were trucked to the demolition area, a pit 400–700 ft deep. The drums were stacked in pyramid fashion and detonated using supplementary charges of TNT, composition C, and nonelectrolytic blasting caps. Tests were made on 150–7400 lb, and it was found that, as the quantity of lead azide increased, the lead retention of the pit rose to a maximum of 75%. The tests were monitored by the U.S. Army Environmental Hygiene Agency and included analysis of the soil, atmosphere, and nearby groundwater for lead. Monitoring stations 4000 ft from the center of the explosion recorded airborne lead concentrations below 0.8 mg/m^3. No evidence was found for submicron lead particles at high altitudes. In an arid area with a clay–sandstone and gravel bed leaching is not a factor. However, the topography of the explosion site must not permit lead run-off by sudden rains. The major problem with this method of destruction is shipping the lead azide to the demolition site in accordance with safe practice [30], which in the United States requires the carrier to submit a routing agenda conforming to regulations of the Interstate Commerce Commission by avoiding stopping in populated areas.

b. *Thermal Destruction of Lead Azide*

Heat will decompose all solid inorganic azides (see for example Chapter 6, Volume 1), but care must be taken to avoid too high temperatures of bulky samples if detonation is not to result from self-heating. Thermal methods for the destruction of lead azide are, nevertheless, authorized by military manuals [30]. However, although dry lead azide may be decomposed at 245–250°C without detonation, Graybush *et al.* [31] cite detonations with as little as 4 mg when confined. Valentine *et al.* [26] studied a combination of a high-boiling solvent dispersant and a counterflow of inert gas at temperatures ranging from 170 to 260°C. A serious drawback is the possible self-heating because of the exothermicity of the decomposition reaction. Effective heat exchange with the dispersant fluid at a temperature less than 200°C, and a nonreactive, cheap dispersing medium, could make this an efficient disposal method because only lead metal and nitrogen are the products. However, the method is not in practice.

c. *Oxidation–Reduction Method*

Chemical destruction by oxidation–reduction methods in aqueous media is well documented [32]. Destruction of small quantities of lead azide with ceric ammonium nitrate is fast but expensive. The electrolytic oxidation of cerous ammonium nitrate can be done without sophisticated apparatus, and it may be possible to set up a continuous process for destruction of lead azide. Destruction of lead azide with nitric acid and sodium nitrite [28] is rapid, but small quantities of nitric oxides and hydrazoic acid are given off even under optimum conditions. The method is used for destruction of 1–5-lb lots. The net reaction is

$$Pb(N_3)_2(s) + 4HNO_3(aq) + 2NaNO_2(aq) \longrightarrow 2N_2(g) + Pb(NO_3)_2(aq) + 2N_2O(g) + 2NaNO_3(aq) + 2H_2O(l)$$

The reaction is pH dependent with a half-life of 10^{-3} sec at a pH of 0–1, after solution of lead azide. The reaction rate is not diffusion controlled, even though a solid–liquid reaction is involved. Substitution of acetic acid for nitric acid is not advised because the solution becomes buffered, and the destruction of hydrazoic acid slows. This results in boil-off of hydrazoic acid.

The procedure for the destruction of 15 lb of lead azide follows:

1. Add 200 gallons of water to a 300-gallon aluminum or iron container.
2. Add 4 gallons of 25% (w/v) sodium nitrite solution.
3. Agitate and add 15 lb of lead azide, preferably in slurry form.
4. Add 4 gallons of 60% nitric acid; allow the gas surge to subside, and add an additional 5 gallons of nitric acid.
5. The pH at the end of the reaction will be 0.7. Add caustic to a pH of 8 and filter off the hydrated lead oxide residue. The aqueous solution contains sodium nitrate and any excess caustic.
6. The heat of reaction is 163.5 kcal/mole, which means a temperature rise of 4.5°C in the 200 gallons of solution. The heat of neutralization will result in another 1-degree rise in temperature.
7. Tests for hydrazoic acid in the gas phase are made with silver chromate paper, prepared by immersing filter paper in 0.1 M K_2CrO_4, washing with water, drying, immersing in 0.1 M $AgNO_3$, and drying. The paper is cut into small strips and stored in a brown bottle. Hydrazoic acid at a pressure of 10 mm will turn the wet paper white within 3–5 sec, if held at the lip of the test vessel at ambient temperatures. At a pressure of 1 mm of hydrazoic it will require 1 min to turn white. The test is a simple monitoring device for experimental work with azides only if large concentrations of chlorides are absent.

d. An Electrolytic Method

The electrolytic destruction of lead azide at 90°C in excess 20% sodium hydroxide to produce nitrogen at the anode and discharge lead at the cathode was also investigated [24]. The pH of this solution is about 13.4 and Stout [33] claimed that, at a pH > 10, the anodic reaction is oxygen discharge. The discharge of nitrogen as claimed is attended by an enormous overpotential of 3.3 V on a platinum anode, whereas the overpotential for oxygen is less than 1 V. This favors discharge of oxygen.

e. *Reconversion to Starting Materials*

Drugmand [29] proposed conversion of lead azide to the starting materials, as suggested by Kramer, by the following reaction:

$$Pb(N_3)_2(s) + 2NaOH(aq) \longrightarrow Pb(OH)_2(s) + 2NaN_3(aq)$$

Lead hydroxide is filtered off, converted to lead nitrate, and stored. The sodium azide in solution is recovered by evaporation and recrystallization. The reaction of lead azide with sodium hydroxide is diffusion controlled with basic lead salts forming on the crystal surface and slowing the reaction rate. Within days the reaction goes to completion. A flow diagram for the method is given in Figure 1.

f. *Storage of Lead Azide in Alkaline Solution*

Drugmand [29] also proposed storage of lead azide in excess 20% sodium hydroxide solution, suggested by Richter as follows:

$$Pb(N_3)_2(s) + 4NaOH(aq) \longrightarrow Na_2PbO_2(aq) + 2NaN_3(aq)$$

In this method the lead azide is stored in solution form. To re-form lead azide, the solution is neutralized with nitric acid to destroy the plumbate ion complex, the phlegmatizing agent is added, and the pH is lowered until lead azide precipitates. A flow diagram for Richter's proposal is presented in Figure 2.

g. *Destruction in Situ*

Ideally, lead azide destruction should be done with little manipulation and no physical movement of the storage drums. This can be accomplished by siphoning off the alcohol-water desensitizing solution from each drum, flushing out the sawdust with water, and filling each drum with a concentrated solution of sodium chloride containing an excess of sodium carbonate. The products are lead azidochloride, basic lead carbonate, and sodium azide. This procedure was investigated at the bench scale [45], and it was found that the lead azide was completely destroyed in three months.

D. LONG-TERM STORAGE OF LEAD AZIDE

The problems of long-term storage arise both from a concern with safety, such as has been alluded to throughout this chapter, and with the deterioration of products such that they will not function satisfactorily. What constitutes "long term" in this context depends on the chemical and physical conditions of storage. In general, any lead azide which has remained in a storage magazine for 3-4 months after manufacture should be considered to be in long-term storage.

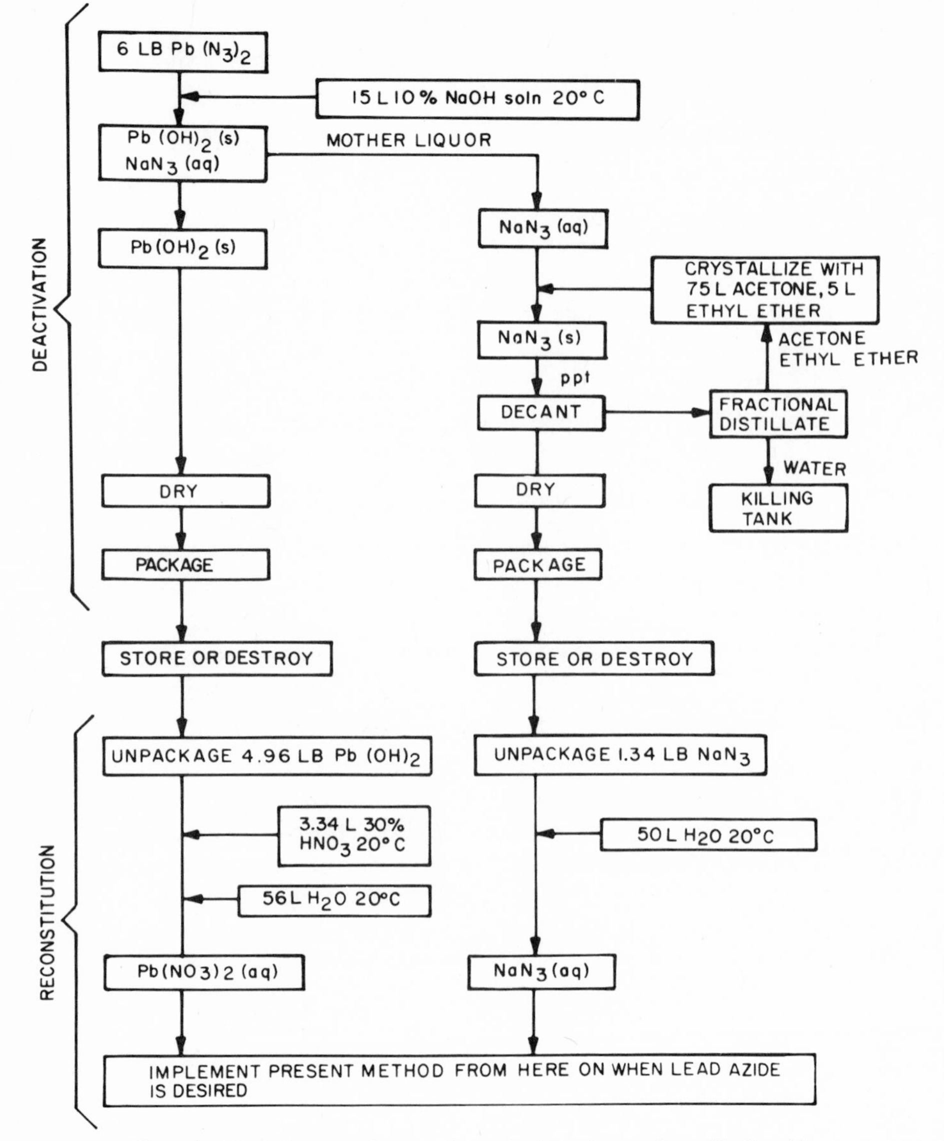

Figure 1. Flow diagram of Kramer's method for destruction of bulk azide.

Lead azide may be stored either as bulk raw material or as pressed pellets in detonators or other explosive-train components. In most European countries the bulk transportation of lead azide is restricted. In such situations it is most common for lead azide to be manufactured and pressed into detonators at the same

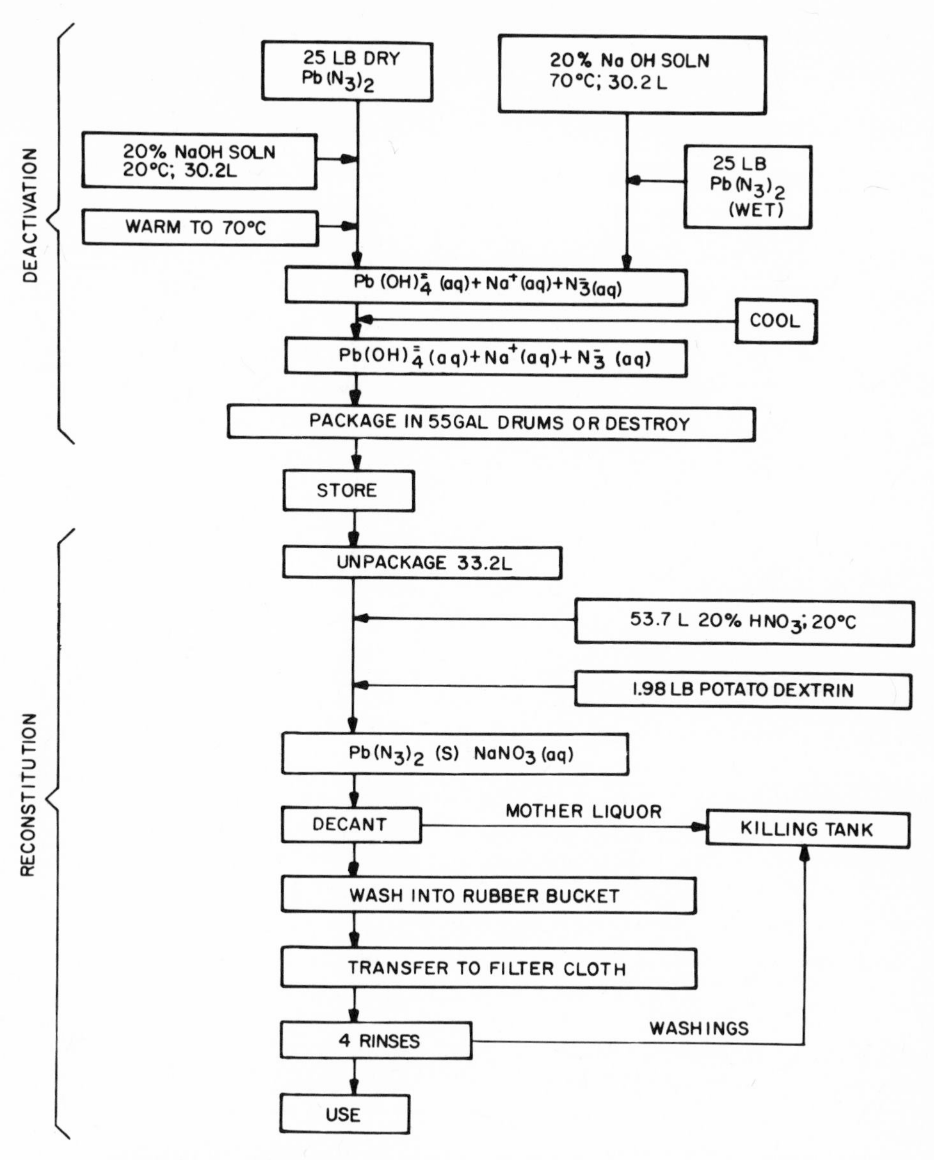

Figure 2. Flow diagram of Richter's method for destruction of bulk azide.

plant. Only limited stocks are then required and are used promptly. Thus, long-term bulk storage as discussed in this section is not a common practice, but has been encountered as a result of special circumstances in the U.S.A.

The long-term storage of lead azide in detonators, particularly military items, is a more general problem, and numerous facets of compatibility with other materials can be envisaged, some of which were discussed in the section on indirect hazards and later in the subsection on chemistry of long-term storage. Careful design and sealing of the items can minimize such problems, as noted, for example, by Lamnevik [34].

In general, problems arise from the presence of moisture, and in the United Kingdom, for example, some products are stored dry. Atmospheric moisture may, nevertheless, also cause deterioration, and the spontaneous detonation of detonators in stores has been attributed to hydrolysis reactions of this sort as discussed later in this chapter. Most commonly, however, bulk lead azide is stored under a "desensitizing liquid," usually water or water–alcohol mixtures, and thus the potential for deterioration is always present.

1. Storage Studies

Historically, the practice of storing lead azide under liquids, and the media used, are continuations of procedures used to desensitize mercury fulminate in storage. In fact, an early investigation by Hopper [35] showed lead azide to be more stable under prolonged storage than mercury fulminate. It was also shown that dextrinated lead azide stored for 25 months under 50:50 water–ethanol mixture showed no change in sensitivity, whereas material stored dry under the same climatic conditions exhibited a marked increase in sensitivity. The investigators believed this anomaly to be due to time-hardened dextrin particles acting as an abrasive in the impact tests.

The solubility of the stored material in the medium selected is extremely important, as is the cost of the medium. The selection of water–alcohol as a medium was based on its comparative cheapness, the low solubility of lead azide in the mixture, and its excellent "antifreeze" properties. The solubility of lead azide in water, although small, is a function of the temperature, and solvation and recrystallization occur continuously under normal storage conditions. Service lead azide formed large agglomerates when stored under water for an appreciable length of time. An accidental explosion with this material was attributed to one or both of these occurrences, although neither was positively established [32]. What was then described as the unpredictable behavior of lead azide was believed to be brought about by the energy released when large crystals or agglomerates were fractured. However, early work carried out in the United States showed that no changes occurred when a duPont lead azide, containing 10% lead carbonate (similar to Service lead azide), was stored for long periods under ethanol–water mixtures. At this time investigations were begun in various countries to find "crystal modifiers" that would allow precipitation of uniform material. It was believed that this would accomplish two things i.e., make the material "more predictable" by eliminating large crystals and also result in a more "free-flowing" product that would be less troublesome during the detonator loading process. Studies conducted in the early 1930s at Picatinny Arsenal in the United States showed that Commercial lead azide withstood storage for three years with no change in appearance or behavior [36]. The belief was that lead azide exhibiting the normal crystalline morphology did not withstand storage, whereas lead azides showing no definite crystal form would store satisfactorily. However, tests conducted later on polyvinyl alcohol lead azide, which has definite crystal form, showed no crystal growth or agglomeration after a year of storage in 50:50 water–ethanol solution [37].

Avrami and Jackson [38] investigating dextrinated lead azide manufactured during World War II and repackaged in 1953, concluded that 25-year-old azide did not differ from recently manufactured dextrinated lead azide with respect to its sensitivity to impact.

Present storage practices resulted from long-term tests conducted using laboratory techniques, i.e., the material tested was stored in laboratory-type containers, usually glass bottles. Samples were withdrawn at selected intervals and tested for their ability to function in detonators and for crystal changes. These tests omitted an evaluation of any interaction of the lead azide with the materials present under the storage conditions of a magazine. Thus separate investigations were conducted to evaluate the interaction of lead azide with selected metals under wet and dry conditions over long periods of time.

The only known *in situ* study [39] of material stored in bulk was conducted on Special Purpose lead azide (Chapter 1), as described in more detail in Section D.4, below. In addition, dextrinated lead azide that had been in storage for 25 years and had been repackaged was tested for suspected sensitivity changes. However, the sensitivity was found to be almost identical to that of a fresh preparation that was tested concurrently [38].

2. Current Packaging Practice in the U.S.A.

In the United States, lead azide manufactured for military use is packaged in steel drums under 50:50 water–ethanol solution; however, isopropanol may be substituted for the ethanol. The drums are stored in above-ground magazines, and surveillance consists of periodic checks of the drum exterior and a yearly inspection of the liquid level. The top of each drum is fitted with a clear plastic plug to allow a more stringent control of the liquid level.

Typical packaging in the United States that meets military specifications is shown in Figure 3. The packaging consists of an outer container (usually a single-trip, type 17H steel drum), a jute or Osnaburg-cloth liner to retain sawdust used as a shock absorber, and a separate inner package containing the lead azide. Different manufactureres use different batch sizes. Each batch of lead azide is wrapped in the "diaper" cloth filter (used during the manufacturing) which is

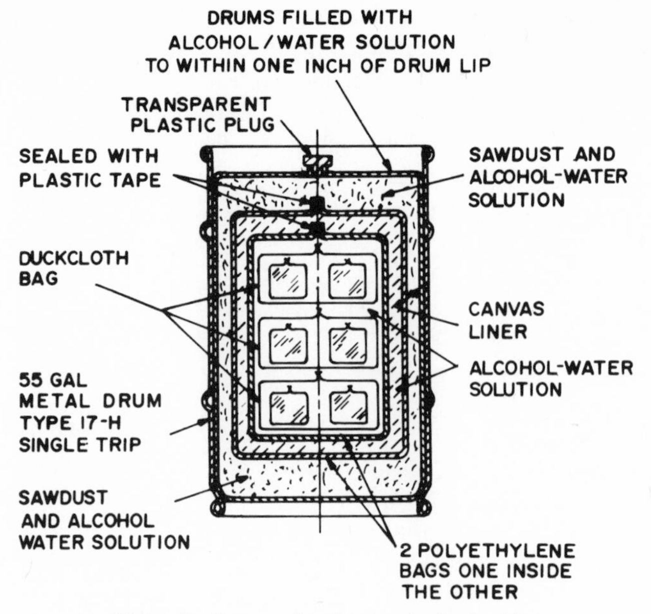

Figure 3. Approved packaging for lead azide.

tied off with string to form a bag. Two or three of the diaper bags are placed in duck-cloth bags, which are also tied off with string. In turn three or four of the duck bags are placed inside a polyethylene bag that is sealed off with black plastic electrical tape. Successively, depending on the manufacturer, the plastic bag is placed in other plastic bags each individually sealed with tape.

The lead azide is wetted with at least 20% of its weight of 50% water–50% denatured alcohol mixture when placed in the duck bag. A duck cap is usually placed on top of the diaper cloth bags before the duck bag is sealed off to prevent escape of lead azide through the orifice.

Previously, packaging specifications allowed the use of a rubberized cloth bag. However, during the long-term storage study discussed below, it was shown that the use of the rubberized bags results in deterioration of the lead azide and necessitates the destruction of the lead azide.

3. Assessment of Current Practice

An investigation was conducted during 1968–1972 to evaluate the adequacy of the packaging practice described above. The investigation consisted of a diurnal temperature cycling study and a concurrent surveillance and analysis of lead and storage media. The temperature cycling study was conducted using laboratory procedures and laboratory containers.

The surveillance study involved analyses of the lead azide and of the media in various regions of the drums, together with paramagnetic resonance and impact-sensitivity measurements on the lead azide. No detectable interaction with the packaging materials or identifiable changes in crystal morphology or size occurred.

In the case where the use of rubberized cloth bags had been permitted in lieu of polyethylene material, analyses showed the presence of high concentrations of hydrolysis products and products of reaction with the drum in all regions of the drum. Electron paramagnetic resonance measurements showed the presence of iron and manganese in the lead azide lattice, although similar measurements on recently manufactured azide showed only trace amounts of iron and no manganese. The lead azide assay was also low.

In the cases where polyethylene waterproof bags were used, the lead azide had survived a storage time of approximately six years, although some difference in impact sensitivity and corresponding lead azide content between manufacturers was found and attributed to differences in permeability of the packaging. In each case the azide remained within specifications.

Notwithstanding the evident ability of commercial products to perform satisfactorily after extended storage, potentially adverse chemical reactions may proceed within the packaging materials. This is discussed in more detail in the following subsection.

4. Chemistry of Long-Term Storage

This section attempts an understanding of the chemical problems of long-term storage; however, complete understanding of the problems is difficult because conditions are generally complex, and numerous side reactions may occur. There is evidence that the long-term stability of lead azide is largely governed by its reactivity with water and heat and the influence which other substances in its environment have on this reactivity. McLaren [40] noted that if lead azide is heated at 250°C for 5 min, basic lead azide is formed, and the detonation velocity is reduced to half its original value. Reitzner [41] found that traces of water affected the thermal decomposition of lead azide at 184–200°C, but larger quantities resulted in hydrolysis with the formation of hydrazoic acid and ammonium azide. Although such temperatures are unlikely to be encountered in normal storage, the data suggest that abnormal high-temperature excursions during diurnal thermal cycling can lead to the deterioration of products. At ambient temperatures Todd *et al.* [42] showed that lead azide is stable when stored dry, but hydrolyzes in water, and properties such as the critical temperature for ignition, the ignition delay time, and detonation velocity are modified. Thornley [43], using electron microscopy to study the surface chemistry of lead azide crystals, found that high humidity was essential for a reaction with carbon dioxide. Blay

Table I. Hydrolysis Constants for Lead Azide (25°C) [48]

Type of lead azide	Solvent	Experimental	pHN_3 (mm)[a]	Hydrazoic acid (moles/liter)
Special purpose	Alcohol–water	3.3×10^{-8}	0.057	5.9×10^{-4}
Special purpose	Alcohol–water	4.8×10^{-8}	0.061	6.4×10^{-4}
Special purpose	Alcohol–water	9.9×10^{-8}	0.068	7.0×10^{-4}
Dextrinated	Alcohol–water	6.3×10^{-9}	0.043	5.2×10^{-4}

[a] At a partial pressure of 1 mm CO_2.

[44] showed lead azide to be stable when dry but that it is decomposed rapidly in the presence of water and carbon dioxide. Kramer [45] demonstrated that in the reaction of lead azide with lithium chloride to form lead azidochloride, the destruction of lead azide was dependent on the diffusion-controlled reaction of lead azide with water and carbon dioxide.

The hydrolysis reactions were studied by Feitknecht and Sahli [46], Lamnevik [47], Todd *et al.* [42], and Blay [44]. The equation for the hydrolysis of lead azide in water was given by Feitknecht and Sahli as

$$Pb(N_3)_2(s) + H_2O(l) \longrightarrow Pb(OH)N_3(s) + HN_3(aq)$$

$$K_A = P[HN_3]_g$$

The equilibrium constant for the reaction is equal to the partial pressure of the hydrazoic acid and was calculated by Feitknecht and Sahli to be 4.5×10^{-2} mm at 25°C [46]. From the distribution coefficient of hydrazoic acid between air and water [48], the concentration of hydrazoic acid in the water is calculated 5.4×10^{-4} M at 25°C. It is speculated that the $Pb(N_3)_2$–H_2O reaction does not proceed beyond the basic lead azide because of the high nucleation energy for the formation of lead hydroxide.

With carbon dioxide present the reaction can be considered in the same manner as the basic hydrolysis reaction. Lamnevik [47] considered the overall reaction as

$$3Pb(N_3)_2(s) + 4H_2O(l) + 2CO_2(g) \longrightarrow 2PbCO_3Pb(OH)_2(s) + 6HN_3(aq)$$

$$K_B = \frac{P^6[HN_3]_g}{P^2[CO_2]_g}$$

and calculated the equilibrium constant K_B to be 8.7×10^{-11} at 25°C. At a partial pressure of 1 mm of carbon dioxide the pressure of hydrazoic acid is 0.021 mm. This pressure will rise as the partial pressure of carbon dioxide rises. From the distribution coefficient at 25°C, the concentration of hydrazoic acid in the liquid phase is 2.5×10^{-4} M. In the presence of large quantities of carbon dioxide, the reaction with water can be represented as

$$Pb(N_3)_2(s) + H_2O(l) + CO_2(g) \longrightarrow PbCO_3(s) + 2HN_3(aq)$$

and

$$K_C = \frac{P^2\,[HN_3]_g}{P[CO_2]_g}$$

Kramer [48] calculated the equilibrium constant to be 3.5×10^{-4} and the pressure of hydrazoic acid is 1.9×10^{-2} mm at a carbon dioxide pressure of 1 mm. Aqueous hydrazoic acid in equilibrium with the gas phase at 25°C would be 2.3×10^{-4} M with respect to hydrazoic acid. The equilibrium constants were measured [48] for military-grade azide in 50:50 alcohol–water solutions and are summarized in Table I.

In these experiments, carried out statically under one atmosphere of carbon dioxide, only basic lead carbonate was detected. Basic lead azide and normal lead carbonate were not observed by X-ray diffraction [48]. This is probably due to a "cocoon effect," with basic lead azide surrounding the kernel center, basic lead carbonate as an intermediate layer, and a thin outer skin of the normal carbonate. It seems likely that in preparation of the sample for X-ray diffraction analysis the basic lead carbonate masked the normal lead carbonate below the X-ray diffraction detection threshold, as noted by Todd [49].

Although the basic lead carbonate may be the result of a secondary hydrolysis, the hydrolysis constants were calculated by measurement of the hydrazoic acid and unreacted carbon dioxide, after equilibrium had been reached and assuming the reaction to proceed to the basic lead carbonate as the primary product. The data are in reasonably good agreement with the calculated values. This suggests a three-tiered hydrolysis:

Kernel hydrolysis:

$$Pb(N_3)_2(s) + H_2O(l) \rightleftharpoons Pb(OH)N_3(s) + HN_3(aq)$$

Intermediate layer:

$$Pb(OH)N_3(s) + CO_2(aq) \longrightarrow PbCO_3(s) + HN_3(aq)$$

$$3PbCO_3(s) + 2H_2O(l) \rightleftharpoons 2PbCO_3 \cdot Pb(OH)_2(s) + H_2CO_3(aq)$$

Outer layer:

$$2PbCO_3 \cdot Pb(OH)_2(s) + 6HN_3(aq) \rightleftharpoons 3Pb(N_3)_2(s) + 4H_2O(l)$$

$$Pb(N_3)_2(s) + CO_2(g) + H_2O(l) \rightleftharpoons PbCO_3(s) + 2HN_3(aq)$$

All calculations are based on intermediate layer hydrolysis, neglecting both the kernel and outer layer hydrolysis.

Basic lead azide dehydrates, when heated, to several polymorphs as described by Todd [49]. Later work by Todd [50] indicates the most usual form to be

$$PbO_{0.66}Pb(N_3)_{0.96}$$

The formation of the normal carbonate was reported by Todd when the carbon dioxide content was in excess of 4%. At ambient temperatures, water in equilibrium with air has a pH of about 5.9, which drops to 3.9 as the carbon dioxide is enriched to 100%. Neglecting the bicarbonate ion, the pH figures mean that water in contact with CO_2 contains 10^{-6}-10^{-4} g ions of carbonate per liter. A saturated solution of lead azide with a solubility of 0.02% (w/v) contains 6.9×10^{-4} g ions of lead per liter. Since the solubility product of lead carbonate is 3.5×10^{-14}, it follows that the carbonate ion concentration has only to rise about 10^{-10} g ions/liter for lead carbonate to precipitate.

In storage the release of about 0.02-0.04 mm of hydrazoic acid into the gas phase and 5×10^{-4} M hydrazoic acid into the liquid phase results in recombination of the basic lead compounds with the hydrazoic acid to reform lead azide or in reaction of the acid with the other contents of the storage containers. Numerous reactions of hydrazoic acid have been reported [51, 52] which could effect the hydrolysis equilibrium. The most important of these occur with most common oxidizing and reducing agents (metals), mineral acids, alkalis, alcohol, aldehydes, ketones, sulfur, and sulfides. The reported reactivity is so considerable that probability of the acid remaining unreacted during storage appears small.

It should be borne in mind that the reactions of lead azide and water–carbon dioxide are reversible, and the extent of lead azide deterioration will be influenced by a number of factors which include temperature, the partial pressure of reactants, diffusion rates, container dead space, and leakage from the container.

Forsyth *et al.* [39] described corrosion of some drums in long-term storage, observing dark blue and brown residues inside the containers. Laboratory studies [48] of lead azide in alcohol–water were conducted in sealed glass capsules which were outgassed and pressurized to one atmosphere with carbon dioxide. Iron, sawdust, and polyethylene were added individually, and the capsules were analyzed at intervals to determine the decomposition rate of lead azide. For iron it was found that the overall reaction is

$$Fe(s) + 3H_2O(l) + HN_3(aq) \longrightarrow NH_4OH(aq) + Fe(OH)_2(s) + N_2(g)$$

The hydrazoic acid resulted from the hydrolysis reaction. Iron in contact with the slurry resulted in 40% decomposition of the lead azide within 90 days, and iron held in the gas phase above the lead azide slurry resulted in 70% destruction of the lead azide within 90 days. The alcohol–water solution turned dark red-brown, and the iron oxidized when the capsule was opened. The hydrazoic acid and ammonia concentrations inside and outside the polyethylene bags were the same. There was spectrophotometric evidence for the $[Fe(N_3)_2]^+$ complex.

The iron from the capsule experiments was brittle and fell apart when rubbed. In this investigation iron azide (ferric or ferrous azide) and hydrazine could not be detected, contrary to the work of Franklin [51] and Curtius and Risson [52]. Polyethylene becomes brittle, opaque, and porous when exposed to hydrazoic acid, and hydrazoic acid diffused through 0.008 in polyethylene bags within 90 days. Blay and Dunstan [53] reported little azide interaction with polyethylene, but noted a marked drop in azide value when Service lead azide was in contact with various rubbers, plastics, and other synthetic packaging materials. A reduction in azide content as high as 70% was shown (Figure 4). This was attributed to the slow release of carbon dioxide from the test material, followed by further hydrolysis of the lead azide.

Complexed iron, discussed later, was found in the rubberized and canvas lead azide bags, but not generally in the polyethylene bags in the work reported by Forsyth *et al.* [39]. This was most likely due to the size of the iron tannate complex. Forsyth also reported considerable quantities of rust in some storage drums [39]. This was possibly due to electrolytic corrosion, after the initial attack on the iron by hydrazoic acid. The air space at the drum top is richer in oxygen than the alcohol-water solution, and this oxygen gradient gives rise to a corrosion cell at the solution-air interface.

Corrosion of the storage drum could lead to several problems. The drum could leak or breathe hydrazoic acid, which would result in the decomposition of more lead azide. Alcohol vapors could be released in the magazine, posing a severe hazard because of its low flash point and detonation limits. A point may also be reached when the magazine has a partial pressure of about 0.04 mm of hydrazoic acid. Although the quantity is small, it is conceivable that the detonation limits for alcohol might be lowered significantly by the presence of hydrazoic acid. During the surveillance tests described above, monitoring indicated

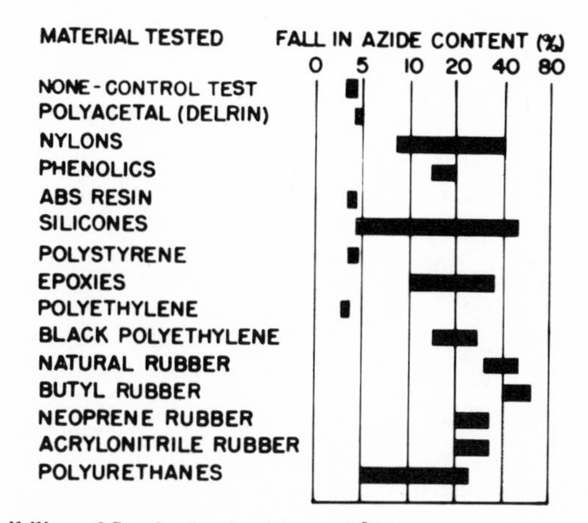

Figure 4. Compatibility of Service lead azide at 60°C, 95% relative humidity, 4-week test.

alcohol fumes in at least one magazine. Later another magazine was destroyed when one drum of lead azide exploded spontaneously. Hydrazoic acid attacks not only the iron drum but also the sawdust filler used in the lead azide packaging. Sawdust contains a polysaccharide easily hydrolyzed in dilute acid solution [54] to form polyphenols and tannic acid. Tannic acids are commonly used as flocculating agents for certain hydrous oxide solutions [55] which pass through fine filter paper. Such an effect was observed in many of the lead azide storage drums under surveillance. In storage the reaction sequence can be summed up as follows:

$$Pb(N_3)_2(s) + H_2O(l) \longrightarrow Pb(OH)N_3(s) + HN_3(aq)$$

$$\text{Sawdust}(s) + HN_3(aq) \longrightarrow \text{glucose}(aq) + \text{tannic acid}(aq)$$

$$Fe(s) + 3H_2O(l) + HN_3(aq) \longrightarrow Fe(OH)_2(s) + N_2(g) + NH_4OH(aq)$$

followed by

$$4Fe(OH)_2(s) + O_2(g) + 2H_2O(l) \longrightarrow 4Fe(OH)_3(s)$$

$$Fe(OH)_3(s) + \text{tannic acid}(aq) \longrightarrow Fe(\text{tannates})(aq) + H_2O(l)$$

The reaction products were isolated and identified by infrared or X-ray diffraction and analysis of the test solutions and residues [48].

The complexity of the storage chemistry makes the complete characterization difficult. There are solid-solid, solid-gas, solid-liquid interactions which may be so slow that complete comprehension of the storage problem is beyond current grasp. The reaction with water should be avoided if possible and consideration given to alternative methods of storage. Seavey and Kerone [56] showed dichloroethylether, a nonsolvent easily removed by drying, to be a cheap, satisfactory desensitizing liquid.

E. ELECTROSTATIC CHARGE GENERATION DURING HANDLING

This section deals with mechanisms of electrostatic charging and charge relaxation in powders and with the important variables, such as powder size, quantity, temperature, humidity, etc., which affect these phenomena. Also considered are experimental techniques for investigating the electrostatic characteristics of sensitive materials, with particular reference to lead azide. Finally, the application of such information to the safe handling of sensitive substances is discussed.

Although the section is directed to the technology of azides primarily, nothing unique electrostatically arises from the fact that they are explosives. Accordingly, reference will be made to other explosive and inert substances, where appropriate, to illustrate a point or to provide supplementary information. Charges generated on other substances and on individuals in proximity to azides can also result in explosions. The approach to the subject will be in terms of classical physical and chemical concepts rather than those of quantum mechanics.

1. General Principles

For broad descriptions of the many ways in which electrostatic charges can be generated in nature, one may refer to Loeb's work on electrostatic charging [57], and in particular to Harper [58] for contact charging and to Montgomery [59] for a general review. The important mechanisms by which dry powders are charged during handling are contact and induction, and they are the ones considered here.

The phenomenon underlying contact charging is the "electrical double layer." When different substances are brought into contact, differences in electronic properties of the surfaces may cause a separation of charges to occur across the interface, and the resulting layers of opposite charge constitute the electric double layer. When the surfaces are separated, charges tend to return to their original conditions, but if one or both of the surfaces are poor conductors, the return is not complete and a net electrification of the two surfaces occurs. It is possible also for the charge carriers to be ions, which may adhere preferentially to one surface.

Although there is a body of theory to account for contact electrification and a large literature on the subject, it is sufficient here to emphasize that contact charging is a common phenomenon, probably impossible to avoid, and that it is intimately related to composition, electrical conductivity, and the mechanics of contact of surfaces. Because even monomolecular contamination can markedly affect both the amount and the sign of the charge, experimental investigations are notoriously erratic in their observations and conclusions.

The second mechanism considered here, inductive charging, typically occurs when a conductive powder, lying on a metal or other conductive surface, is subjected to an electric field. In such cases the field induces a charge on the powder, the sign of which depends on the polarity of the field. If the contact between the powder and the surface is broken and the powder removed from the field, as by pouring, the charge remains on the powder. In handling sensitive materials the significant points to keep in mind concerning inductive charging are that the degree of conductivity required is quite small, and may result from adsorption of moisture under ambient conditions, and that the required electric field can result from stray charges on nearby insulators or even on electrically charged personnel.

The curves of Figure 5, taken from an experiment by Pollock [60] illustrate the two mechanisms of charging. RDX, an insulator, charges by contact, the graphite-coated propellant charges by induction, and lead azide charges by both mechanisms. At this point a word of caution may be in order. At sufficiently high field strengths, electrostatic attraction may cause the charged particles to be picked up and subsequently scattered indiscriminantly to create an indirect hazard. The effect was noted by Pollock *et al.* [61] in their experiments with lead azide and is most apparent with the finer particle sizes.

The charge generated on powders is most commonly measured by means of a Faraday cage and either an electrometer or electrostatic voltmeter. In experiments with sensitive explosives it is preferable to use the electrometer, because this instrument permits using very small quantities of powder. The Faraday cage is essentially a capacitor, and in its classical form consists of two concentric cups insulated from each other. The inner cup is one "plate" of the capacitor and is connected to the "high" side of the electrometer or electrostatic voltmeter. The outer cup serves as the second plate and also shields the inner one from extraneous fields that would otherwise influence the measurement. When a charged object eneters the inner cup, an equal charge of the same sign is induced on the outside of that cup, and the cup assumes a potential difference with respect to the outer cup given by the relationship

$$V = q/C$$

where V is the potential in volts, q is the charge in coulombs, and C is the combined capacitance of the Faraday cage, leads, and electrometer input. Once the

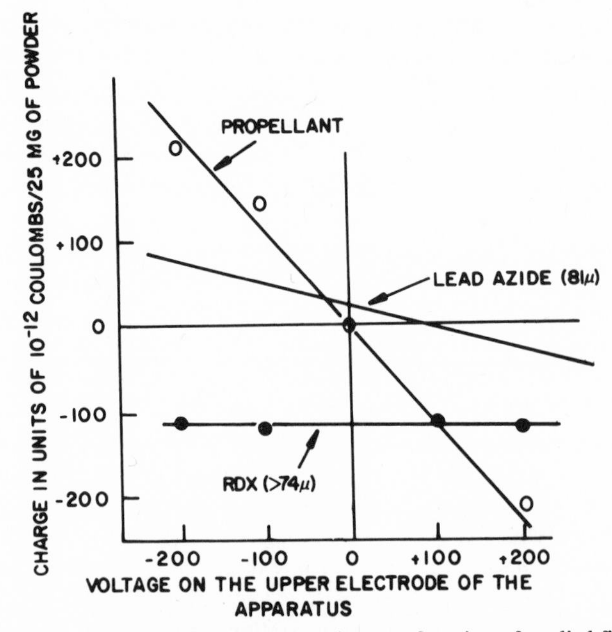

Figure 5. Electric charges induced on powders as a function of applied field in a vibrating-trough apparatus.

powder enters the inner container, the measurement is independent of the behavior of the charge. The charge that was induced on the outside of the cup will remain until it relaxes through the normal leakage, which is a function of the circuit capacitance and the leakage resistance of the measuring circuit.

The experimental studies are reviewed in Sections E.2 and E.4.

2. Charging Studies

The comments concerning charging behavior later in the section were derived from the investigations outlined briefly here. Rathsburg and Schmitz [62], using commercial equipment and commercial quantities of several primary and main-charge explosives and inert powders, investigated the effect of air flow, substrate, grounding of the substrate, humidity (over a limited range), and quantity. They also made observations of the effects of radiation, such as from radioactive sources and ultraviolet light, on the discharge of their apparatus. Braid *et al.* [63] used a pouring technique to investigate means of controlling charging and found that by coating a tray or other device, such as sieve, with tetryl, subsequent batches of tetryl would not charge during the usual handling operations. They concluded that the tetryl eliminated contact charging by preventing metal-to-particle contact. Cleves *et al.* [64] also used a pouring technique to study the charging of lead azide, lead styphnate, and barium styphnate as a function of temperature and relative humidity. Significant findings were that at a given temperature, the charge decreased, although not drastically, in a linear manner with increasing relative humidity, and their curves were displaced downward somewhat, with increasing temperature. Relative humidities ranged from 20 to 80% and experimental temperatures from 10 to 40°C. They also derived recommended combinations of temperature and relative humidity to limit charging to acceptable levels.

Nakano and Mizushima [65] also studied charging of lead azide, lead styphnate, DDNP, mercury fulminate, and a number of inert inorganic substances when caused to flow down chutes of aluminum, brass, and polyvinyl chloride. They presented data for temperatures in the range 10–40°C and relative humidities of 40–89%. They plotted families of curves of coulombs/g vs. temperature at fixed humidities. The data showed some irregularities and some overlaps among curves, probably due to uncontrolled pouring; however, the general trend was in agreement with Cleves *et al.* The effect of temperature was irregular and not marked. Their overall range of charge density was $2\text{-}20 \times 10^{-9}$ coulomb/g, and the range for a given powder over their experimental range was only about fourfold. The results of Nakano and Mizushima were in qualitative agreement with those of Cleves *et al.*

Pollock *et al.* [61] used a vibrating trough apparatus, fitted with an insulated electrode that permitted study of both contact and induction charging, to investigate the charging of commercial lead azides as a function of quantity, particle size, electric field, and rate or manner of feeding of the powder. Significant findings were that charge per gram is inversely proportional to particle diameter, from which they inferred that charging is dependent on the number of particle-metal contacts. Charging of azides was also markedly affected by an applied electric field, which could reverse the sign of the charge at strengths of only a few hundred volts per cm. The effect was self-limiting in that, as the field increased, particles were picked up and could assume an opposite charge when they contacted the upper electrode. Charging was also found to increase with quantity at less than a linear rate. Results in this respect were in agreement with those of Rathsburg and Schmitz [62]. The effect was attributed to mutual repulsion of charges on the particles in the case of small amounts of powder and to impeded contact in the case of larger amounts.

3. Electrostatic Hazards Estimation

The approach taken in here is to present the data of Rathsburg and Schmitz for commercial quantities of various explosives under normal ambient conditions and then to discuss the effects of the various process variables in order to provide an estimate of uncertainties due to these effects. Because of the dearth of data on charging of large amounts of explosives, data for materials in addition to lead azide are given. Unfortunately they obtained data for lead azide only up to 50 g, stating that its charging paralleled that of nitrosoguanidine. The data for lead azide quantities above 50 g in Table II are extrapolated values based on those for nitrosoguanidine. The data are for relative humidities in the range 35-50% and tray contact areas of 200 cm^2. The values are the maximum obtained,

Table II. Maximum Charge on Explosives due to Handling as a Function of Quantity[a]

	Charge (μcoulombs)				
Explosive	50 g	100 g	300 g	500 g	1000 g
Mercury fulminate	0.39	0.49	0.53	0.55	–
Lead styphnate	0.29	0.37	0.39	0.40	–
Nitrosoguanidine	0.19	0.24	0.29	0.32	–
Lead azide[b]	(0.20)	(0.25)	(0.30)	(0.34)	–
Tetracene	0.10	0.17	0.22	0.34	–
PETN	0.13	0.17	0.26	0.29	0.33
Tetryl	0.04	0.06	0.10	0.12	–
RDX	0.04	0.06	0.10	0.12	0.16
TNT	0.02	0.03	0.05	0.06	0.07

[a] Data are taken from Figures 2 and 3 of reference [62].
[b] Data for lead azide are based on those for nitrosoguanidine.

and the design of the trays was such as to tend to maximize readings, according to Cleve *et al.*

A comparison of the data [61,62,64] suggests that Rathsburg and Schmitz's data may be taken as a good basis for extrapolation to other conditions. Thus a linear extrapolation of the average of Pollock *et al.* for five different lots gives 2.3×10^{-6} coulombs for 50 g, and extrapolation of Nakano and Mizushima's data gives $0.1\text{-}1.0 \times 10^{-6}$ coulomb, compared to 0.2×10^{-6} coulomb for Rathsburg and Schmitz. Pollock *et al.* worked with quantities of only 0.3–35 mg, and Nakano and Mizushima worked with 0.2 g. In view of the nonlinear dependence of charge on quantity, the agreement is good. It should be mentioned here that the sign of the charge can be affected by the substrate and immediate previous treatments; the magnitudes given, therefore, are for either sign, although as a rule Pollock *et al.* found fairly consistent behavior.

a. Particle Size

Pollock *et al.* [61] investigated the dependence of charge on particle size of a number of azides against aluminum and stainless steel, using well-characterized size fractions, and found consistently that for a given set of conditions, charge per unit weight was inversely proportional to the first power of the particle diameter in the range 37–88 μm. Since contact charging is due to intimate contact between the particle and substrate, it was inferred that for a given contact pair, charging depends on the number of contact points per particle and the area of the contact. Thus differences between lots of azides (or other powders) and substrate were attributed to a combination of particle shape or geometry and mechanical factors of the substrate material. In support of this inference is the experience of Braid *et al.* [63], who noted that roughening their trough decreased charging and suggested that such roughening interfered with particle-to-metal contact.

It is not possible to assess the effect of the substrate surface independently; however, the powder can be characterized with respect to effective size by means of a simple sieve analysis that can be used to get a cumulative size distribution. If the total fraction that passes each size sieve is plotted against the sieve opening on normal probability paper, the mean weight-diameter from the 50% point is obtained and the standard deviation from the sizes corresponding to the 13% and/or 83% diameter. The effective (surface mean) particle diameter can then be calculated by means of the equation given by Orr and Dallavalle [66]:

$$\log d_s = \log M' - 4.605(\log g)^2$$

where d_s is the surface mean particle diameter, M' is the weight-mean diameter from the plot, and g is the ratio of the "83%" diameter to M'.

Except for Pollock *et al.*, none of the other investigators provided size data;

however, since commercial materials were used by Rathsburg and Schmitz and the range of sizes used by Pollock *et al.* represented those characteristic of commercial types, no great error is likely in assuming the validity of Rathsburg and Schmitz's values.

b. Relative Humitidy and Temperature

The effects of humidity and temperature are interrelated in that adsorption of moisture on surfaces depends on these two variables, and the effect of electrification is most likely due not only to the effect of electronic (i.e., quantum mechanical) properties but also on surface conduction. The data of Cleves *et al.* on humidity show the same trend as those of Nakano and Mizushima's, but the curves of Cleves *et al.* are more regular. The minimum charge occurred at the higher temperatures and humidity, and the greatest charging (by about an order of magnitude) occurred at low temperature and low humidity. Cleves *et al.* state that the curves for lead azide and barium styphnate were similar, but that they were about 1/10 and 1/3, respectively, the values for lead styphnate at 65% relative humidity and 20°C. On the basis of the foregoing, they derived recommended values of humidity and temperature for safe operation (Table III).

c. Electric Field

The effect of an electric field in inducing an electric charge on a powder that is nominally nonconducting is illustrated in Figure 6, but has been known from

Table III. Recommended Values of Temperature and Relative Humidity for Safe Working[a]

	Temperature (°C)					
Material	10	15	20	25	30	40
Cotton drill		68	65	61	58	54
Teryline		71	65	58	53	
Nylon		70	65	60	55	
Khaki serge		70	65	59	56	
Viscose rayon		70	65	61	57	52
Woolen flannel		71	65	59	52	
Lead styphnate	72		65		57	46
Lead azide	72		65		56	49
Barium styphnate			65		58	49
Average value	72	70	65	60	56	50
Recommended value	76	70	65	61	57	52

[a]Taken from reference [64].

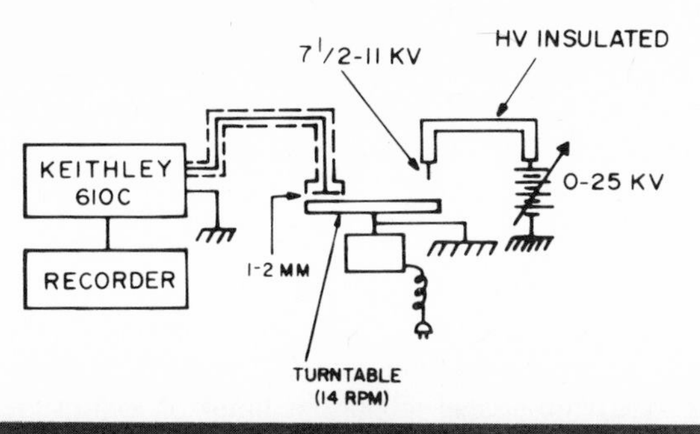

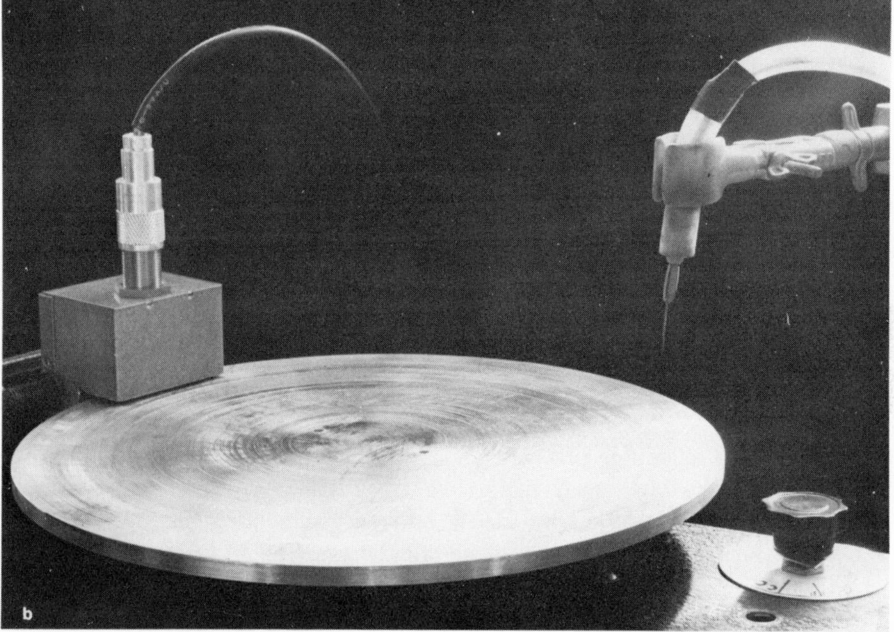

Figure 6. Corona charging apparatus for measuring electrostatic relaxation times on explosives: (top) schematic, (bottom) prototype turntable.

previous work. For example, Petersen [67] found the effect to be linear with field strength in the case of borosilicate spheres, provided that some water vapor was first introduced into his system; the effect was negligible immediately after bakeout and before the addition of the water vapor. Electrification follows an equation of the form

$$Q = Q_0 + AE_0$$

where Q is the total electrification, Q_0 is the electrification due to contact, A is a proportionality constant, and E_0 is the field strength.

The principal property determining the relative importance of contact and induction is the conductivity due to the combined conduction of the surface and bulk. It is inferred from the foregoing and from the work of Cleves *et al.* that electrification by contact is reduced, while induction is increased, with increasing humidity and temperature.

4. Charge Relaxation on Explosives

Pollock [60] adapted a corona charging method described by Zabel and Estcourt [68] to measure charge relaxation times on explosives. The apparatus consisted of a turntable, a corona charging needle spaced about 10-15 mm above the turntable, a 0-25-kV power supply, a field-sensing electrode about 1 mm above and near the edge of the turntable, an electrometer, and a recorder. The photograph of Figure 6b shows the prototype used for earlier work to investigate charge relaxation under ambient conditions and the effect of antistatic additives and of "doping." A later model was provided with an enclosure to permit investigating the effect of humidity and of illumination by sunlight.

The procedure used was to spread 10-20 mg of powder as evenly as possible over about 1 cm^2 near the periphery of the turntable, and after turning on the turntable motor, power supply, and electrometer and recorder, to raise the voltage until a corona discharge occurred. When a steady-state charge was reached, the power supply was turned off and the decay followed by the electrometer and recorder. Figure 7 shows a typical record for lead azide, lead styphnate, and tetracene.

Figure 8 is a semilog plot of charge against time. The curve for lead azide shows an initially high decay rate. This high initial rate was also observed by Lawver and Wright [69], who used a different technique and attributed the effect to polarization. Inasmuch as the effect was found to be most noticeable

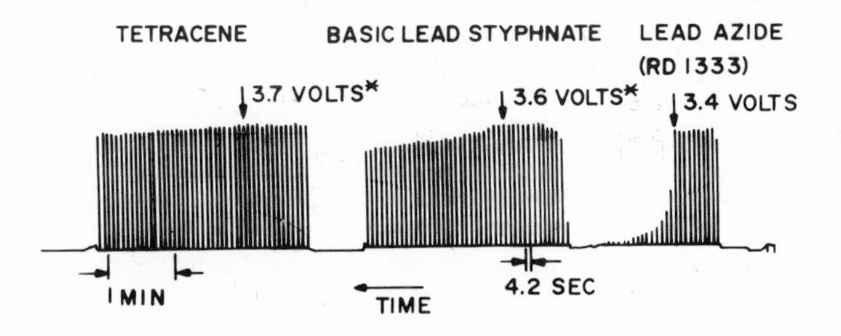

Figure 7. Electrostatic charge relaxation on three primary explosives.

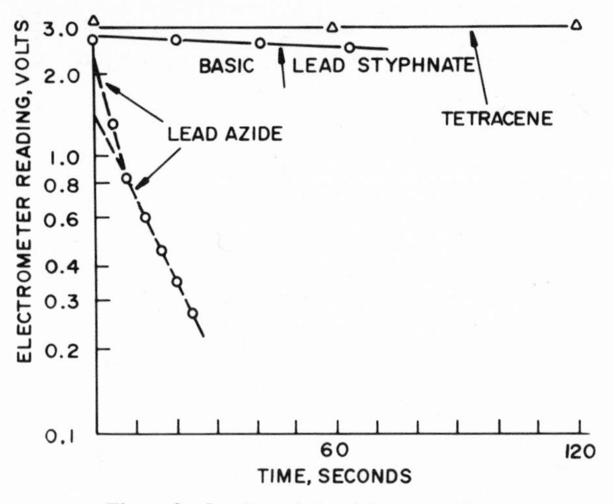

Figure 8. Semilog plots of charge vs. time.

on those samples having fast decay times (i.e., high surface conductivity associated with high humidity), this explanation is plausible.

Table IV summarizes representative data for a number of explosives. The first three entries are for the explosives used to obtain the records in Figure 7. The values in parentheses were obtained at a time when external light could enter the hood in which the experiments were carried out and illustrate the effect of photoconductivity on relaxation of charge on lead azide and tetracene. Lead styphnate did not show the effect.

Table IV. Charge Relaxation Times of Explosives at 25°C and 40–50% Relative Humidity

Sample	Relaxation time (sec)	Resistivity (ohm-cm)
Powders used in electrification studies		
Propellant		10^5
RD1333 lead azide	17 (6)[a]	10^{14}
RDX (Class B, Type A)	1980	10^{16}
Other explosives		
PVA lead azide	29 (13)	10^{14}
Lead azide with 0.1% antistatic agent	1	10^{13}–10^{12}
Basic lead styphnate	275	10^{15}
Tetracene	2400 (150)	10^{16}

[a]Values in parentheses are for samples illuminated by UV light and are indicative of photoconduction.

The values of resistivity were obtained by assuming that relaxation follows similar behavior to that in a uniformly charged body of liquid and are a function of a characteristic time given by the expression

$$\Upsilon = \xi\xi_0\rho$$

where Υ is the relaxation time in seconds, ξ is the dielectric constant, ξ_0 is the permittivity of space, 0.885×10^{-13} farad/cm, and ρ is the resistivity. For convenience, the product $\xi\xi_0$ was taken to be 2×10^{-13} in all cases. It should be noted that the "resistivity" is only an empirical value which is dependent on particle shape, number and nature of contacts with other particles and the substrate and, of course, conduction. In the case of most explosives the latter will be very markedly affected by surface conduction.

Inspection of the resistivity values shows a definite trend, or correlation, with charging behavior. Thus, one may estimate that contact charging drops off for powders below about 10^{11} ohm-cm, while contact charging becomes predominant above about 10^{15}.

Addition of commercial antistatic agents was found to increase relaxation rates, although the effect appeared to be sensitive to small residual amounts of alcohol. Humidity control still appears to be the most effective practical means of controlling charging and increasing charge relaxation. Pollock found, for example, that increasing the relative humidity from about 10% to 66% increased the relaxation rates by about an order of magnitude.

5. Discussion

The chief value of electrostatic property data and knowledge of the manner in which electrostatic charges behave on powders is in prevention of accidents and in avoiding nuisances.

For electrostatic ignition to occur, it is necessary to meet four conditions: An electrostatic charge must have been generated; a means must exist by which the charge, usually dispersed, can be brought to a point and create a localized discharge; an electrostatically ignitable substance must be present; exposure of the substance to the discharge, or coupling between the discharge and the explosive, must exist.

The factors considered in this section have concerned the generation and magnitude of the charge, and in considering charge relaxation or resistivities, there is implicit recognition of the mechanism of localization. The matter of the degree of sensitivity of the explosive, and of the coupling between the discharge and explosive, pertains to the subject of sensitivity and the measurement and design of suitable test methods and is dealt with in Chapter 5. It may, however, be mentioned that a quantitative estimate of hazard could be made by comparing the energy (or possibly the charge) available with that required to cause initiation.

It is difficult to determine the energy possessed by a mass of charged powder, since the energy is a function of the charge and capacity with respect to the surroundings, and this latter is usually indeterminate or at best difficult to approximate. If the resistivity of the powder in question is not too high, any charge on it will soon dissipate to the container. If the latter is grounded, no energy is possessed by the powder or its container. If the resistivity is very high, the charge will tend to remain on the powder, but by the same token will not be likely to be completely available or easily localized, and will, therefore, not be able to sustain a very intense discharge. Insight into the importance of such charges appears to be lacking, and there is a need for experimental and analytical investigation. In the meantime, it seems prudent to follow the rule of keeping charges to a minimum in any locality and equipment connected with the processing or handling of explosives.

A second possible source of difficulty due to electrostatic charging is that powders that tend to retain their charges can give rise to processing difficulties. Either they tend to cling to metal parts and make them difficult to handle, as noted by Braid *et al.*, or if the powder is very fine, mutual repulsion of like charges will tend to disperse them, and this may give rise to a very great potential for ignition of dust films or of dust clouds.

There appears to be little data on commercial quantities of explosives. However, it has been shown that relaxation measurements can be used to correlate with probable electrostatic behavior. Such measurements are easily made with safety, and the method lends itself to a variety of studies, since only a few milligrams need be used. It should be noted that none of the explosives studied exploded due to the corona discharge, although the charge was estimated to be two to three orders of magnitude greater than would be possible due to normal contact or induction.

REFERENCES

1. P. Guenther, R. Meyer, F. Mueller-Skjold, *Z. Phys. Chem.*, *A175*, 154 (1935).
2. P. Gray, T. Waddington, *Nature*, *179*, 576 (1951).
3. O. Loew, *Ber. Dtsch. Chem. Ges.*, *24*, 294 (1891).
4. L. Smith, C. Wolf, *J. Med. Res.*, *12*, 451 (1904).
5. R. Stern, *Klin. Wochenschr.*, *6*, 304 (1927).
6. W. Biehler, *Arch. Exp. Pathol. Pharmakol.*, *126*, 1 (1927).
7. F. Hildebrand, K. Schmidt, *Arch. Exp. Pathol. Pharmol.*, *187*, 155 (1937).
8. J. Graham, *Arch. Int. Pharmacodyn.*, *77*, 40 (1948).
9. J. Graham, J. Rogan, D. Robertson, *J. Ind. Hyg. Toxicol.*, *30*, 98 (1948).
10. J. Graham, *Br. J. Pharmacol.*, *4*, 1 (1949).
11. E. Werle, R. Fried, *Biochem. Z.*, *322*, 507 (1952).
12. F. Roth, J. Schurr, E. Moutis, W. Govier, *Arch. Int. Pharmacodyn.*, *108*, 473 (1953).
13. E. Werle, F. Stucker, *Arzneim. Forsch.*, *8*, 28 (1958).

14. L. Schwartz, *Ind. Med.*, *11*, 457 (1942).
15. L. Fairhall, W. Jenrette, S. Jones, E. Prichard, *U.S. Public Health Rep.*, *58*, 607 (1943).
16. I. Sunshine, ed., *Handbook of Analytical Toxicology*, The Chemical Rubber Co., Cleveland, Ohio, 1969, p. 705.
17. E. Bassendowska, Z. Kowalski, K. Knobloch, S. Szendzikowski, *Med. Pracy.*, *16*, 187 (1965).
18. S. Lamnevik, Proceedings of the Symposium on Lead and Copper Azides, Explosives Research and Development Establishment, Waltham Abbey, England, Report WAA/79/0216, 1966, p. 75.
19. A. T. Thomas, Proceedings of the Symposium on Lead and Copper Azides, Explosives Research and Development Establishment, Waltham Abbey, England, 1966, p. 103.
20. S. Lamnevik, Proceedings of the Symposium on Chemical Problems Connected with the Stability of Explosives, Explosives Research and Development Establishment, Waltham Abbey, England, 1968, p. 21.
21. S. Lamnevik, Proceedings of the Symposium on Lead and Copper Azides, Explosives Research and Development Establishment, Waltham Abbey, England, 1966, p. 92.
22. H. Kramer, unpublished results.
23. F. Feigl, *Spot Tests in Inorganic Analysis*, Elsevier, Amsterdam, 1958, p. 331.
24. T. W. Stull, R. E. Stouder, Electrolytic Destruction of Lead Azide, Burlington Iowa Ammunition Depot, Burlington, Iowa, Tech. Rept. 191, 1970.
25. T. A. E. Richter, W. Fisco, unpublished results.
26. R. S. Valentine, H. Berzof, M. F. King, J. F. Kitchens, Lead Azide Disposal Process Development, Atlantic Research Corporation, Alexandria, Va., Tech. Rept. 4743, 1974.
27. J. Williams, U.S. Army Armaments Command, Rock Island Illinois, personal communications, 1973.
28. H. Kramer, M. Warman, unpublished results.
29. J. R. Drugmand, A Comparison of Two Alternate Methods of Storing Lead Azide to the Present Method, M.S. thesis, Texas A&M University (1972).
30. Military Explosives, U.S. Department of the Army Technical Manual TM9-1300-214, 1967.
31. R. J. Graybush, F. G. May, A. C. Forsyth, *Thermochim.* Acta, *2*, 153 (1971).
32. B. T. Fedoroff, H. A. Aaronson, E. F. Reese, O. E. Sheffield, G. D. Cliff, *Encyclopedia of Explosives and Related Items*, Vol. I, pp. A574–575, Picatinny Arsenal, Dover, N.J., 1960.
33. R. Stout, *Trans. Faraday Soc.*, *41*, 87–95 (1945).
34. S. Lamnevik, Proceedings of the International Conference on Research in Primary Explosives, Paper 4, Explosives Research and Development Establishment, Waltham Abbey, England, March 17–19, 1975.
35. J. D. Hopper, unpublished results.
36. W. H. Rinkenbach, J. D. Hopper, A Study of the Explosive Characteristics of Commercially Prepared Lead Azide, Tech. Rept. No. 852, Picatinny Arsenal, Dover, N.J., 1937.
37. A. C. Forsyth, Laboratory Investigations under Project TA 3-5101, Inter-Arsenal Communication, Picatinny Arsenal, Dover, N.J. 1955.
38. L. Avrami, H. J. Jackson, Results of Laboratory Studies on TRIEX-7 Dextrinated Lead Azide, Tech. Memo. No. 1877, Picatinny Arsenal, Dover, N.J., 1972.
39. A. C. Forsyth, J. R. Smith, D. S. Downs, H. D. Fair, The Effects of Long Term Storage on Special Purpose Lead Azide, Tech. Rept. No. 4357, Picatinny Arsenal, Dover, N.J., 1972.
40. A. C. McLaren, *Research*, *10*, 409 (1957).
41. B. Reitzner, *J. Phys. Chem.*, *65*, 948 (1961).

42. G. Todd, R. Eather, T. Heron, Proceedings of the Symposium on Lead and Copper Azides, Explosives Research and Development Establishment, Waltham Abbey, England, Report WAA/79/0216, 1966.
43. G. M. Thornley, The Reactions of Lead Azide with Carbon Dioxide and Water Vapor, University of Utah Report, 1963.
44. N. J. Blay, Proceedings of the Symposium on Lead and Copper Azides, Explosives Research and Development Establishment, Waltham Abbey, England, Report WAA/79/25–26, 1966.
45. H. Kramer, The Kinetics of the Reaction of Lithium Chloride and Lead Azide Exposed to the Atmosphere, Tech. Rept. No. 3842, Picatinny Arsenal, Dover, N.J., 1969.
46. M. Feitknecht, M. Sahli, *Helv. Chim. Acta*, *37*, 1433 (1954).
47. S. Lamnevik, Proceedings of the Symposium on Lead and Copper Azides, Explosives Research and Development Establishment, Waltham Abbey, England, Report WAA/79/0215, 1966.
48. H. Kramer, unpublished results.
49. G. Todd, *Helv. Chim. Acta*, *54*, 2210 (1971).
50. G. Todd, personal communication, 1973.
51. E. C. Franklin, *J. Am. Chem. Soc.*, *56*, 568 (1934).
52. T. Curtius, J. Rissom, *J. Prakt. Chem.*, *58*, 291 (1898); *61*, 408 (1901).
53. N. J. Blay, I. Dunstan, Compatability Testing of Primary Explosives and Pyrotechnics, Ministry of Defense, Explosives Research and Development Establishment, Tech. Rept. 115, 1973.
54. P. Karrer, *Organic Chemistry*, Elsevier, New York, 1939, pp. 351–352.
55. F. J. Welcher, *Organic Analytical Reagents*, Van Nostrand, New York, 1946.
56. F. R. Seavey, E. B. Kerone, U.S. Pat. 2000959 (1935).
57. L. Loeb, *Static Electrification*, Springer Verlag, Berlin, 1958.
58. W. R. Harper, *Contact Electrification*, Clarendon Press, Oxford, 1967.
59. D. J. Montgomery, in: *Solid State Physics: Advances in Research and Application*, Vol. IX, F. Seitz, D. Trumbull, eds., Academic Press, New York, 1959.
60. B. D. Pollock, Charging and Charge Relaxation in Explosives, Proceedings of the 16th Explosives Safety Seminar, Hollywood, Fla., 1974.
61. B. D. Pollock, W. B. Zimny, C. R. Westgate, Electrification of Lead Azide Powders under Ambient Conditions, Tech. Rept. No. 4214, Picatinny Arsenal, Dover, N.J., 1971.
62. H. Rathsburg, L. Schmitz, *Chem. Eng. Tech.*, *21*, 386 (1949).
63. P. E. Braid, R. C. Langille, A. M. Armstrong, *Can. J. Technol.*, *34*, 45 (1956).
64. A. C. Cleves, J. F. Sumner, R. M. H. Wyatt, in: *Proceedings of the Third Conference on Static Electrification*, D. K. Davies, ed., Static Electrification Group, Institute of Physics, London, 1971.
65. Y. Nakano, Y. Mizushima, *J. Ind. Explos. Soc., Jpn.*, *32*, 305 (1971).
66. C. Orr, J. M. Dallavalle, *Fine Particle Measurement*, Macmillan, New York, 1959, pp. 42–52.
67. J. W. Petersen, *J. Appl. Phys.*, *25*, 501 (1954).
68. L. W. Zabel, A. R. Estcourt, *J. Appl. Polym. Sci.*, *13*, 1909 (1969).
69. J. E. Lawver, J. L. Wright, *Proc. AIME*, *244*, 78–82 (1969).

4

The Sensitivity to Impact and Friction

L. Avrami and R. Hutchinson

A. INTRODUCTION

The sensitivity of an explosive is a measure of the ease or minimum amount of energy required to initiate fast decomposition in the material [1-5]. It is not a quantity which can be defined precisely, and as indicated in the introduction to Volume 1, it is often expressed as a probability that an explosion will occur following the application of a given stimulus at a specified level. Very often the nature of the decomposition regime initiated is not determined with any precision, although in assessing the potential consequences it is clearly important to know whether the regime is stable and whether detonation is achieved.

Sensitivity plays a pivotal role in explosives technology, because on the one hand it is indicative of the hazards associated with handling a material, and on the other hand it is a key parameter determining the effectiveness of an explosive in an explosive train. In the former case occurrences of low probability are of interest, while in the latter case reliable functioning of an azide demands certainty that it will detonate in response to a given stimulus.

The term sensitiveness is sometimes used, particularly in British literature, to define the quality of being sensitive, when it connotes the hazard level associated with an explosive. In this usage sensitivity is a term reserved for the effectiveness or performance level of the explosive. The distinction is by no means universally accepted, and as will be clear from the content of this chapter, both measures are often based on explosions which occur with a relatively high probability. When testing sensitive explosives, such as azides, the range of stimuli directed at samples does not vary greatly, and the probability ranges encountered are such that the measurable hazard and performance levels approach each other closely. Thus, the single term sensitivity will be used here, but it should be noted that in general neither measurement techniques nor the statistical theory are well enough developed to provide satisfactory data on events which occur with low probability, unless a corresponding large number of trials are made.

1. Development of Sensitivity Tests

The lack of a quantitative understanding of the mechanism of initiation of solid explosives, as reflected in the discussions of Chapters 6–9 of Volume 1, led to the development of empirical tests to relate the sensitivities of different materials to significant stimuli, such as shock, impact, friction, electric discharge, and heat.

Since the turn of the century, the development of empirical tests has been directed primarily toward simulating a hazardous situation in the manufacture or use of explosives, and only comparatively recently have tests been developed specifically to quantify performance levels. Some of the latter are discussed in Chapter 7. Many variations of tests were developed to determine the sensitivity of explosives to each type of stimulus, and it was soon apparent that consistent values are not obtained even with ostensibly the same apparatus. In general, relative values obtained for one stimulus do not necessarily rank materials with respect to another stimulus.

The nonreproducibility of data implies that the tests and the meaning of the term "sensitivity" need further study and understanding [6]. Questions arise as to whether certain tests actually measure the effect of the specified stimulus, even as one among several stimuli directed at the test sample by the apparatus. Kistiakowsky and Connor [1] concluded that a definite mechanical sensitivity of an explosive was not found. Koenen *et al.* [2] and Afanas'ev and Bobolev [5] also concluded that an intrinsic impact sensitivity of an explosive does not exist, but is a function of the test method, container, compression, arrangement of the sample, state of the sample tested, its method of preparation, particle size and shape, and possibly the climactic conditions during testing. By extension these factors can also be expected to affect sensitivities to other stimuli.

This chapter deals explicitly with the sensitivity of the explosive azides to impact and friction. Chapters 5 and 6 discuss, respectively, their sensitivity to electrostatic discharge and to heat and radiation. Shock sensitivity, although a form of impact sensitivity, is discussed in detail in Chapter 7. Although these are not the only types of stimulus that may be applied to an explosive, they are representative of the technology.

2. Sensitivity Relative to Other Explosives

The tests developed for azides are, in general, not different from those used for other solid explosives. However, in common with other primary explosives,

Table I. Relative Sensitivities of Explosives According to Laboratory Tests

Explosive	Impact (cm)		Friction[a]	Static electricity (J) [7]		Heat stability[b]		Minimum
	Picatinny Arsenal	Bureau of Mines	Picatinny Arsenal	Unconfined	Confined	Explosion temp. (°C)	Vacuum stability (ml)	detonating charge[c] (g)
Ammonium picrate	43.2	>100	U	0.025	6.0	318[d]	0.4	0.4
Black powder	40.6	32	S	>12.5	0.8	427[e]	0.9	
TNT	35.6	>95	U	0.06	0.44	475[d]	0.23	0.27
HMX	22.9	60	E	0.40		327	0.45	0.30
Tetryl	20.3	26	C	0.007	0.44	257	1.0	0.10
PETN	15.2		C	0.06	0.21	225[d]	0.5[f]	0.03
Lead styphnate	20.3		E	0.0009	0.0009	282	0.3	0.001
Lead azide	7.6	75	E	0.007	0.007	340	0.07	

[a] Qualitative results: E, explodes; C, crackles; S, snaps; U, unaffected.
[b] Explosion temperature test: 5-sec value in open cup. Vacuum stability test: Amount of gas evolved after 40 hr. at 120°C.
[c] Amount of lead azide required to shock-initiate explosive.
[d] Decomposes.
[e] Ignites.
[f] Value at 100°C, value at 120°C is >11 ml.

the level and range of stimulus values applied are considerably smaller than for secondary explosives, and this has led to the development of apparatus specifically for azides. In addition, because of their great sensitivity, azides are handled using special procedures, and tests have occasionally been developed to simulate the procedures and to detect significant differences in the parameters of the explosives being handled.

For purposes of comparison the sensitivities of lead azide to various stimuli as measured in some common laboratory tests are given in Table I. The values given are not definitive except with respect to the particular apparatus and samples used, but they do indicate azide sensitivities in relation to those of other explosives measured with the same apparatus. The tests referred to in Table I are described later in the chapter.

B. IMPACT SENSITIVITY

1. Character and Variety of Tests

The sensitivity of azides to the impact of the falling hammer or weight, or to collisions with other massive objects, is among the properties commonly determined when a compound or artifact is first prepared [5,8]. A crude impact test is easy to devise, but it is only one of several tests that assess the effect of comparatively weak mechanical action on explosives (the effect of strong shocks on azides is discussed in Chapter 8, Volume 1, and in Chapter 5). However, the impact sensitivity is not easy to quantify precisely [5] because the mechanical action may be separated into the effect on a confined explosive and the effect on an unconfined explosive. In practice it is difficult to distinguish the degree to which the two kinds of action occur simultaneously in a porous sample. In Chapter 8, Volume 1, Chaudhri and Field summarize some of the general principles involved in the design and application of apparatus that quantify the response of explosives to this type of stimulus.

Basically, the apparatus drops a weight from a known height onto a steel striker resting on the explosive, which, in turn, rests on a steel anvil. Impact tests differ from shock sensitivity tests, such as gap tests, flyer-plate tests (Chapter 7) both in the amplitude and the duration of the impulse imparted. Shocks characteristically deliver peak stresses of the order of tens of kilobars over periods of about a microsecond. The characteristic response times in falling-weight impact tests are at least two orders of magnitude greater (depending on the dimensions of the hammer), while peak stresses are at least two orders of magnitude less [4,9].

The ordering of explosives in terms of their response to impact was started at the beginning of the century with the work by Kast [10] and Rotter [11]. Since then, the procedure has been adopted as a standard test by laboratories in many countries [5,12]. Various modifications to particular apparatus have been introduced to make them suitable for different explosives. Even apparatus of similar design frequently do not give consistent data, and there are investigators who believe that the structural foundations and nature of the subsoil have a significant effect on data given by the larger machines. Apparatus is commonly named for the establishment where it was developed, such as the U.S. Bureau of Mines, the Explosive Research Laboratory (ERL), Bruceton, Picatinny Arsenal, or for the designer, such as the Rotter machine, or simply for the type of weight used, such as the ball-drop or drop-weight machines.

A brief description of representative apparatus follows. Test results on lead azide are available for all these machines for analysis later in this chapter.

a. Bureau of Mines Impact Apparatus

The Bureau of Mines apparatus (Figure 1) subjects the explosive sample to an impact of a free-falling 5-kg weight. The weight may be raised to any desired height up to 330 cm, where it is held by an electromagnet until allowed to fall on to a hardened steel anvil and a plunger. The plunger transmits the impact to a striking pin that fits into a steel cup containing the test sample (approximately 35 mg). The stainless-steel striking pins (0.5 in. diam by 0.75 in. long) slide freely in the sample cups (0.5 in. diam and 0.25 in. deep).

The weight is raised to a given height and allowed to drop on the plunger. If

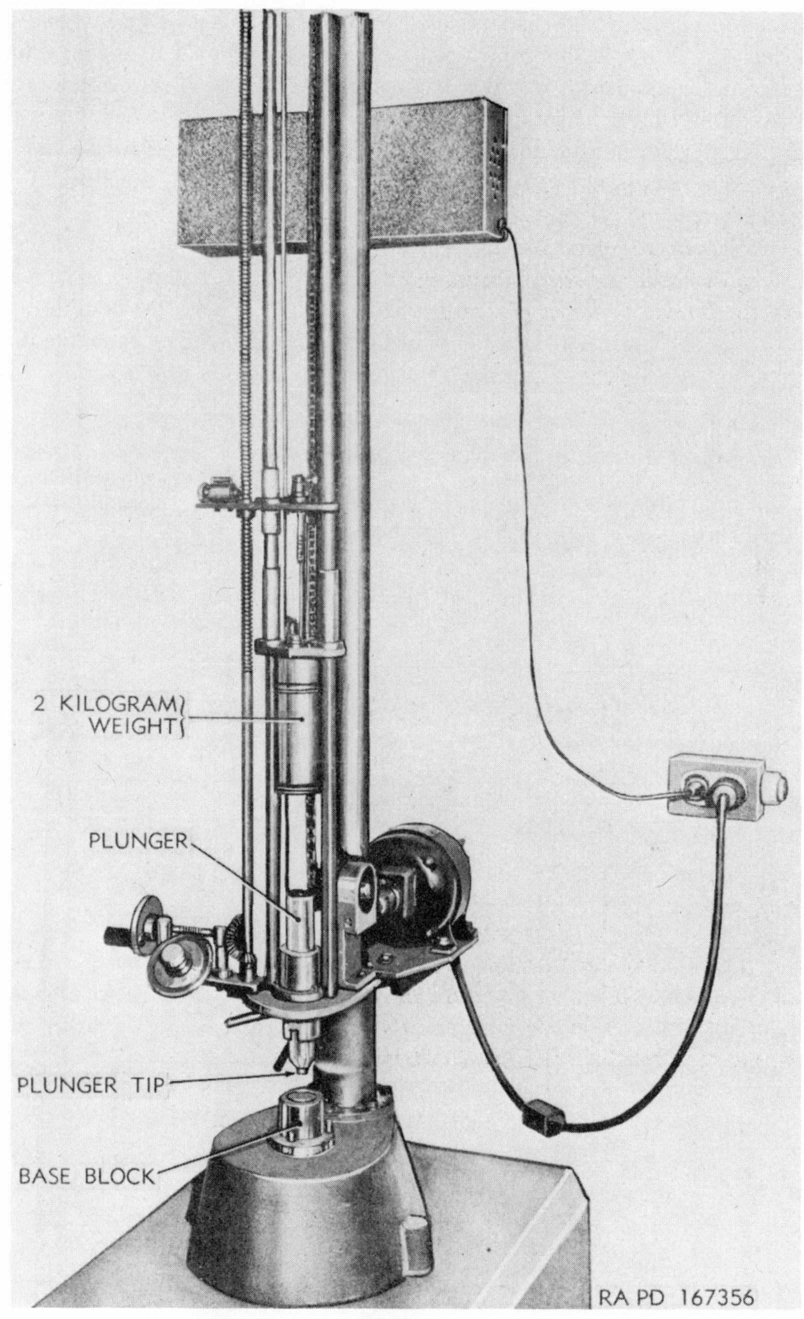

Figure 1. U.S. Bureau of Mines impact apparatus [12].

no reaction occurs, the weight is raised to successively greater heights and the test repeated with a fresh sample until a reaction occurs or until the maximum range of the equipment is reached. If a reaction occurs, fresh samples are tested at successively lower heights until a point of no reaction is reached. Once this height is determined, a sample is tested at a given increment below the level, if the previous sample reacted, and at a given increment above the level, if the previous sample did not react.

Two laboratories that use this apparatus are the U.S. Bureau of Mines at Bruceton, Pennsylvania, and Picatinny Arsenal, Dover, New Jersey. The apparatus at the Bureau of Mines has been instrumented [13] with the addition of a force gauge and a reaction detector, which make it possible to examine the force exerted on the sample and to make estimates of the reaction times.

b. ERL Impact Apparatus

This apparatus was developed during World War II by the Explosives Research Laboratory (ERL) of the U.S. National Defense Research Committee then also located at Bruceton, Pennsylvania. Figure 2 depicts its principal features, which are similar to those of the foregoing Bureau of Mines apparatus.

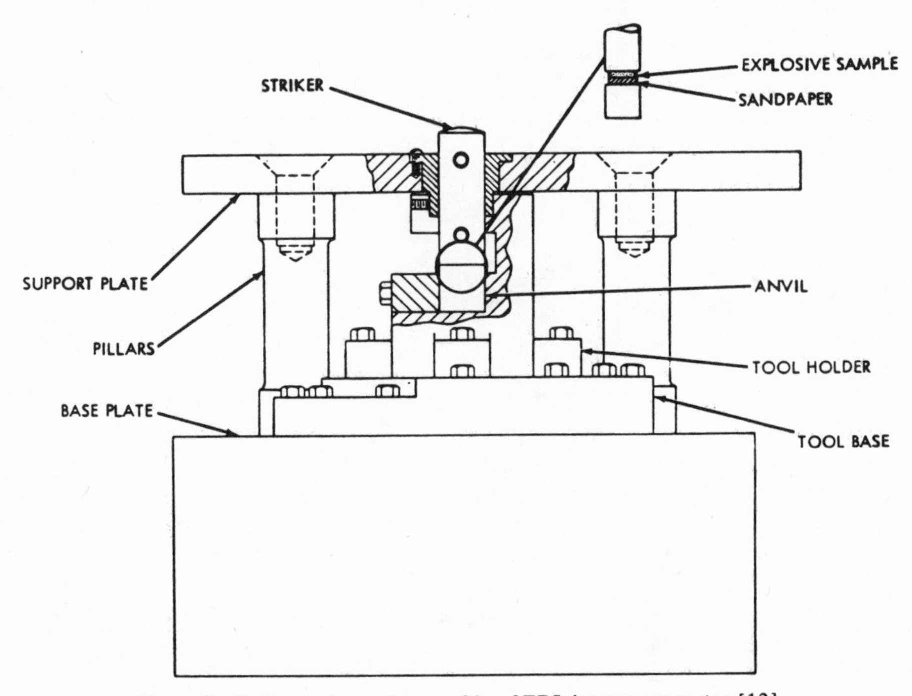

Figure 2. Striker and sample assembly of ERL impact apparatus [12].

The weight is dropped from heights up to 320 cm and is usually 2.5 kg although 2- and 5-kg weights are alternatives. The 35-mg sample rests as a loose pile in the center of a 1-in.-square piece of sandpaper, placed on a hardened steel anvil, 1.250 in. diam. The 1.25-in.-diam striker rests on the explosive and slides freely within a guide. A "noisemeter," an audiovisual instrument, is used to assist the operator in his judgment [4].

The ERL machine is used at the Naval Surface Weapon Center, White Oak, Maryland, and also at the Los Alamos Scientific Laboratory (LASL) and Lawrence Livermore Laboratory (LLL) with modifications.

At LASL the original hollow drop weight has the same geometry and weight but with a hardened steel core surrounded by a solid case of aluminum alloy. The design concentrates more of the mass along the line of impact. The anvil and holder are of one piece and instrumented with strain gauges. No sandpaper is used, but the striker and anvil are roughened by sandblasting.

c. *Picatinny Arsenal Impact Apparatus*

This apparatus (Figure 3) consists of an anvil, two guide bars equipped with an adjustable support for a weight, a supply of hardened steel vented plugs and die cups, and brass die cup covers. The steel anvil is supported by a steel sheet which in turn is imbedded in a large block of reinforced concrete. Two guide bars are approximately 6 ft high, and their inner sides have tongues which guide the falling weight. A yoke equipped with a release pin is attached to the bars so that the weight can be moved to any desired height and held in place by hand screws. The apparatus is used with weights of $\frac{1}{2}$, 1, or 2 kg. The lighter weights are used with the more sensitive materials, such as lead azide. Samples are passed through No. 50 and retained on No. 100 US standard sieves. A brass cup is pressed down until in contact with the top rim of the die cup, and a vented plug is placed on top of the cap. The falling weight drives the plug into the explosive. An explosion is defined as any audible or visual evidence of decomposition, such as a crack, flash, smoke, or charring.

In comparison with the Bureau of Mines apparatus, the Picatinny Arsenal apparatus confines the sample more completely, distributes the translational impulse over a smaller area, and is often believed to introduce more friction between the parts of the apparatus and the explosive.

d. *Rotter Impact Apparatus*

The Rotter machine (Figure 4) contains a steel anvil on a steel base. A short striker passes through a gland, and a second striker, resting upon the first, receives the blow from the falling weight. The 5-kg weight can be raised to a desired height by means of a winch and released by an electromagnet; a device

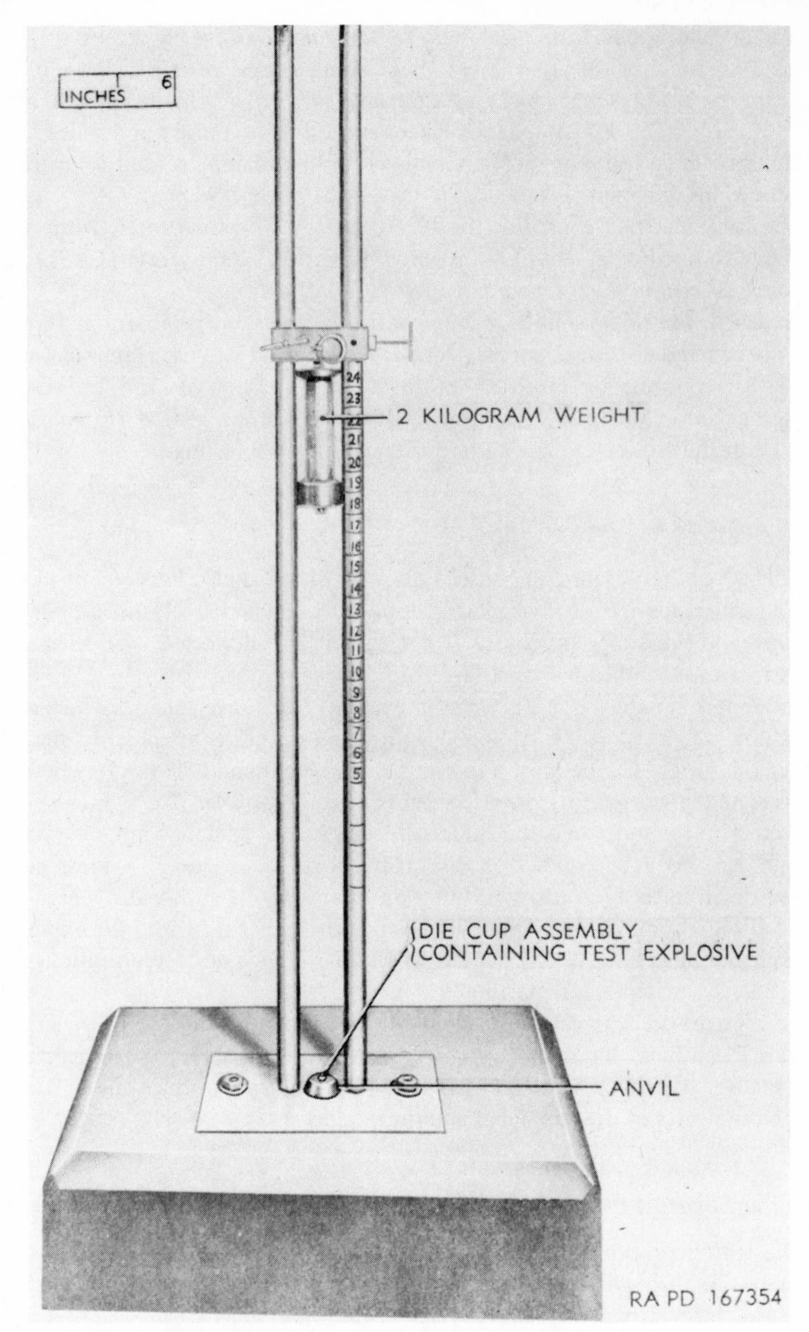

Figure 3. Picatinny Arsenal impact test apparatus.

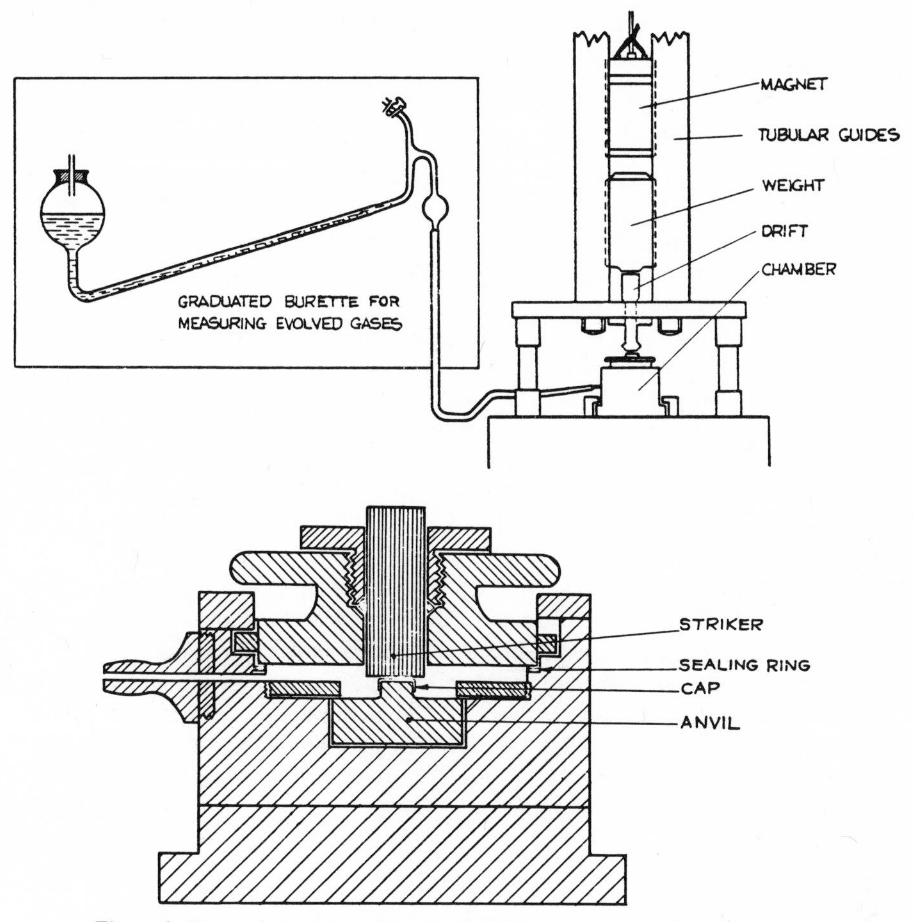

Figure 4. Rotter impact apparatus (top) [12]; explosion chamber (bottom).

for arresting the weight after rebound is provided. Brass caps contain the 40-mg sample, usually in the form of a powder, and are fitted over the spigot of the anvil, thus confining the explosive layer between a steel and a brass surface.

The housing is sealed so that on impact the gas evolved can be recorded by a gas-measuring buret. If the volume of gas evolved by a 0.03-mg sample is more than 1 ml, then the test is considered an explosion; a smaller gas evolution is considered a nonexplosion.

The Rotter test is used throughout Great Britain and in Canada, where it has been further instrumented [11]. The volume of gas is measured by a photocell attached to the manometer, the height for subsequent drops is selected automatically depending on the results of the previous drop, a series of samples is accommodated, and the results are automatically recorded on an electric typewriter.

e. Ball-Drop Apparatus

The ball-drop test (Figure 5) is the simplest and has been used for many years to determine the impact sensitivity of lead azide, particularly in industrial laboratories, such as those of the duPont de Nemours Company in the U.S.A. and the Nobel Explosives Company in Scotland. A $\frac{1}{2}$-in.-diam steel ball weighing 8.35 g is dropped from heights varying by 1-in. increments onto azide powder spread in a 0.03-in layer on a hardened and polished steel block. After each explosion, the lead deposited on the block is cleaned off and the ball and block are replaced whenever their surfaces become noticeably affected.

Factors which affect results include the mass per unit area of the explosive, as determined by its bulk density and thickness. Higher masses decrease the height for a 50% probability of explosion by about 25 in. per g/ml. The surface finish, lead contamination on the block and ball, and the moisture content of the sample also affect results.

In some instances the ball is dropped by allowing it to roll off an inclined track and is, therefore, rotating when it hits the sample. In the case of a no-fire, the ball bounces away from the block and only impacts the sample once, thus overcoming one of the inherent problems of multiple impacts produced by drop hammers. In other cases the ball is retained at the drop height and is released electromagnetically. The ease of sample testing, the low cost of the expendable

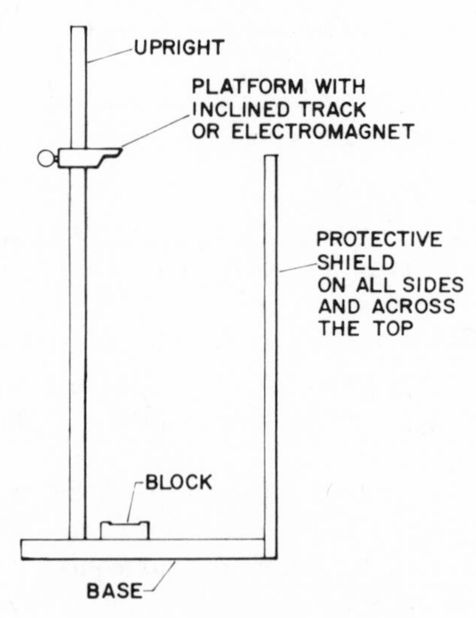

Figure 5. Ball-drop apparatus.

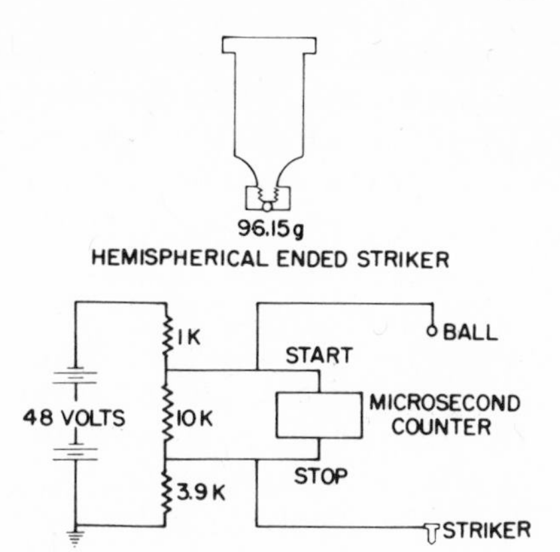

Figure 6. Striker and circuit for ball and disk impact apparatus [15].

parts, and the spread of data make this test desirable for explosives as sensitive as lead azide. Because of the impractical drop heights required, it is not convenient for less-sensitive explosives.

f. Ball and Disk Impact Apparatus

This apparatus is used in British laboratories for testing primary explosives. The description by Wyatt [15] is based on a technique first described by Copp *et al.* [16]. The apparatus employs a 95-g steel ball as the falling weight and a striker of 96.15 g to which is fitted a $\frac{5}{32}$-in. hardened steel ball as the impacting surface (Figure 6). The sample is placed on a $\frac{3}{4}$-in.-diam steel anvil by means of a filling plate 0.018 in. thick having a hole 0.157 in. in diam. A brass disk, 0.025 ± 0.005 in. thick, is placed on top of the explosive. The procedure normally follows that described for the Bureau of Mines apparatus, but uses a logarithmic distribution of heights with an increment of 0.075 in. A circuit for measuring the time of drop to impact is shown in Figure 6.

2. Design and Analysis of Impact Sensitivity Experiments

In principle a sample should remain unaffected, undergo unsustained decomposition, deflagrate, or detonate under the stimulus of a drop weight, and identical samples should react in the same way. In practice, no two samples are identical as to packing density, particle arrangement, etc., and no two stimuli are identical, due to variations in hammer alignment, guide friction, etc., so experimental results are statistical in nature. From low elevations the weights produce no reactions even in extended trials and from high elevations always produce deflagrations or detonations; intermediate elevations result in reactions in some of the trials, with the percent reacting being a function of the elevation of the weight.

Variations in procedure revolve principally around the number of drops, required to make a valid determination of the "sensitivity," and the sequence in which the successive heights are varied. The sequence of height selection depends on the subsequent analysis of the results and the level of probability at which the test is being conducted. The two most common sequences are the "up-and-down" and the "run-down" methods.

The up-and-down method, also called the Bruceton or staircase method, is a widely used statistical procedure for the determination of the height at which 50% of the samples detonate or react and its standard deviation [17,18]. The advantages and disadvantages of the method are described elsewhere [18-20]; basically the test gives a reliable 50% value but not a reliable standard deviation. The number of tests required at each height also has been investigated [21].

Investigators who use the Rotter machine often carry the data analysis further. After obtaining the height at which 50% of the samples react using the up-and-down method, explosives are compared by a "figure of insensitiveness." The height for the sample is compared to the corresponding height for a standard explosive and multiplied by the FI of the standard. Therefore, the FI of a sample is as follows:

$$\frac{\text{Median 50\% height of sample}}{\text{Median 50\% height of standard}} \times \text{FI of standard}$$

The standard normally employed is a specially prepared RDX having an assigned FI of 80.

The run-down method was developed at Frankford Arsenal, Philadelphia, and produces a more complete frequency and probability of reaction curve [22]. Although a larger number of tests is used than in the up-and-down procedure, the method makes possible a better evaluation of the distribution of the population. Starting at a level expected to be between 0% and 100% reactions levels, 20–25 tests are made at each of several levels above and below the starting level until the 0% and 100% levels are reached. The increments between the levels normally are equal to or less than the expected standard deviation. A cumulative probability curve is then plotted from the results of the entire test which is considered to be the frequency distribution of the population.

With the ball-drop apparatus, two procedures are followed. These are the up-and-down method, where at least 20 drops are made following the standard procedure, and the group-data procedure where 20 drops are made at five consecutive levels differing by 1 in., one near the 50% point, two above, and two below it. The up-and-down method determines the height required to produce an explosion in 50% of the trials with a 95% confidence interval of 1.4 in., but does not accurately determine the standard deviation or the probability of firing at low stimulus levels. Data from the group-data method are plotted on probability graph paper and values of both the 50% point and the standard deviation are determined from the line that best fits the data points. For lead azide, the typical standard deviation is 2.3 in. The height to produce an explosion in 10% of the trials can also be determined from the group data, and this is frequently of more interest from a safety standpoint.

With respect to the 10% point, a method used at Picatinny Arsenal determines the minimum height at which at least one of ten trials results in a reaction.

The level corresponding to a 50% probability of initiation is the value reported by most in the field of impact sensitivity. However, as indicated in the introduction and in the previous paragraph, for greater assurance with respect to hazards, the 10% probability or, as close as possible, the zero probability levels are more appropriate.

Normally to determine the latter points with confidence requires many more tests than the 50% value. However, a method developed by Einbinder [23] describes a technique using a sequential sensitivity test strategy and estimation methodology to determine extreme percentage points of a response function. This method appears to be more efficient for the extreme points, and also no loss of accuracy results in the estimation of the 50% point. The Einbinder method–sometimes designated as the one-shot transformed response (OSTR) strategy test–is robust to many forms of the underlying response distribution, does not require a possibly limiting assumption of normality and stimulus step size, and is insensitive to the choice of initial stimulus level. This leads to a significant reduction in the number of tests at the extreme points without any apparent loss in reliability [24].

Other statistical methods include the probit, normit, and logit procedures. However, these are not data-collecting but analytical procedures for the estimation of the distribution. They may be used with data collected by the up-and-down or the run-down methods [25].

The need often arises to determine whether the sensitivity of a product has changed as, for example, during storage or through variations in a manufacturing process or to determine whether any difference exists between two lots made by the same method. To achieve this, Kemmey [26] combined the χ^2 or goodness-of-fit test with the Karber test to rate the materials in terms of relative sensitivity. When the χ^2 test indicates that the two samples are significantly different, the Karber method is useful in arranging them according to sensitivity. In the χ^2 test, any level of confidence can be chosen, and the degrees of freedom correspond to the number of drop heights used.

The various parameters and characteristics of the impact tests, and of the materials tested, should be included and interpreted with the data. Levy [27] derived a probability equation based on the Picatinny Arsenal machine and the explosive tested which required that only a few points be obtained on a % fire curve in order to determine the sensitivity of a sample.

The definition of a reaction affects the results. At many laboratories, any visual observation of burning, smoke, flash, or flame, or any audible noise, such as a crackle, pop, or bang, is considered a reaction. Audio devices are also used to detect reaction, or in the Rotter test the volume of gas produced defines reaction. With primary explosives, such as lead azide, the reaction is usually well-defined since a detonation is the normal consequence.

3. Impact Sensitivity of Azides

Initiation by impact is generally considered to be a thermal process (see Chapters 8 and 9, Volume 1). Although the probability that a sample will initiate as a result of impact is a function of the energy absorbed by the sample and of its rate of application, sensitivities are normally reported as a function of the potential or kinetic energy of the weight dropped or as a function of the height from which it was dropped. This is done usually without stipulating the test conditions and because no current apparatus permits the energy absorbed by the sample to be determined. Ling and Hess [28] interpreted sensitivity data on the assumption that the initiation probability is controlled by the distribution of active sites among the crystal affected by the external energy source. Smith and Richardson [29] emphasized the importance of the rate of energy delivery, but the data are seldom available.

It is thus not surprising that, even neglecting differences in apparatus, data that relate the impact sensitivities of explosives are generally not consistent, a comment which applies even to the sensitivity of lead azide relative to that of secondary explosives. For example, there exist data by Kondrikov [30] showing that in a comparatively well-instrumented apparatus lead azide appeared less sensitive to impact than TNT, PETN, and TNB. Tables II and III illustrate the degrees of consistency obtainable with present apparatus.

Table II suggests that the ranking, or qualitative ordering, for the various explosives shows some agreement. However, the relative sensitivities of RDX and TNT are, for example, reversed when comparing the impact sensitivities with the Rotter apparatus [11] (Table III).

Another parameter which may be assessed in comparing data is the time delay for the initiation following impact. Cook [32] found the delay in primary explosives to be virtually independent of the potential energy of the drop weight (Table IV).

The data presented in this chapter for the sensitivities of azides are subject to the same kinds of inconsistency and are reported as a function of the potential energy of the drop weight or as the distance of fall of the weight. Matsuguma *et al.* [33] determined the sensitivity of colloidal lead azide using five weights between 0.5 and 3.0 kg (Figure 7). The same data, combining the weights and distances as potential energy, gave an envelope of curves (Figure 8), a presenta-

Table II. Impact Sensitivities of Explosives [31]

Explosive	ERL apparatus							
	NSWC 50%		LASL 50%		B Mines 50% (cm)	PA 10%[b] (cm)	PA apparatus (cm)
	cm	σ^a	cm	σ^a			
Lead azide	4	0.12				17	12.7
PETN	12	0.13	12	0.05	43	17	15.2
RDX	24	0.11	22	0.01	79	32	20.3
HMX	26	0.10	26	0.02		32	22.9
Tetryl	38	0.07	42	0.02	94	26	20.3
RDX/TNT/wax	60	0.13	59	0.02		75	35.6
TNT	157	0.10	154	0.03	183	98	38.1
Ammonium picrate	254	0.05	190				43.2

[a]Standard deviation of the 50% height.
[b]Heights at which at least one explosion was observed in 10 trials; 2.54 cm below these heights there were no explosions in 10 trials.

Table III. Impact Sensitivities with the Rotter Apparatus [11]

Explosive	Median height	Figure of insensitiveness
Lead 2 : 4 dinitroresorcinate[a] (standard)	61	11
Lead styphnate[a]	69	12
Lead azide[a]	113	20
PETN[b]	66	51
HMX[b]	73	56
RDX, military grade[b]	98	75
RDX, standard[b]	104	80
Tetryl[b]	112	86
Black powder[b]	117	90
RDX/wax, 91/9[b]	127	98
RDX/TNT, 60/40[b]	137	105
RDX/TNT/wax[b]	152	117
RDX/TNT/Al/wax[b]	190	146
TNT[b]	197	152
Black powder[b] (w/o sulfur)	253	194

[a]2-kg weight; lead 2 : 4 dinitroresorcinate = 11, standard.
[b]5-kg weight; 30-mg samples; RDX = 80, standards.

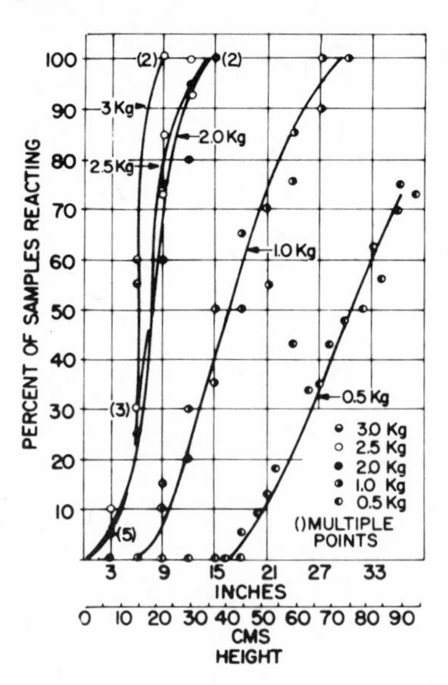

Figure 7. Impact sensitivity of lead azide as function of wieght and height [33]; Picatinny Arsenal apparatus; 20 samples per data point.

tion which may indicate some effect due to the different rates at which energy is applied by the weights.

a. Barium Azide

This inorganic azide has been investigated as a constituent of a binary explosive. The impact sensitivity data obtained using the Picatinny Arsenal apparatus (Figure 9) include that for barium azide made with 1-2% rhodamine B dye.

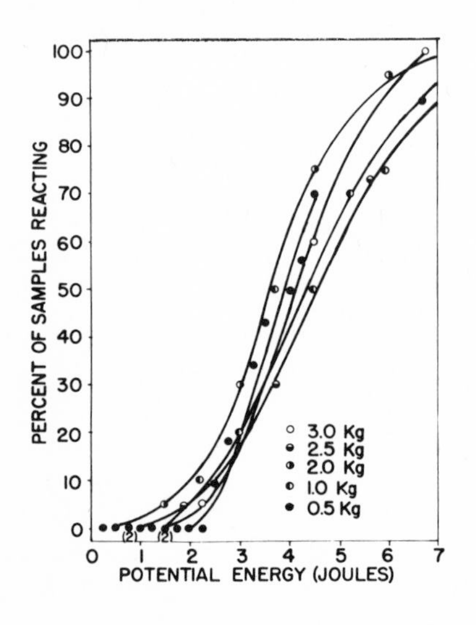

Figure 8. Impact sensitivity of lead azide as function of potential energy [33]; Picatinny Arsenal apparatus; 20 samples per data point.

Table IV. Impact Initiation Delay in Primary Explosives [32]

Impact energy (J)	Average delay (μsec)	Delay range[a] (μsec)
Mercury fulminate		
6.34	45	22–102
4.85	38	24–104
3.40	85	24–177
1.94	112	40–300
1.29	114	50–350
0.47	196	50–630
0.20	97	40–200
Lead azide		
6.34	106	40–295
4.85	164	40–380
3.40	296	91–450
1.94	318	50–620
0.98	197	40–810
0.47	114	40–300
Lead styphnate		
6.34	133	60–200
4.80	193	80–990
3.40	191	70–420
Diazodinitrophenol		
6.34	277	80–235
4.85	361	80–620
3.40	425	180–1335
1.94	494	240–680
0.98	274	100–620

[a]For approximately 20 initiations.

Pai-Verneker and Avrami [34] studied factors that affect the sensitivity of barium azide. An increase in surface area of the barium azide crystals increased the impact sensitivity. The effects of time, temperature, and of additions of about 1% sodium as an impurity suggest that aging of the material initially brings about a desensitization followed by a sensitization to the original level. The impurity also decreased the sensitivity, decreases from whatever cause generally being reductions up to 50% in the probability of reaction for a given drop height.

b. Copper Azides

The impact sensitivity of three forms of cuprous azide, obtained by a ball-drop technique, are shown in Figure 10. Singh [35] found that cuprous azide

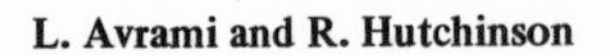

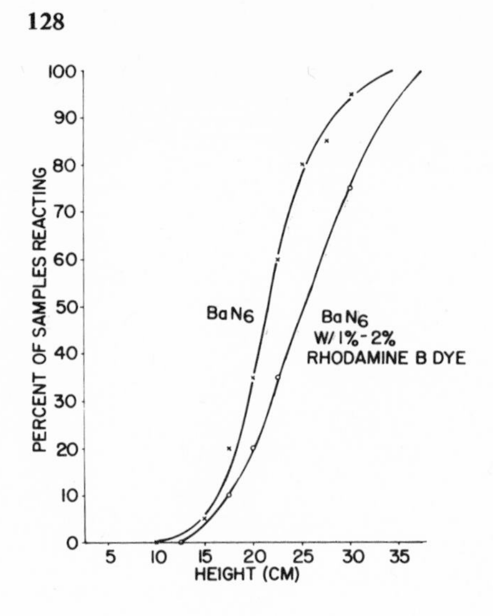

Figure 9. Barium azide impact sensitivity [34]; Picatinny Arsenal apparatus; 2-kg weight; 70°F; relative humidity, 55%; 20 samples per data point.

is more sensitive to impact than lead azide and that an increase in the crystal size of cuprous azide increases the sensitivity of that azide to impact. Data for the basic cupric azide are shown in Table V where they are compared with data for lead azides.

c. *Thallous Azide*

The effect of temperature on the impact sensitivity of thallous azide is shown in Figure 11. At −100°C a decrease was obtained when tested with the Picatinny Arsenal apparatus. The results at room temperature indicate that

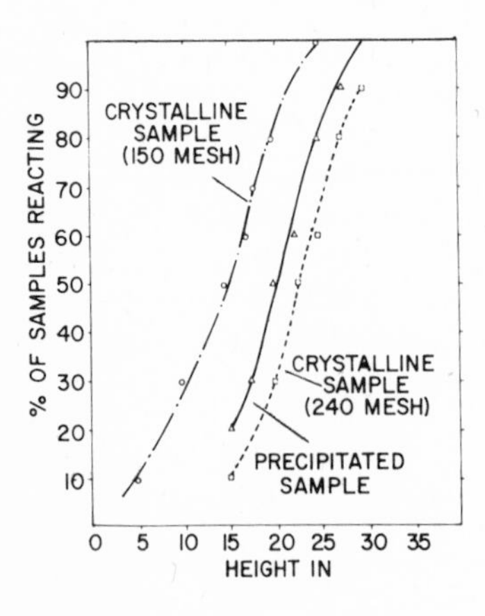

Figure 10. Impact sensitivity of cuprous azide on a ball-drop apparatus [35]; $\frac{7}{8}$-in. ball; 15 trials at each height.

Table V. Ball and Disk Impact Sensitivities of Azides [41, 42]

Explosive	Height for 50% reactions (cm)	Standard deviation (cm)
Service lead azide	15.24	0.106
Beta lead azide	15.04	0.078
Gamma lead azide	7.35	0.121
90% Service lead azide/10% beta lead azide	11.91	0.081
90% Service lead azide/10% gamma lead azide	10.31	0.149
Monobasic cupric azide, $Cu(N_3)_2 \cdot Cu(OH)_2$	7.26	0.170
Dibasic cupric azide, $Cu(N_3)_2 \cdot 2Cu(OH)_2$	11.04	0.057

thallous azide is less sensitive to impact than barium and lead azide; but more sensitive to changes in temperature than lead azide [36].

d. Silver Azide

In Chapter 1 different processes of preparing silver azide are described, and the impact sensitivities of samples made by processes are presented in Figure 12. They were tested with the Picatinny Arsenal apparatus and also with the ball-drop apparatus.

With the Picatinny Arsenal apparatus, one batch of silver azide made by the Costain process (Chapter 1) was almost as sensitive as the RD1333 lead azide (Figure 12). However, two other samples (one made by the Taylor process and the other by the Costain process) were significantly less sensitive. A difference between the two groups of over 25 in. was evident in heights that caused 50% of the samples to react under the action of a 2-kg weight.

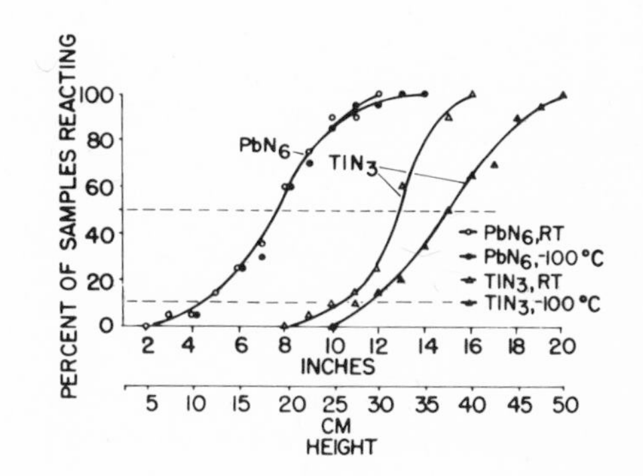

Figure 11. Effect of temperature on impact sensitivity of thallous and lead azides [36]; Picatinny Arsenal apparatus; 2-kg weight; 70°F; relative humidity, 60%; 20 samples per data point.

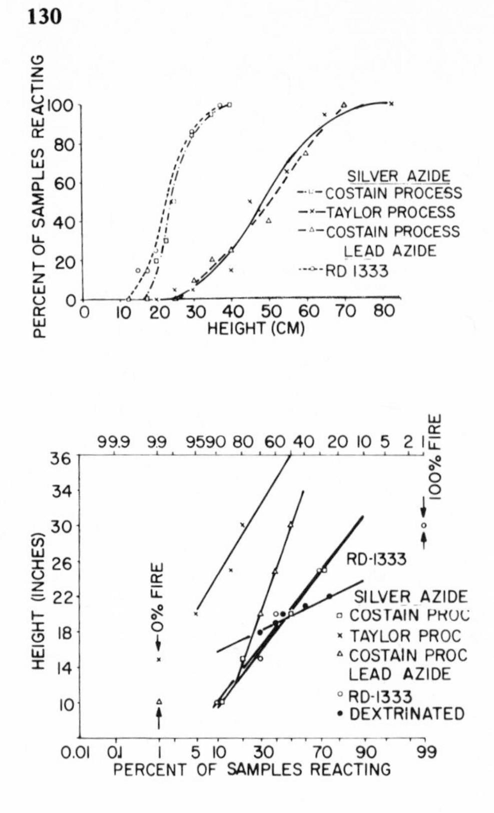

Figure 12. Impact sensitivities of silver and lead azides; Picatinny Arsenal apparatus; 2-kg weight; 68° F; relative humidity 60%; 20 samples per data point.

Figure 13. Ball-drop sensitivity of silver and lead azides [37]; 69°F; relative humidity 61%; 20 samples per data point.

In the ball-drop test the RD1333 lead azide and the Costain process silver azide had practically the same impact sensitivities, but the other two silver azides, although still less sensitive, showed a wider disparity (Figure 13).

Silver azide produced in England and known as composition RD1336 has an FI of 13-16 as obtained with the Rotter apparatus. On the ball and disk apparatus the following results were obtained:

Distance of fall (cm)	12	14	16	19	21
Percent reactions	4	18	38	76	88

e. Lead Azide

Of the many types of lead azide—pure or basic, dextrinated, colloidal, Service, polyvinyl alcohol, Special Purpose, RD1333, RD1343, RD1352, and dextrinated colloidal—the types which have been used for military or commercial

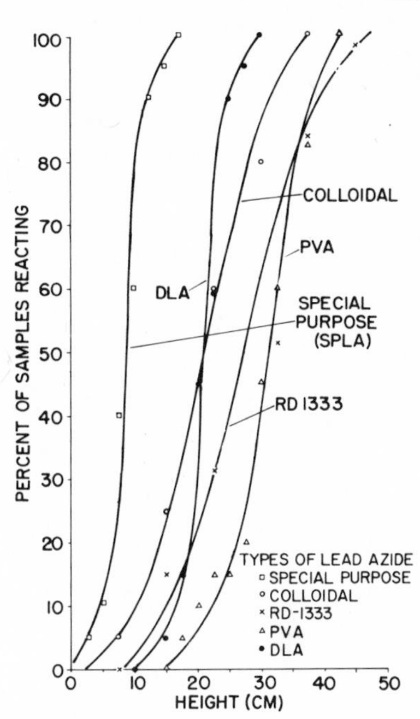

Figure 14. Impact sensitivities of lead azides; Picatinny Arsenal apparatus; 2-kg weight dropped at indicated heights; 70°F; relative humidity 55%; 20 samples per data point.

purposes have been dextrinated (DLA), British Service (SLA), polyvinyl alcohol (PVA), RD1333, Special Purpose (SPLA), and RD 1343.

Figure 14 displays the impact sensitivities of several types of lead azide using a 2-kg weight in the Picatinny Arsenal apparatus. The apparent greater sensitivity of the Special Purpose product has been attributed to the method of drying the material (see discussion below). Each of the data points for RD1333 lead azide is the average of 100 trials, while the points for SPLA are the average for 140 trials.

With the ball-drop apparatus (Figure 15) the sensitivity rank PVA > RD1333 was the reverse of that found with Picatinny Arsenal apparatus [38].

The impact sensitivity of DLA was determined by Hollies *et al.* [39] in the early 1950s with a ball-drop test. Three sizes of steel ball were used, and the kinetic energy delivered to the explosive was determined by subtracting the kinetic energy of the rebounding ball. It was concluded, using dextrin lead azide (Table VI), that change in momentum was a more important factor than kinetic energy in determining the probability of explosion.

It has been indicated that the method of drying lead azide can affect its impact sensitivity. This observation is based on the work of Siele *et al.* [40] (Figure 16). Mass spectrometric analysis of gas samples taken during vacuum

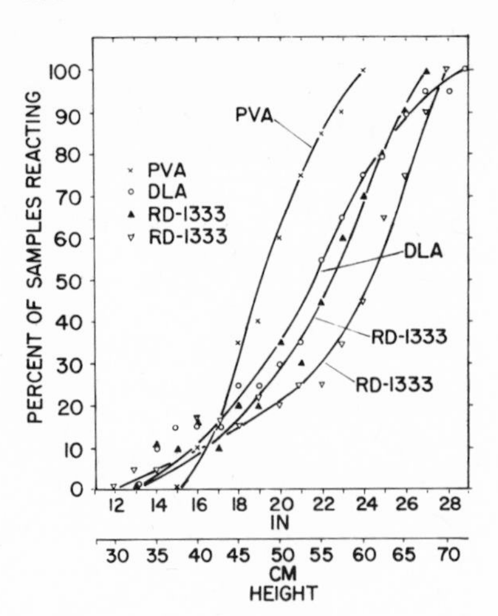

Figure 15. Ball-drop impact sensitivities of different types of lead azide [38]; $\frac{1}{2}$-in. steel ball; 70°F; relative humidity 55%; 20 samples per data point.

drying at 60°C showed that water, ethanol, Freon TF, and CO_2 were liberated under this more rigorous drying condition. (Water, ethanol, and Freon TF were used to wash the SPLA before drying.) It was suspected, therefore, that the dispersion in the percentage of samples reacting above the 50% level in curve A, and the decrease in sensitivity, was due to the absorption of the volatile con-

Table VI. Sensitivity of Lead Azide to Kinetic Energy and Impulse [39]

Ball diam (in.)	Ball mass (g)	Height (cm)	e	Detonations (%)	Potential energy[a] (J)	Net kinetic energy[b] (J)	Impulse[c] ($J - s \times 10^{-4}$)
$\frac{7}{8}$	44.66	90	0.615	10.9 (20)	0.394	0.230	0.310
	44.66	100	0.644	15.0 (20)	0.437	0.256	0.326
	44.66	110	0.613	17.5 (40)	0.482	0.281	0.342
	44.66	120	0.617	20.0 (20)	0.525	0.307	0.358
1	66.68	60	0.510	17.5 (40)	0.392	0.290	0.346
	66.68	80	0.513	35.0 (40)	0.523	0.386	0.403
	66.68	100	0.514	47.5 (78)	0.654	0.483	0.451
	66.68	120	0.509	57.0 (100)	0.784	0.579	0.493
$1\frac{1}{8}$	95.04	100	0.372	70.0 (20)	0.933	0.804	0.580
	95.04	120	0.368	87.5 (40)	1.120	0.965	0.632

[a]Potential energy = mgh.
[b]Net kinetic energy = $mg(h_1 - h_2) = 980\, mh_1(1 - e^2)$.
[c]Impulse or change in momentum = $Ft = m(v_1 - v_2) = 44.3m(1 + e)h_1^{1/2}$, where m = ball mass, g = gravitational constant, h_1 = initial fall height, h_2 = rebound height, $e = v_2/v_1 = (h_2/h_1)^{1/2}$ = coefficient of restitution, F = force, t = time over which ball goes from v_1 to v_2, v_1 = falling velocity before duration of contact, v_2 = rebounding velocity after duration of contact.

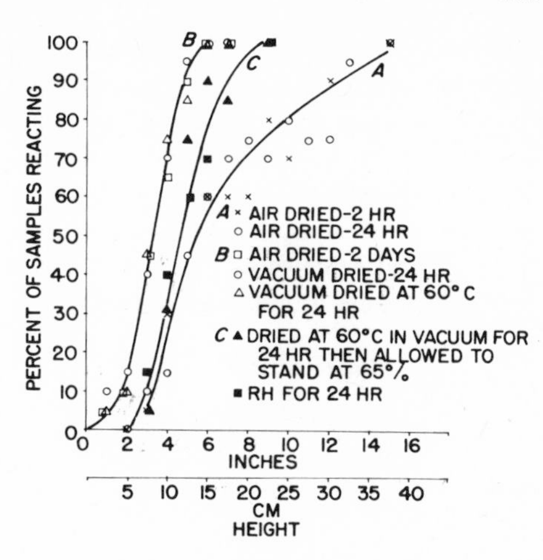

Figure 16. Impact sensitivity of special purpose lead azide: effect of drying and humidity [40]; Picatinny Arsenal apparatus; 2-kg weight dropped at indicated heights; 20 samples per data point.

taminants. Once the contaminants were removed, reproducible impact test results were attainable.

Curve C in Figure 16 indicates that after standing for 24 hr at 65% relative humidity the lead azide reabsorbed moisture and the sensitivity to impact decreased. Current practice is to vacuum dry primary explosives at 60°C for 24 hr prior to impact testing with the Picatinny Arsenal apparatus.

Using this drying procedure, it becomes clear (Figure 17) that the potential energy of the weights and the distances through which they fall are not equiva-

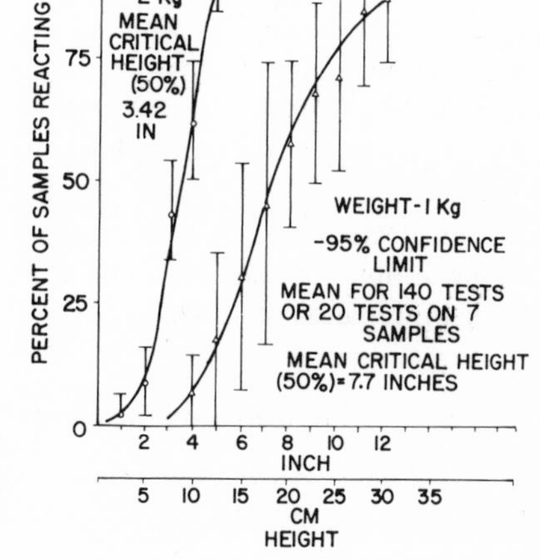

Figure 17. Impact sensitivity of vacuum-dried, Special Purpose lead azide [40]; Picatinny Arsenal apparatus; weights dropped at indicated heights, 20 samples per data point.

lent, as the 2-kg weight is a more effective initiator of SPLA for a given energy delivered and there are much larger confidence limits in the 1-kg impact data.

Copp *et al.* [16] described the results with Service and dextrinated lead azides at length, so only some of the more recent work will be discussed here. Wyatt [41] and Taylor [42] reported ball and disk impact sensitivities for various forms of lead azide. The extreme sensitivity (Table V) of the gamma form of lead azide should be noted. Ball and disk data for RD1333 and RD1343 lead azides were found to be the same:

Distance of fall (cm):	10	12	14	16	18	20
Percent of samples reacting:	2	18	34	64	78	94

The Rotter machine produced a value of 18–20 for the FI of RD1333 lead azide and a value of 21 for RD1343 lead azide.

The following 10% points were obtained for RD1343 with the Picatinny Arsenal apparatus: 2-kg weight, 12.7 cm; 1 kg, 33 cm; 0.45 kg, 50.8 cm; and 0.23 kg, 150 cm. On the same material ball-drop values of 83.8, 76.2, 58.4 and 55.3 cm were obtained for the 10% point in four separate tests. The 10% point is the value 2.54 cm above the height where 10 trials did not fire [9].

The effects of storing lead azide for long periods in 50:50 water–ethyl alcohol mixtures have been investigated by impact sensitivity tests. In a specific instance [43] dextrinated lead azide which had been stored for about 25 years was tested, and no significant change in sensitivity to impact or heat was detected when compared to freshly made material.

Because of the slight solubility of lead azide in 50:50 water–ethyl alcohol and the possibility of recrystallization during diurnal temperature changes, Forsyth *et al.* [44] conducted impact tests on various lots of Special Purpose and RD1333 lead azides that had been subjected to storage for up to 14.75 months. The heights at which 10 and 50% of samples reacted were obtained with the Picatinny Arsenal apparatus with a 2-kg weight (Table VII). The variations in sensitivity shown in Table VII are not considered to be significant within the normal limits of experimental error for these materials.

Ball-drop tests were also conducted on different SPLA lots after periods up to four years. No significant changes were detected and the heights causing 50% of the samples to react were consistent within normal measurement uncertainties using both pieces of apparatus [44].

4. Effect of Impurities on the Impact Sensitivity of Azides

It is well known that impurities, whether chemically combined or in the form of mixtures, can significantly affect the sensitivities of explosives, as has been indicated to some degree earlier in this volume. The effects can be purely

Table VII. Changes and Impact Sensitivity of Lead Azide due to Storage [44]

	Storage time (months)							
	As received		6.5		10.5		14.75	
Sample	10%	50%	10%	50%	10%	50%	10%	50%
SPLA								
Lot 1	6.50	8.20	5.10	6.80	5.19	6.95	5.10	6.82
Lot 2	6.77	9.41	4.48	7.20	5.22	7.02	5.05	7.00
Lot 3	6.39	8.66	4.78	7.45	5.13	6.61	7.60	9.70
Lot 4	6.20	9.64	6.20	8.78	5.25	7.25	6.38	7.78
Lot 5	6.29	8.37	4.89	7.61	4.81	6.78	5.27	7.23
Lot 6	3.48	6.72	3.35	5.95	4.58	6.38	5.21	6.53
RD1333								
Lot 1	6.96	8.76	3.02	5.74	4.76	6.44	3.93	6.29
Lot 2	5.12	7.02					5.14	7.62

mechanical, as in the sensitization caused by gritty particles which has very hazardous consequences for the handling of azides. Soft, low-melting impurities are believed to significantly reduce sensitivities to impact and shock, as in the cases of waxes present in amounts of just 1-5%. The precise relationship between such changes of sensitivity and the mechanical properties of additives is, however, not well defined, and for this reason empirical tests are necessary to determine the magnitude of changes possible.

The effects can also be thermal or such as to significantly affect the rate of the chemical reactions that underlie the initiation process. Aspects of these rate changes are discussed in Chapters 6 and 8, Volume 1, and in Chapter 6 of this volume. As heat sinks, impurities are also believed to affect the sensitivity of explosives, particularly azides, even when present as mixtures. In this sense both solids and liquids have been combined with azides to decrease their sensitivity. If sufficient quantities of impurity are incorporated, the reaction regime may change from a detonation to a deflagration or ignitions may fail to become self-sustaining. Again the approach is largely empirical, and this section presents a somewhat random selection of data to illustrate the effects on impact sensitivity. Some other effects on impact sensitivity specific to the chemical doping of azides are discussed in Section B.6 below.

a. Effect of Grit

It was known that the inclusion or addition of grit (a hard inert substance) enhances the impact sensitivity of solid explosives, especially the primary explosives, long before any systematic effort was made to determine all the effects.

It was not until 1932 that Taylor and Weale [45] conducted experiments on the impact sensitivity of mercury fulminate and ground glass. Their findings led them to the conclusion that "traces of a hard gritty material cause a sharp increase in sensitiveness, while addition of oils, waxes, and lubricants, or substances like water which wet the surfaces of the (explosive) particles and have a chemical quenching action, cause rapid decrease in sensitiveness" [46]. The initiation of explosion was believed to be tribochemical, based on the hardness of the grit particle, with the rise in surface temperature limited by the melting point of the grit.

Copp *et al.* [16] stressed the importance of the hardness of the grit when using the Rotter impact test to determine the sensitivity of high explosives to grit as a function of percentage and size. With a grit of Moh's hardness more than 4 the sensitivity increases with an increase in the weight percentage of grit. With the ball and disk apparatus mercury fulminate is much less sensitive to grit than Service lead azide. Urbanski [47] also showed that when lead azide is mixed with pulverized sand it is more sensitive to impact than mercury fulminate. Although he mentioned that lead azide does not necessarily explode when rubbed in a porcelain mortar, he gave no explanation.

Copp *et al.* believed that percussion sensitiveness (with the ball and disk apparatus) was very complex and involved, in addition to the formation of hot spots through friction, a tribochemical reaction in which there was a more direct transfer of mechanical energy to activation energy than was the case when the mechanical energy is first converted to heat. An example of this is shown with experiments using nickel and tin disks to confine Service lead azide in which the surfaces of the low melting point lessened the grit sensitivity but enhanced the impact sensitivity.

Bowden and coworkers [48-50] pursued the studies on the effect of grit and found that the melting point of the grit is the determining factor and that all grits which sensitize explosives to impact or friction have melting points above 400-550°C. (The lower limit for primaries is about 500°C.)

It was found that silver nitrate, silver bromide, and lead chloride had no appreciable sensitizing effect on lead azide when the impact was performed with a 240-g striker falling 29 cm. However, borax and chalcocite gave 100% explosion efficiency. Bismuthinite and galena had only small sensitizing effects in spite of their high melting points, but both of these materials were soft compared to the lead azide.

Besides the hardness, the size of the grit particles is important, and particles larger than 20 μm diam are more effective. However, particles of diamond as small as 0.5μm still had a sensitizing effect on a number of solid explosives. Particles as small as this are present in most explosives, no matter how carefully they are prepared.

Experiments by Bowden and Gurton [48] also revealed the importance of the thermal conductivity of the grit. Glass, which was the hardest grit used and which has a low thermal conductivity, gave the greatest sensitization with most of the explosives. Bowden concluded that the initiation was due to the localized hot spots formed by intergranular friction and not to direct mechanical activation of surface molecules.

Scullion [51] developed an equation to relate the sensitization of an explosive to the amount of grit added. The impact sensitivity (ψ) of an adulterated explosive was related to the insensitivity (ψ_{max}) of the uncontaminated explosive by the equation

$$\psi = \psi_{max} - R\,[1 - \exp(-kg)]$$

where g is the weight fraction of added grit (mixtures contained 0–0.1% grit), and R and k are constants which depend on the explosive-grit system and the method of test.

The equation was found to be in reasonable accord with the results from a series of tests using TNT, RDX, and HMX in various admixtures of fine grits. Whether the equation can be applied to primary explosives is not known.

b. Lead Azide Mixtures with RDX, Ground Glass, Amorphous Silica, and Liquid Halocarbons

Grit has been used deliberately to sensitize explosives [52,53]. The range of effects on impact sensitivity are illustrated by variations detected in compositions that included lead azide, RDX, ground glass, and pyrogenic silica, and they were desensitized for transportation and handling by saturation with a volatile halocarbon.

The data were gathered using the Picatinny Arsenal apparatus (1-kg weight) and the up-and-down method. The nature of each sample's response was noted as well as the height for which 50% of the samples reacted [54,55].

Dry mixtures of Special Purpose lead azide and RDX had impact sensitivities intermediate between those of the ingredients. At azide concentrations above about 10 wt %, the impact sensitivity was essentially that of pure lead azide. If the same mixtures were flooded with a volatile halocarbon (Freon TF), compositions of <25% PbN_6 were desensitized but compositions of >25% PbN_6 were more sensitive than dry mixtures (Figure 18).

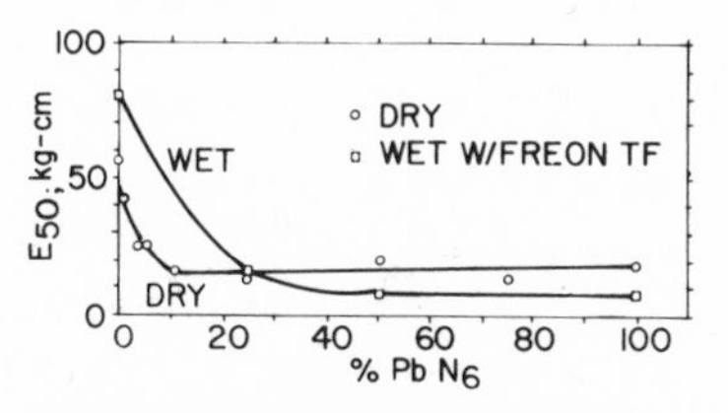

Figure 18. Impact sensitivity of RDX/PbN_6 mixtures [55]; Picatinny Arsenal apparatus; 20 samples per data point.

Liquids are generally incompressible; thus a possible rationalization of the observed sensitization by the liquid is that, in the confined situation represented by the die cup, the liquid aided the efficient transfer of shock when the sample was subjected to impact.

The addition of 6% pyrogenic silica (Cab-O-Sil) to the above mixtures changed the picture only slightly.

In general, wetting may not produce impact desensitization in lead azide or in mixtures that contain more than about 25% lead azide. In the case of pure lead azide wetting may sensitize it.

The addition of pyrogenic silica to wet mixtures had little effect on the height for 50% reactions, but it weakened the character of the reactions. This is termed "moderation" and is to be distinguished from "desensitization." The finding demonstrated that the percentage of reactions is not an adequate measure of handling safety, but it needs to be supplemented by data on the nature of the response to the impact stimulus.

When response character was added to the impact data for the RDX/PbN_6/Cab-O-Sil mixtures, the wet mixtures were as "sensitive" to stimuli as the dry mixtures, but the violence of their reaction was reduced. Initiations did not propagate throughout the samples.

The moderation effect is not universal. In the absence of Cab-O-Sil, wetting does not moderate the reactivity of lead azide although it moderates the behavior of RDX/PbN_6 mixtures [56].

The confined conditions of samples in the Picatinny Arsenal apparatus may not be significantly different from conditions likely to be encountered in shipping and handling of explosives, e.g., a unit volume of lead azide is confined by the other lead azide in a large shipping container. Impact tests were conducted with the liquids normally used in shipping or transporting the lead azide. Figure 19 shows that lead azide with water, water–alcohol, or Freon was more sensitive than lead azide alone. The density of the medium surrounding the lead azide affects the sensitivity of the mixture.

However, temperature affects the sensitivity of Cab-O-Sil–lead azide and RDX–lead azide mixtures. The sensitivity of the Cab-O-Sil mixture is decreased and approaches that of lead azide, while the RDX–lead azide mixture becomes less sensitive than lead azide. It was concluded that Cab-O-Sil–lead azide and RDX–lead azide mixtures are not truly compatible [40].

c. *Effect of Liquid Nitrogen on the Impact Sensitivity of Lead Azide*

If liquids act as thermal sinks to desensitize explosives, liquid nitrogen might be presumed to maximize the effect. Tests using a Picatinny Arsenal apparatus modified with the addition of a holder for the nitrogen indicate that of the primary explosives tested only dextrinated lead azide displays a decrease in impact sensitivity (Table VIII) [57].

Table VIII. Effect of Liquid Nitrogen on the Impact Sensitivity of Primary Explosives [57][a]

	Dext. PbN_6	PVA PbN_6	RD1333 PbN_6	Lead styphnate	KDNBF	DDNP
A. Mean height for 50% probability of reaction[b]						
Control (dry)						
Height (cm)	32.13	28.75	31.75	33.43	27.31	9.65
σ	5.05	3.33	12.29	13.39	1.24	4.10
LN_2 Test						
Height (cm)	43.31	33.02	31.75	32.16	27.61	9.47
σ	6.58	9.32	5.18	4.52	4.24	2.72
B. Height for 10% probability of reaction[c]						
Control (dry)						
Height (cm)	15.24 (12.70)[d]	12.70(7.62)	12.70(7.62)	15.24(7.62)	5.08(7.62)	5.08(2.54)
LN_2 Test						
Height (cm)	25.4	15.24	12.70	17.78	12.50	5.08
C. Effect of temperature cycling at 50% height[e]						
In LN_2						
% Fire[f]	55%	35%	45%	60%	75%	45%
Dry						
% Fire[f]	45%	50%	60%	30%	60%	35%

[a] 2-kg weight in modified PA impact machine.
[b] 50% point determined by up-and-down technique.
[c] 2-kg weight in modified PA impact machine for 10% point.
[d] Values in ( ) indicate 10% point in regular PA test.
[e] Temperature cycling consisted of soaking sample in LN_2 for 15 min and then letting stand at ambient for 1 hr for five cycles; 20 samples tested dry and 20 in LN_2.
[f] % fire in cycled tests indicates percentage of samples fired of 20 samples tested at control 50% fire height.

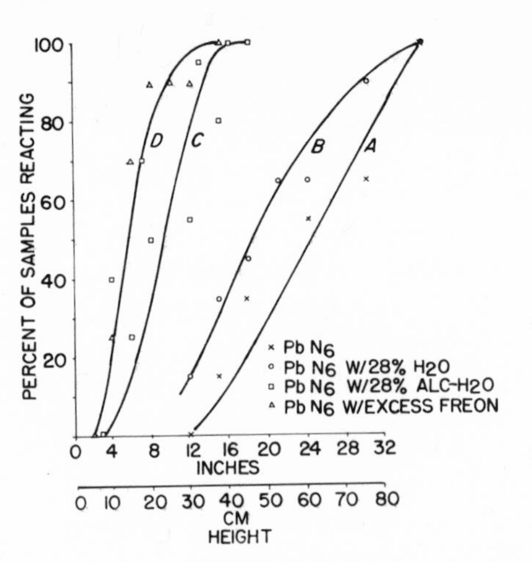

Figure 19. Impact tests of lead azide and lead azide with different liquids [56]; Picatinny Arsenal apparatus; 1-kg weight dropped at indicated heights; 20 samples per data point.

d. Impact Sensitivity of Unconfined Mixtures

The increased sensitivity of confined, wet mixtures of lead azide led to tests with unconfined samples [57]. An earlier investigation [58] indicated that wet lead azide is more sensitive to impact than the dry substance when measured on the Naval Surface Weapons Center ERL machine. The Picatinny Arsenal apparatus was modified to reduce the confinement of the sample by introducing a well to retain the sample in relatively large amounts of liquid.

Dry, unconfined lead azide was initiated by impact from lower heights than the confined material. Unconfined mixtures of lead azide and water, lead azide and alcohol, and lead azide and Freon TF were less sensitive than dry, unconfined lead azide.

e. Effect of Coprecipitants on Impact Sensitivity

Minute amounts (0.01% to 2%) of impurities significantly affect the rate of thermal decomposition of lead and silver azides (Chapter 6, Volume 1). For example, the ionic impurities Ag^{+}, Fe^{2+}, $[FeN_3]^{2+}$, and $[BiN_3]^{2+}$ increase [59-61] and Cu^{2+} decreases [100] the rate of thermal decomposition of lead azide; Cu^{2+} increases the rate for silver azide. In contrast to ionic impurities incorporated in the azide lattice, semiconductors [62,63] in contact with the surface and proton- or electron-donor vapor adsorbed on the surface of azide (Chapter 4, Volume 2) affect its decomposition properties.

Presently, the relationship between thermal decomposition and sensitivity is not clear, and it is not possible to extrapolate thermal decomposition properties to explosive properties. However, in view of the influence of impurities on thermal decomposition, the question arises whether impurities within the crystal lattice affect the mechanical sensitivities of azides. Only a limited amount of data bears directly on the important question.

The influence of impurities on explosive properties is relevant for both theoretical and practical reasons. Subjecting doped azides to thermal decomposition and explosive studies enables a comparison of effects across the two regimes, thereby helping to elucidate the relationship between thermal and explosive decomposition and the mechanisms of explosion. Of practical interest are the effects of impurities on sensitivity. For instance, there was concern about batches of "pink" lead azide obtained from the U.S. suppliers during the period 1966–1972 [64]. The pink color was tentatively related to iron impurities in the sodium azide used to produce the lead azide. Because of fear that the pink material might be more sensitive than normal lead azide, it was disposed of without further testing. The need to know the influence of impurities on sensitivity before the situation arises is indicated by this example. Furthermore, selected doping of azides offers the possibility of improving their characteristics.

Observed Effects of Dopants. Sukhushin and Zakharov investigated the effects of impurities on detonation velocity and impact sensitivity of lead and silver azide [65]. Their results are presented in Figures 20–22 for lead azide-silver azide, cupric azide-lead azide, and cupric azide-silver azide, respectively. The compositions indicated in the figures are based on the composition of the metal nitrate solutions prior to precipitation of PbN_6 with sodium azide. However, Pichugina and Ryabykh determined by analysis that azide crystals generally contain less impurity than the starting material [66]. Therefore, the curves in the figures should be compressed toward lower impurity levels.

The change in detonation velocity at the low impurity levels is unexpected based on hydrodynamic theory. Impact sensitivity and detonation velocity tend to vary inversely.

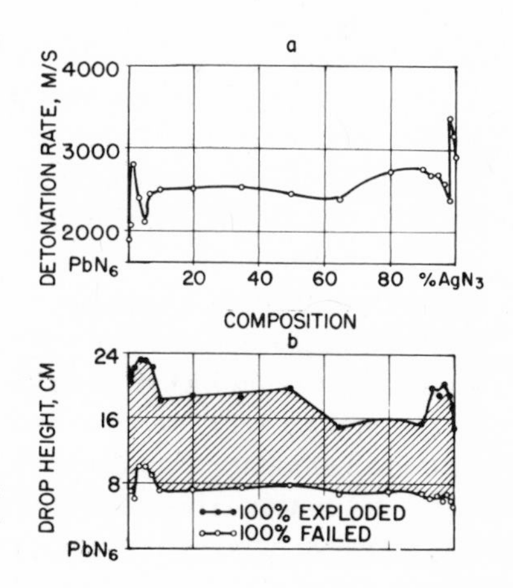

Figure 20. Dependence of detonation rate (a) and impact sensitivity (b) in $Pb(N_3)_2$–AgN_3 composition [65].

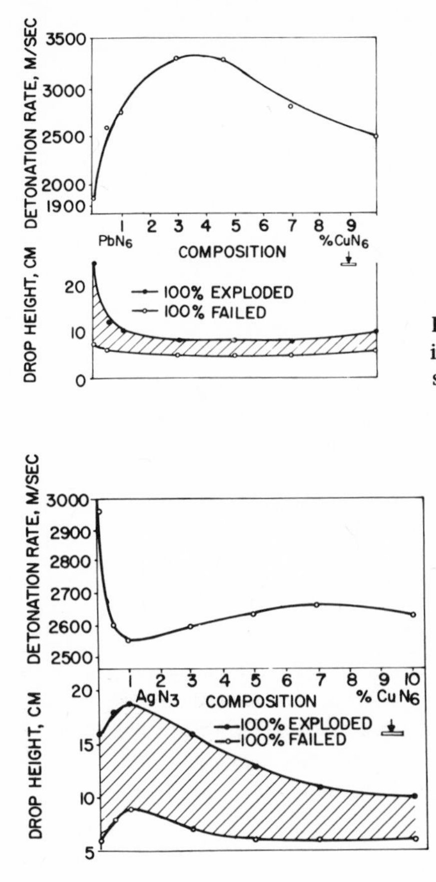

Figure 21. Dependence of detonation rate and impact sensitivity in $Pb(N_3)_2$–$Cu(N_3)_2$ composition [65].

Figure 22. Dependence of detonation rate and impact sensitivity in AgN_3–$Cu(N_3)_2$ composition [65].

It should be mentioned that the samples were compressed to only 1.8 g/ml; in military applications lead and silver azides are pressed to 3-3.5 g/ml. Whether or not the same trends would be observed at higher loading densities is subject to question.

In spite of these restrictions, the relative effects of impurities appear real and offer intriguing possibilities: Based on Figure 20 lead azide containing a very small amount of silver has essentially the same detonation velocity and impact sensitivity as pure silver azide. Could silver-doped lead azide be produced and used in place of the more expensive silver azide? The lead-doped silver azide appears to have a higher detonation velocity than pure silver azide and, therefore, could be very useful in small detonator applications. The results indicate that small amounts of ionic impurities can increase or decrease sensitivity of the azides to impact. This observation is most important from safety and quality-control viewpoints.

The importance of impurity on azide sensitivity was also demonstrated by Singh, who observed that $[BiN_3]^{2+}$ (0.24 wt %) increased impact sensitivity (Figure 23) [61]. Shorter times to explosion for the more thermally reactive $[BiN_3]^{2+}$-doped azide is expected based on current theory of thermal sensitivity. Increased impact sensitivity along with increased thermal sensitivity and thermal decomposition rate give indirect indication that impact sensitivity is thermal in nature.

The question of the effects of iron impurities on the explosive properties, of lead azide was addressed by Hutchinson, who subjected pure and Fe^{2+}-doped lead azide (0.016 mole % iron) to thermal decomposition and explosive tests [67]. Both samples pressed to the same density had identical detonation velocities, 4650 m/sec at a 3.5 g/ml. The iron-doped material was slightly more sensitive to impact (ball-drop test) and to heat (time to explosion tests) than was the pure material. However, the differences in results were close to the scatter in the data.

The undramatic effects of the iron impurity could have been caused by the extremely low concentration of the impurity. Only an 8% increase in rate of thermal decomposition would be expected at 0.016 mole % Fe^{2+} impurity (Chapter 6, Volume 1). Further testing at higher iron concentrations (0.01–1%) is needed to resolve the question.

Discussion. The limited amount of work concerning the effects of ionic impurities on the explosive properties of the azides is more tantalizing than conclusive. It appears that small amounts of impurities can increase or decrease both the sensitivity and the detonation velocity of the azides. The degree to which the effects will hold up under military loading conditions remains to be seen.

The influence of impurities on sensitivity has practical significance from a safety viewpoint, and the ability to reduce the sensitivity of azides with certain impurities appears possible. Increasing the detonation velocity of the azides by adding impurities could produce more powerful detonations for critical applications.

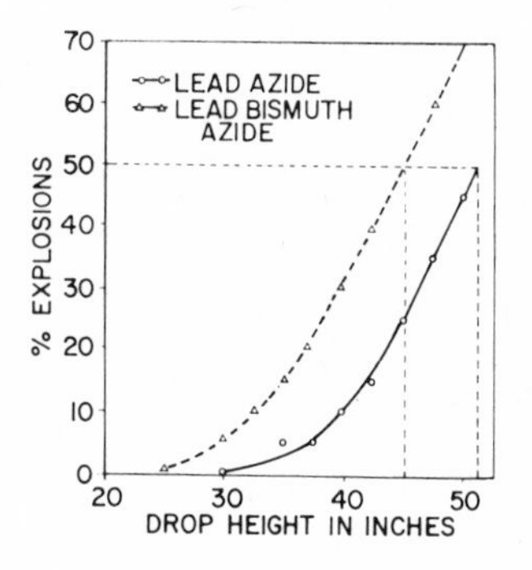

Figure 23. Percentage explosions vs. drop height for lead azide and $(BiN_3)^2$-doped lead azide [61].

In addition to direct practical applications, it appears that doped azides are a potential tool for testing the theories of explosion and sensitivity, and for bridging the gap between thermal decomposition effects and explosive characteristics. For instance, the observed effect of minute impurities on detonation velocity challenges classical hydrodynamic theory and offers a basis for refining that theory. Impurity doping is especially useful in such studies because impurities can be added without altering other physical properties such as particle size and loading density. The use of doped explosives as a tool in understanding and unraveling the mysteries of the explosion process would seem to warrant vigorous future investigation in view of the positive preliminary results.

Progress towards standardization has been slow since all investigators recognize the limitations of impact sensitivity assessments.

The need for more refined tests, a better understanding of what is going on during an impact in an explosive, and improve statistical approaches to the treatment of low-probability events are the paramount requirements for future studies in impact sensitivity.

C. FRICTION SENSITIVITY

Of all the hazardous stimuli which can be encountered in the handling of explosives, friction is probably the most universal, the most difficult to eliminate, and the least understood. With most stimuli, e.g., impact, shock, temperature, and electrostatic discharge, steps can be taken to guard against or eliminate their consequences, but friction in some form is usually present in handling, pouring, mixing, or packaging, or is associated with the movement of handling devices which include sliding, rotating, pressing, and scraping. The problem is further complicated because friction *per se* as an adhesion and energy-dissipation process cannot be quantitatively interpreted [68].

The classical laws of friction were noted in the works of da Vinci [69, 70], Amontons [71], Coulomb [72], and Euler [73]. In simplest terms friction is the resistance to motion which occurs whenever an object slides across another surface. The laws of sliding friction may be summarized as:

1. Frictional force is directly proportional to the load or to the total force acting normal to the sliding surface.

2. Frictional force for a constant load is independent of the contact area.

Friction between unlubricated surfaces is due to a combination of adhesion and plastic deformation [74]. Adhesion occurs at the regions of contact. For friction to occur, the junctions or welds have to be sheared. This part of friction can be written as

$$F_{\text{adhesion}} = AS$$

where A is the true area of contact and S the average shear strength of the junctions.

Plastic deformation is caused by the ploughing, grooving, or cracking of surface asperities. If this is designed as P, and there is negligible interaction between the two, friction can be considered the sum

$$F_{\text{adhesion}} + F_{\text{deformation}} = AS + P$$

Research on friction has concentrated on the quantitative description of the areas of actual contact, the strength of adhesion between surfaces, the shear strength of the interface, interactions between A and S, the deformation component P, and interaction between the deformation and adhesion.

Other frictional phenomena of interest in explosives sensitivity include the temperature rise during sliding, stick-slip (friction-induced oscillations), friction of rolling bodies, fretting (a severe form of wear), and internal friction [74–80]. Fracturing can also occur in sliding and grinding operations, and friction is interrelated with electrostatics. Those topics are discussed in Chapter 9, Volume 1.

Although major advances in the understanding of frictional phenomena have occurred during the past two decades, explosives have not normally been the subject of investigation. Brown [77] reiterates that the treatment of the friction sensitivity of explosives is still a "black art," and that the testing and safe handling techniques applied to explosives are highly empirical.

1. Friction Sensitivity Apparatus

The friction testing of explosives is not as common as impact testing in the U.S. It is difficult to devise a test which delivers a simple frictional stimulus without also impacting the sample and indirectly heating it by contact with other sliding components of the apparatus. Friction tests even then commonly do not produce reactions in any but the most sensitive explosives. Brown [77] found that there were many varieties of friction sensitivity test, but that they could be grouped into three categories:

1. Those which shear a thin layer of explosive between two rigid plates of steel or other materials of construction; some impart linear, or single-pass, motions and some impart rotary and continuous motion;

2. Those which rub a block of explosive violently on a hard or abrasive surface;

Those which subject a sample to extreme deformation in an impact or extrusion event.

Only a few tests fit the last two categories; the only test that fits the last description is the Susan test [12], in which one pound of explosive in a special projectile is driven against a hard target at selected velocities. It is really a form of impact test and is used only with secondary explosives.

Friction tests that utilize violent rubbing against a hard or abrasive surface are the Pantex skid test, the oblique impact test, and also the friction-impact torpedo test [12]. These tests also are for secondary explosives and involve the shear strength and penetration hardness of the softer materials.

Most friction tests are those in the first category. The explosive shear between two surfaces, which may be smooth, rough, or covered with abrasive. If the two surfaces come in contact before or at the same time as acting on the explosive, they can be heated sufficiently to cause ignition by simple thermal mechanisms. This possibility has led some to question whether the tests correctly characterize friction sensitivity; in this view the apparatus are merely crude heat machines.

Among many shear-type tests the following are representative: the mallet friction test, the sliding-block friction test, and the emery-paper friction test of the Explosives Research and Development Establishment, England, the U.S. Bureau of Mines pendulum friction test, and the Julius Peters friction test. Each of these can be used to test primary explosives, and some may be used also on secondary explosives.

a. Mallet Friction Test

In this simple test [12] a small sample is spread on an anvil at floor level, the operator stands astride the anvil, and delivers the explosive a glancing blow as he swings a mallet gently between his legs.

Anvils of softwood, hardwood, and York stone are used. Softwood is Norwegian yellow pine, hardwood is English oak, and the York stone is from Yorkshire quarries.

The mallets are made of boxwood and mild steel. The boxwood mallet weighs 25 oz. and the mild steel 34.5 oz. Moving pictures taken of the operator swinging the mallet lead to the claim that a velocity of 60 ft/sec is the average swing.

Anvils are also made of mild steel, aluminum bronze, and naval brass. When no ignitions are obtained with the boxwood mallet on the softwood, hardwood, and York stone, a steel mallet is used on metal anvils. The metal anvils and mallets represent materials that may be utilized in the manufacture or use of the explosive.

The results are variable, and a reaction is recorded for any crackle, spark, or flash. The records are based on 10 swings with 1-6 ignitions indicating a 50% probability of reaction, and above 6 being a 100% probability.

The mallet test provides data required to classify explosives into three categories. When combined with the FI of the explosive from the Rotter impact test, the mallet test yields the categories:

1. Very sensitive–those explosives with a FI of 30 or less which also give ignitions with the standard wood mallet on softwood.

2. Sensitive—those explosives with a FI of 90 or less which do not give ignitions with the standard wood mallet.

3. Comparatively insensitive—those explosives with a FI greater than 90 which do not give ignitions with the standard wood mallet.

Handling of "very sensitive" and "sensitive" compositions requires extreme care. Friction and impact between materials harder than softwood must be avoided. For "comparatively insensitive" materials it is usual to recommend that blows between hard surfaces be avoided, but simple transfers between metal containers (i.e., pipes, valves) are permitted.

b. Sliding-Block Friction Test

In the sliding-block friction test (Figure 24), a sample is subjected to frictional forces when clamped between two prepared steel surfaces [12]. The upper sliding block is constrained to move horizontally over the lower fixed block by guides. A pedal-operated mechanism permits the load to be released so that the sliding block assembly can be inserted.

A 5-ft pendulum with striker is released from a raised position by means of an electromagnet. After being struck, the upper block is ejected through the guides into a rubber bin which is in line with the swing of the pendulum. The pendulum continues upward and is arrested by means of a spring-loaded braking system. The velocity of strike for the maximum height is 22 ft/sec. The applied load can be varied in the range from ounces to 150 lb.

The surfaces of the blocks are ground, and the explosive sample is dispensed onto the bottom plate in the center by means of a filling plate, 0.015 in. thick with a hole 0.095 in. diam.

Ten trials at the same load are performed, then the load is either increased or decreased depending upon the result. Reactions are recorded as audible reports, visible flashes, flash marks on the blocks, smoke, or fumes. The sensitivity is defined in terms of the load required to give one ignition in ten trials.

A sliding-block friction test apparatus is located at Picatinny Arsenal.

c. Emery-Paper Friction Test

In the emery-paper friction test [12] a sliding surface is also caused to move relative to a fixed block by means of a blow from a pendulum (Figure 25). The stationary surface is a steel cylinder to which Oakey Grade 0 emery paper (English designation equivalent to 400 grit) is glued on its underside. Emery paper is also glued toward one end of the sliding block. The explosive is dispensed in the same volume as in the ball and disk impact test and is spread on the sliding block in a 2-cm circle toward the end of the emery paper.

A 50-trial up-and-down test is carried out using a logarithmic distribution of

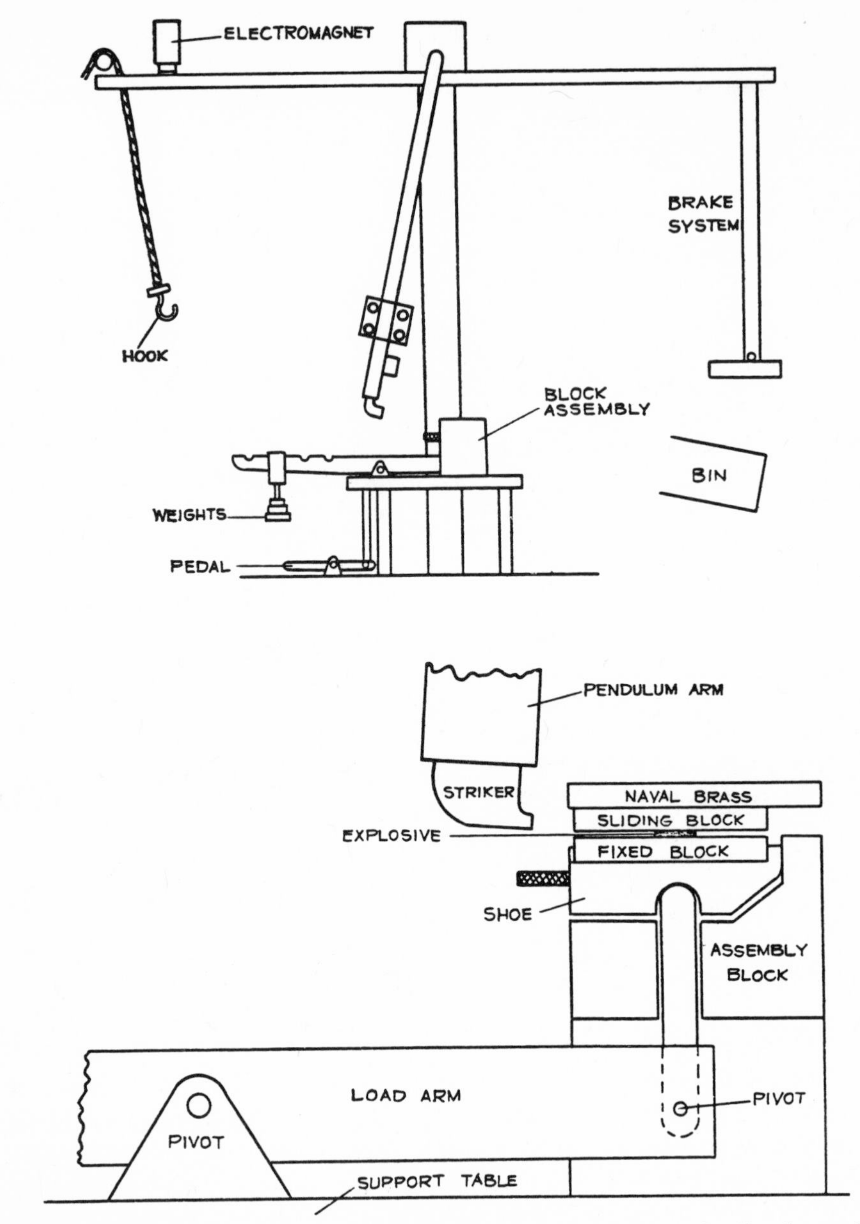

Figure 24. Sliding-block friction apparatus (top) and block assembly and striker (bottom) [12].

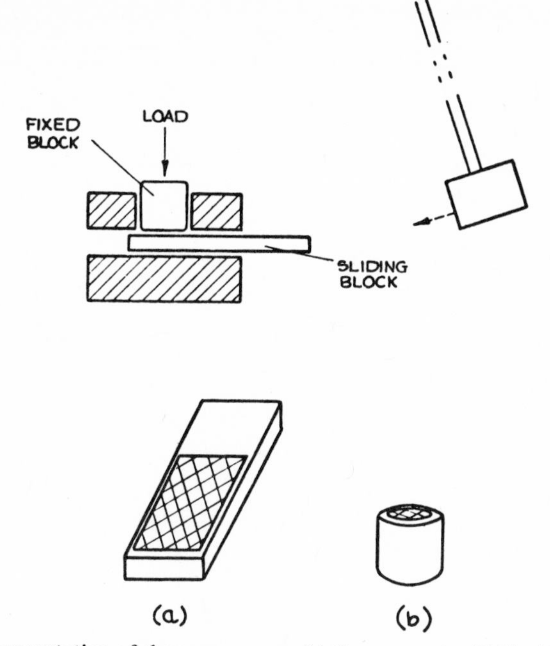

Figure 25. Representation of the emery-paper friction apparatus [12]: (a) sliding block; (b) fixed block.

velocity with an increment of 0.100. The results also are listed as the number of ignitions in ten trials at selected strike velocities.

The emery-paper friction test determines the sensitiveness of primary explosives with a higher degree of precision than the mallet test. One of the major hazards with primary explosives is their sensitization by grit. In general, any sample which can be initiated with velocities of strike less than 10 ft/sec is very sensitive and proper handling precautions must be taken.

d. Pendulum Friction Test

The pendulum friction apparatus [12] was developed by the U.S. Bureau of Mines and is used also by Picatinny Arsenal (Figure 26). To the lower end of the pendulum is attached a 20-kg shoe with an interchangeable face of steel or fiber. The shoe falls from a height of 1 m and sweeps back and forth across a grooved steel anvil. With no explosive present the pendulum is adjusted to pass across the anvil 18 ± 1 times before coming to rest. The adjustment is made by raising or lowering the A-frame by means of an eccentric shaft. A 7-g sample is spread evenly in and about the grooved portion of the anvil, and the shoe is allowed to

Table IX. Mallet Friction Sensitivity [12]

	Mild steel mallet on anvils of			Standard wood mallet on anvils of		
Explosive	Mild steel	Naval brass	Aluminum bronze	York stone	Hardwood	Softwood
RD1333 lead azide				100	100	100[a]
Dextrinated lead azide				100	100	100
Service lead azide				100	100	100
RD1343 lead azide				100	100	100
Silver azide				100	100	100
PETN	50–100	50–100	50–100	100	0	0
RDX	50–100	50–100	50–100	0–50	0	0
RDX/TNT	50–100	0–50	0–50	0	0	0
TNT	0	0	0	0	0	0

[a]Percent "ignitions."

sweep back and forth until it comes to rest. Ten samples are tested, and the number of snaps, crackles, ignitions, and/or detonations is noted. The fiber shoe tends to show differences between explosives which are not detectable with steel shoes.

An explosive passes the friction pendulum test if in ten trials with the fiber-faced shoe, there is no more than local crackling, regardless of the behavior under the action of the steel shoe.

e. Julius Peters (BAM) Friction Test*

The Julius Peters friction apparatus [81, 82], which is extensively used in Europe, originates in a mortar-and-pestle test. It consists of a porcelain plate upon which a sample is located, and a pestle under load rests on the sample. Loads in the range 10–1000 g are applied for primary explosives. By means of an electric mortar the porcelain plate is rotated back and forth in arcs of about 10 mm, the maximum velocity of the pestle relative to the plate being about 7 cm/sec.

Three grades of surface roughness are used, each pestle is used once only, and six tests are conducted for each weight. Sensitivity is defined in terms of no reactions, partial reactions, inflammations, crackles, and explosions.

*The apparatus was built according to the directions of the Bundesanstalt für Materialprüfung (BAM).

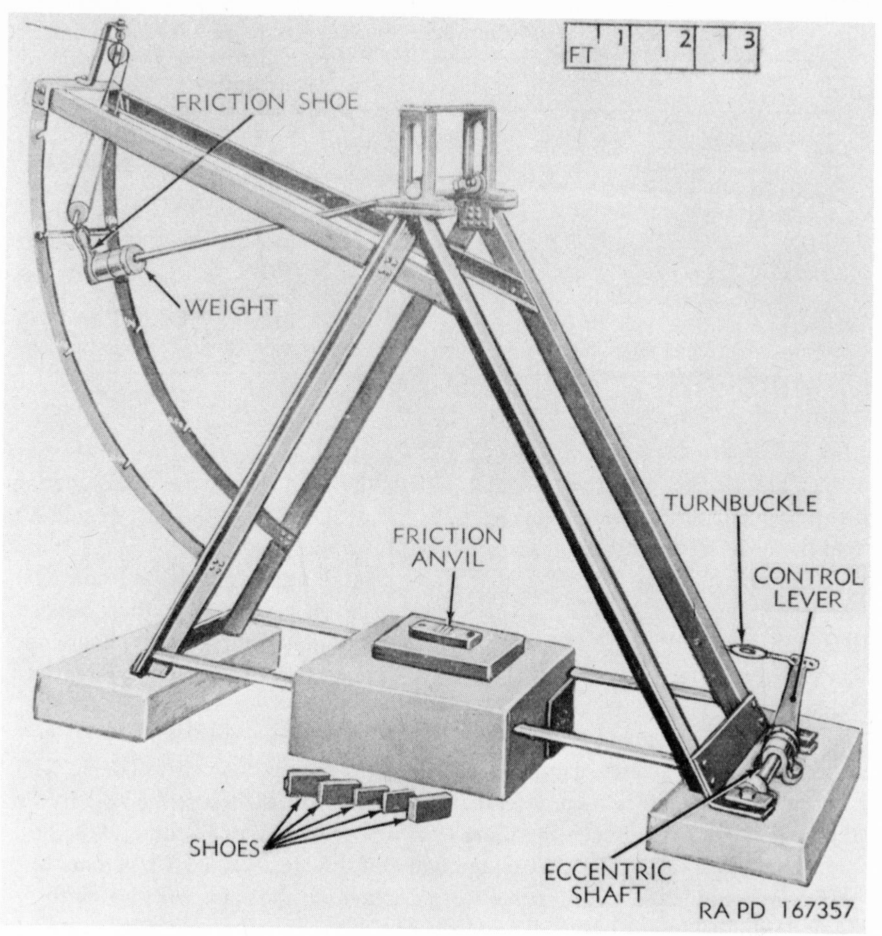

Figure 26. U.S. Bureau of Mines pendulum friction apparatus [12].

2. Friction Sensitivity of Azides

The lack of understanding of the mechanism of frictional initiation of azides limits the operational usefulness of the data obtained. There is little intrinsic merit in the isolated data for any explosive taken alone, but when the data are compared with those for other explosives for which there is also laboratory or industrial handling experience, friction sensitivity data can serve as guides to potential hazards.

This is shown by the results obtained with the mallet friction test (Table IX). On a relative basis it can be seen that according to the mallet friction test the

Table X. Pendulum Friction Sensitivity [12]

Explosive	Fiber shoe	Steel shoe
Lead azide	Failed	Failed
Dextrinated lead azide	Failed	Failed
Silver azide	Failed	Failed
PETN	Passed	Failed
RDX	Passed	Failed
Tetryl	Passed	Failed
Composition B (RDX 59.4/TNT 39.6/wax 1.0)	Passed	Passed
TNT	Passed	Passed
Explosive D (ammonium picrate)	Passed	Passed

azides are all very sensitive and that TNT is the least sensitive of the materials listed. Similar results were obtained with the pendulum friction test (Table X), and the order of sensitivity was practically the same.

The emery-paper friction test is more discriminating (Tables XI and XII), the test revealing that dextrinated lead azide is less sensitive than Service, RD1333, and RD1343 lead azides. The values obtained for lead styphnate and barium styphnate suggest that they are as friction sensitive as dextrinated lead azide.

Using the up-and-down methodology, 90% Service lead azide mixed with either 10% beta or 10% gamma lead azide was the most sensitive (Table XII). Among the lead azides, gamma lead azide required the highest strike velocity for the level of sensitivity corresponding to a 50% probability of ignition. When the same materials were subjected to the ball and disk test (Table V), gamma lead azide and monobasic cupric azide were also among the least sensitive, with dibasic cupric azide not even falling in the category of a "very sensitive" material.

Table XIII illustrates the role surfaces play in determining friction sensitivity. The effect of emergy-papered surfaces is very apparent with Service lead azide, dextrinated lead azide, and lead styphnate.

Tests were also conducted to determine the effect of temperature on the friction sensitivity of the same materials [16]. None of the primary explosives revealed any marked temperature effect; however, the fact that no large temperature effect was observed led Copp *et al.* to conclude that the action of the apparatus was mechanical rather than thermal.

The sensitivity of primary explosives to mechanical action as distinct from the sensitivity to hot spots produced by mechanical action was proposed by Copp *et al.* to be associated with their relatively high lattice energies.

The tests were extended to determine the effect of liquids in grit sensitization (Table XIV). Fluids minimized the effect of grit, although the effect was

Table XI. Emery-Paper Friction Sensitivities [12]

Explosive	Ignitions/trial at strike velocity (ft/sec)											Type of ignition
	2.5	3	4	5	6	7	8	9	10	11	12	
Service lead azide	1/20	3/20	4/20	15/20	19/20							Complete
RD1333 lead azide	1/20	2/20	5/20	8/20	17/20							Complete
R1343 lead azide	1/20	2/20	5/20	8/20	17/20							Complete
Dextrinated lead azide							0/10	2/10	2/10	4/10		Complete
Lead styphnate							0/10	3/10	7/10	9/10		Complete
Lead 2,4-dinitroresorcinate				1/10	4/10	7/10	7/10	8/10				Complete
Barium styphnate									0/20	2/20	2/20	Complete and partial
Silver azide		0/10	4/10	5/10		5/5						Complete

Table XII. Emery-Paper Friction Sensitivity (50% Initiation Values) [41, 42]

Explosive	Velocity of stroke for 50% reactions (ft/sec)	Standard deviation, σ (ft/s)
Service lead azide	4.46	0.184
Beta lead azide	4.39	0.185
Gamma lead azide	6.29	0.194
90% Service lead azide/10% beta lead azide	3.61	0.206
90% Service lead azide/10% gamma lead azide	3.57	0.242
Monobasic cupric azide, $Cu(N_3)_2 \cdot Cu(OH)_2$	5.25	0.124
Dibasic cupric azide, $Cu(N_3)_2 \cdot 2Cu(OH)_2$	>10	

less with Service lead azide than with lead styphnate or mercury fulminate. Desensitization was attributed to lubrication, fewer fractures, or to quenching marginal initiations.

With the Julius Peters apparatus lead azide and mercury fulminate ranked in the same order (Table XV) as those obtained with grit or the sliding-block friction test.

A modification of the sliding-block technique was developed by Bowden and Yoffe [83], the apparatus subjecting the explosive to rapid shear while under a

Table XIII. Sliding-Block Friction Sensitivities [16]

Explosive and surface[a]	Ignitions/trial at velocity strike (ft/sec)									
	5	6	7	8	9	10	11	12	13	14
Service lead azide	0/10	0/10	0/10	0/10	0/10	0/10	0/10	0/10	0/10	0/10
Steel on steel	0/10	1/10	0/10	0/10	0/10	0/10	0/10	0/10	0/10	0/10
Emery	4/10	4/10	9/10	10/10						
Dextrainated lead azide										
Steel on steel	0/10	0/10	0/10	0/10	0/10	0/10	0/10	0/10	0/10	0/10
Emery					0/10	1/10	1/10	5/10	6/10	10/10
Lead styphnate										
Steel on steel	0/10	0/10	0/10	0/10	0/10	0/10	0/10	0/10	0/10	0/10
Emery					0/10	1/10	1/10	5/10	5/10	9/10
Mercury fulminate										
Steel on steel		0/10	1/10	5/10	8/10	8/10	7/10	9/10	10/10	10/10
Emery			0/10	1/10	4/10	4/10	7/10	9/10	10/10	10/10
Lead dinitroresorcinate										
Steel on steel	0/10	3/10	5/10	8/10	10/10	10/10				
Emery				0/10	1/10	5/10	7/10	5/10	9/10	9/10

[a] With steel on steel surfaces, substantially the same results were obtained with a dead load of 40 lbs instead of 6 lbs on the tilting table.

Table XIV. Effect of Liquids on Sliding-Block Friction Sensitivity [16]

Desensitizer	Ignitions/trial[a,b]		
	Service lead azide[c]	Lead styphnate[c]	Mercury fulminate[c]
Without desensitizer	10/10	10/10	10/10
Sprayed 20 sec w/water or w/20% glycerin solution	6/10	6/10	2/10
Same, plus 0.1% perminal[d]	8/10	6/10	2/10
Same, plus 0.1% turkey red oil	5/10	5/10	
One drop ethyl alcohol	1/10	0/10	0/10
20 sec spray butyl alcohol	6/10		
One drop butyl alcohol	0/10		

[a]Sliding block friction tester with 6 lb load.
[b]Number of ignitions at 14 ft/sec.
[c]10% by weight grit.
[d]Perminal is sodium salt of isopropyl naphthalene sulfonic acid.

known load. The amount of explosive sample was 25 mg with or without grit. The effect of adding grit on lead azide sensitivity for a load at 64 kg is shown in Table XVI. Only grits of melting points 500°C and greater were effective sensitizers.

Kinoshita and Arimura [84] utilized a Yamada-type friction sensitivity apparatus [85] to test combinations of explosives and materials in designing manufacturing equipment. Tests were conducted using friction test pieces of emery, ebonite, bakelite, hardened vinyl, copper, brass, iron, lead, aluminum, wood, polyethylene, and hard rubber. Some of the results are listed in Table XVII.

Table XV. BAM Friction Sensitivities [82]

Explosive	Reaction observed	Load on porcelain peg (g)			
		10	20	30	40
Lead azide	No reaction	2	0		
	Ignition	0	0		
	Detonation	4	6		
Cyanidetriazide	No reaction	2	1	1	0
	Ignition	0	0	0	0
	Detonation	4	5	5	6
Mercury azide	No reaction	5	2	0	0
	Ignition	0	0	0	0
	Detonation	1	4	6	6

Table XVI. Friction Initiation of Lead Azide in the Presence of Grit [83]

			Explosion efficiency (%)			
Grit	Hardness (Mohs' scale)	Melting point (°C)	Lead azide (height of fall, 60 cm)	Lead styphnate (height of fall, 40 cm)	PETN (height of fall, 60 cm)	RDX (height of fall, 60 cm)
Nil			0	0	0	0
Silver nitrate	2–3	212	0	0	0	
Silver bromide	2–3	434	0	3	50	40
Lead chloride	2–3	501	30	21	60	20
Silver iodide	2–3	550	100	83	100	100
Borax	3–4	560	100	72	100	100
Bismuthinite	2–2.5	685	100	100	100	
Glass	7	800			100	100
Chalococite	3–3.5	1100	100	100	100	
Galena	2.5–2.7	1114	100	100	100	
Calite	3	1339	100	93	100	

When ebonite, bakelite, and similar materials were used, ignitions were more easily obtained than with the metals. No significant differences occurred in the sensitivity with the different materials. Kinoshita and Arimura believed that ignition with emery was by a different mechanism than by ebonite. They considered the rate of rupture of single crystals, rate of displacement between particles, and heat dissipation by the friction surfaces. When ground glass was added to the explosive, the results obtained with ebonite approximated those obtained with emery.

Yamamoto [86] utilized a similar type of apparatus, but he examined the effect of different metal and resin surfaces on the friction sensitivity of lead styphnate.

Table XVII. Results of Friction Sensitivity Test Utilizing Yamada Friction Apparatus [85]

			Load for 50% initiations (kg)		
Explosive	Particle size (mm)	Additive	Emery	Ebonite	Copper
Service Lead Azide	0.1 ~ 0.3	—	1.7	66.0	>100
		Ground glass 100 mesh	1.2	11.0	8.0
Diazodinitrodiphenol	0.3 ~ 0.5	—	15.0	52.0	>100
		Ground glass 100 mesh	12.5	27.0	32.0
Lead Styphnate		—	20.0	35.0	>100
		Ground glass 100 mesh	15.5	32.0	100

Szulacsik and Lazar [87] demonstrated an effect of particle size and load on the friction sensitivity of mercury fulminate. The sensitivity to friction increased with particle size, but whether a similar trend applies to the azides has apparently not been determined.

D. SUMMARY AND CONCLUSIONS

Numerous problems remain to be solved before it will be possible to quantify the sensitivity of explosives to impact and friction in terms that express an intrinsic property of the chemical substances. The need for more refined tests, a better understanding of the phenomena that occur during the test procedures, and improved statistical approaches to the treatment of low-probability events are paramount requirements for more meaningful data. Even then it will continue to be difficult to relate measurements in laboratory apparatus to the conditions encountered in the industrial handling and field utilization of azides.

Available information on the impact sensitivity of azides, predominately lead azide, shows that the data are affected by the test conditions and the condition of the material itself. The energy and the rate of energy transmitted by impact to the test sample are critical for initiation.

The concept of a standard test with a standard material has been the aim of many investigators. The differences in impact apparatus, as well as in the preparation, testing, and response of samples, have affected the values to a higher degree than recognized by even the most perceptive user. It cannot yet be said that the physical basis for a standard test has emerged.

The trend toward separate apparatus for primary and secondary explosives has become more apparent in recent years. While this approach may permit greater precision in the test measurements, it will attribute little to the basic understanding of initiation by impact unless improved diagnostics of the thermomechanical behavior of the substances are also introduced.

As with the impact sensitivity, one of the major difficulties in assessing friction sensitivity lies in the lack of knowledge of the mechanisms and influences to which the explosive is subjected during "friction" tests. This, combined with limited data on the intrinsic properties of the explosive, results in the crudest measures of sensitivity and potential hazards being available. The various friction tests attempt to simulate specific hazards that may be encountered in practice, but the data obtained are pertinent only in comparison with those for other materials using the same apparatus under the same conditions.

Attempts to correlate the data from various types of apparatus have not been successful, and again a paucity of information on the properties of azides and the action of the stimulus has prevented the development of a universal standard test. To progress from the barely qualitative determinations of friction sensitivity quantitative measures requires a knowledge of the frictional properties of the azides which is currently almost nonexistent.

Both impact and friction were considered by Bowden and Yoffe [8, 83] to be stimuli which lead to initiation via the formation of hot spots. Insofar as no existing test is entirely free of both impulsive loads and the movement of explosive particles with respect to contiguous surfaces, both adhesion and shear may be suspected to play a role in determining sensitivity, and it is not uncommon to consider the brittleness and hardness of azides and other primary explosives when comparing their sensitivities to those of secondary explosives.

On this basis the more brittle the explosive crystal, the more sensitive it is to impact and friction; high rates of shear combined with high pressures ensure that heating occurs [88, 89]. When a brittle substance is impacted, high pressures are developed and transmitted without rapid attentuation. Displacements occur within the material (internal friction), in addition to the movement of, say, the striker relative to the substance (external friction). The most significant difference between friction and impact tests may well be that in the friction tests the normal and tangential forces are applied independently of each other, whereas under impact they are applied simultaneously.

Andreev [88] suggested that explosives which have high "fluidity" (elasticity or plasticity) due to a low coefficient of internal friction, or which have a large critical diameter, are not very sensitive to friction. TNT is then among the materials which are relatively safe to handle. Conversely, lead azide is among the relatively dangerous substances. On this basis one role of phlegmatizers as discussed in Chapter 1 is to increase "fluidity" and thus reduce sensitivity. There is, however, no well-documented evidence that phlegmatizers yield consistently lower sensitivities, the improved "fluidity" in such instances rather being the improve pourability of the powders.

In general, initiation only from the heat generated by friction is very difficult to achieve except under extremely high pressures or sliding velocities [90]. Various investigations have shown that only primary explosives, such as the azides, can be detonated by friction in the absence of grit particles [8, 83]. Secondary explosives melt before ignition occurs, but primary explosives decompose explosively prior to melting. Experimenters generally utilize grit particles, at least to some degree, in friction sensitivity tests.

Frictional heating has a secondary effect on the ignition process in that the temperature rise changes the mechanical properties of the explosive. Friction properties may also change with temperature. Wear also removes material and creates new surfaces, the consequences of which are not well understood, although "fresh" surfaces appear to be more sensitive than surfaces exposed briefly to the normal environment.

Hoge [91] took these factors into consideration in obtaining the frictional behavior of various secondary explosives. He separated friction into components which included wear, as well as adhesion and plastic deformation, because the explosives have low shear strength and high coefficients of friction and are thus prone to wear. Of the explosive materials tested, plastic-bonded explosives were found to have a considerably higher coefficient of friction than self-bonded crystals. The plastic-bonded explosives also showed some viscoelastic or time-dependent effects.

With the ability to grow large crystals of lead azide (Chapter 2, Volume 1), it is feasible to consider that work done with secondary explosives can now be carried over to primary explosives.

REFERENCES

1. G. B. Kistiakowsky, R. Conner, in: *Science in World War II. Chemistry*, W. A. Noyes, Jr., ed., Chapters V and VI, Little, Brown, Boston, 1948.
2. H. Koenen, K. H. Ide, W. Haupt, *Explosivstoffe*, *1958*, 178–189, 202–214, 223–235 (1958).
3. C. Boyars, D. Levine, *Pyrodynamics*, *6*, 53–77 (1968).
4. A. Macek, *Chem. Rev.*, *62*, 41–63 (1962).
5. G. T. Afanas'ev, V. K. Bobolev, *Initiation of Solid Explosives by Impact*, I. Shectman, translator (1971), Izdatelstvo Nauka, Moskva, 1968; NASA Translation Report, U.S. Dept. of Commerce, NTIS, Springfield, Va., TT 70-5074, NASA TT F-623.
6. R. F. Walker, Military Research and Development on High Energy Materials: Explosives and Detonation, Tech. Rept. 4391, Picatinny Arsenal, Dover, N.J., 1972.
7. F. W. Brown, D. J. Kusler, F. C. Gibson, Sensitivity of Explosives to Initiation by Electrostatic Discharges, Report RI-3852, U.S. Bureau of Mines, Pittsburgh, Pa., 1946.
8. F. P. Bowden, A. D. Yoffe, *Fast Reaction in Solids*, Academic Press, New York, 1958.
9. B. T. Federoff, O. E. Sheffield, *Encyclopedia of Explosives and Related Items*, Vol. 7, Tech. Rept. 2700, Picatinny Arsenal, Dover, N.J., 1975.
10. H. Kast, *Schiess Sprengstoffwesen*, *4*, 263 (1909).
11. H. N. Mortlock, J. Wilby, *Explosivstoffe*, *14*, 49 (1966).
12. G. R. Walker, ed., *Manual of Sensitiveness Tests*, Canadian Armanent Research and Development Establishment, Valcartier, Quebec, Canada, 1966.
13. K. R. Becker, C. M. Mason, R. W. Watson, Bureau of Mines Instrumented Impacted Tester–Preliminary Studies, U.S. Bureau of Mines Report 7670, Pittsburgh, Pa., 1972.
14. R. C. Bowers, J. B. Romans, The Acoustical Spectrum as a Tool for Investigating the Impact Sensitivity of Explosives, U.S. Naval Research Laboratory Report 5466, Washington, D.C., 1960.
15. R. M. H. Wyatt, Some Studies on the Dynamics of Impact Sensitiveness of Loose Mercurcy Fuliminate, in: *Proceedings of the International Conference on Sensitivity and Hazards of Explosives*, E. G. Whitbread, ed., Explosives Research and Development Establishment, Waltham Abbey, Essex, England, 1963.
16. J. L. Copp, S. E. Napier, T. Nash, W. J. Powell, H. Skelly, A. R. Ubbelohde, P. Woodward, *Phil. Trans. R. Soc.* (*London*), *A241*, 197 (1948).
17. W. J. Dixon, J. Mood, *Am. Stat. Assoc.*, *43*, 109 (1948); W. J. Dixon, F. J. Massey, Jr., *Introduction to Statistical Analysis*, 2nd ed., McGraw-Hill, New York, 1957, pp. 318–327.
18. K. A. Brownlee, J. L. Hodges, Jr., M. Rosenblatt, *J. Am. Stat. Assoc.*, *43*, 262 (1953).

19. M. G. Natrella, *Experimental Statistics*, National Bureau of Standards Handbook 91, Washington, D.C., 1963.
20. Principles of Explosive Behavior, U.S. Army Material Command Phamphlet 706-108, Washington, D.C., 1972.
21. R. L. Grant, A Combination Statistical Design for Sensitivity Testing, U.S. Bureau of Mines Information Circular 8324, Pittsburgh, Pa., 1967.
22. C. W. Churchman, *Statistical Manual, Methods of Making Experimental Inferences*, Pittman Dunn Laboratories, Frankford Arsenal, Philadelphia, Pa., 1951.
23. S. K. Einbinder, Ph.D. thesis, Polytechnic Institute of Brooklyn, New York, 1973.
24. G. Weintraub, Applications of Sequential Sensitivity Test Strategies and Estimation Using a Werball Response Function for Extreme Probabilities and Percentage Points, Proceedings of the Twentieth Conference on the Design of Experiments in Army Research Development and Testing, Fort Belvoir, Va., October 23–25, 1975, Army Research Office, Washington, D.C., ARO Report 75-2, Part 2, pp. 579–599.
25. D. E. Hartvigsen, J. P. Vanderbeck, Sensitivity Tests for Fuzes, NAVORD Report 3496, Naval Weapons Center, China Lake, Calif., 1955.
26. P. J. Kemmey, Analysis of the Sensitivity of Explosive Materials to Impact; Various Statistical Methods, Tech. Memo. 1812, Picatinny Arsenal, Dover, N.J., 1968.
27. P. W. Levy, *Nature*, *182*, 37 (1958).
28. R. C. Ling, W. R. Hess, An Interpretation of the Sensitivity of Explosives, Tech. Rept. FRL-TR-21, Picatinny Arsenal, Dover, N.J., 1960.
29. D. Smith, R. H. Richardson, Interpretation of Impact Sensitivity Test Data, Alleghany Ballistics Lab. Report ABL/Z-86, 1965.
30. B. N. Kondrikov, Certain Methods of Determination of the Sensitivity of Explosives to Shock, Vzryvnoe Delo, No. 68/25, 168, 1970; U.S. Army Foreign Science and Technology Center, Charlottesville, Va., Translation FSTC-HT-315-72). Also data from informal lecture Princeton, N.J., February 16, 1970.
31. D. C. Hornig, Standardization of High Explosive Sensitiveness Tests, in: *Proceedings of the International Conference on Sensitivity and Hazards of Explosives*, E. G. Whitbread, ed., Explosives Research and Development Establishment, Waltham Abbey, Essex, England, 1963.
32. M. S. Cook, *The Science of High Explosives*, Reinhold Publishing, New York, 1958.
33. H. J. Matsuguma, N. Palmer, J. V. R. Kaufman, Mechanical Parameters of the Picatinny Explosive Sensitivity Test Apparatus, in: *Proceedings of the International Conference on the Sensitivity and Hazards of Explosives*, E. G. Whitbread, ed., Explosives Research and Development Establishment, Waltham Abbey, Essex, England, 1963.
34. V. R. Pai-Verneker, L. Avrami, *J. Chem. Phys.*, *72*, 778 (1968).
35. K. Singh, *Trans. Faraday Soc.*, *55*, 124 (1959).
36. T. Richter, unpublished results.
37. T. Costain, unpublished results.
38. W. L. Shimmin, J. Huntington, L. Avrami, Radiation and Shock Initiation of Lead Azide at Elevated Temperatures, Physics International Co. Report PIFR-308, San Leandro, Calif., AFWL TR-163, June 1972.
39. N. R. S. Hollies, N. R. Legge, J. L. Morrison, *Can. J. Chem.*, *31*, 746 (1953).
40. V. I. Siele, H. A. Bronner, H. J. Jackson, Impact Sensitivity of Lead Azide, RDX, CAB-O-Sil System: Effect of Temperature, Freons, and Water on Sensitivity, Tech. Rept. 3671, Picatinny Arsenal, Dover, N.J., 1968.
41. R. M. H. Wyatt, The Sensitiveness of the Polymorphs of Lead Azide, Proceedings of the Symposium on Lead and Copper Azides, Explosives Research and Development Establishment, Waltham Abbey, England, October 25–26, 1966, Ministry of Technology Report WAA/79/0216.

42. G. W. C. Taylor, The Preparation of Gamma Lead Azide, Proceedings of the Symposium on Lead and Copper Azides, Explosives Research and Development Establishment, Waltham Abbey, England, October 25–26, 1966, Ministry of Technology Report WAA/79/0216.
43. L. Avrami, H. J. Jackson, Results of Laboratory Studies on Triex-8 Dextrinated Lead Azide, Tech. Memo. 1877, Picatinny Arsenal, Dover, N.J., 1969.
44. A. C. Forsyth, J. R. Smith, D. S. Downs, H. D. Fair, Jr., The Effects of Long Term Storage on Special Purpose Lead Azide, Tech. Rept. 4357, Picatinny Arsenal, Dover, N.J., 1972.
45. W. Taylor, A. Weale, *Proc. R. Soc.* (*London*), *A138*, 92 (1932).
46. W. Taylor, A. Weale, *Trans. Faraday Soc.*, *34*, 995 (1938).
47. T. Urbanski, *Chemistry and Technology of Explosives*, Vol. III, Pergamon Press, New York, 1967.
48. F. P. Bowden, O. A. Gurton, *Nature*, *162*, 654 (1948); *161*, 348 (1948); *Proc. R. Soc.* (*London*), *A198*, 337 (1949).
49. F. P. Bowden, A. Yoffe, *Research*, *1*, 581 (1948).
50. F. P. Bowden, *Proc. R. Soc.* (*London*), *A204*, 20 (1950).
51. H. J. Scullion, *J. Appl. Chem.*, *29*, 194 (1970).
52. G. T. Afanas'ev, V. K. Bobolev, *Dokl. Akad. Nauk S.S.R.*, *138*, 886 (1961).
53. K. K. Andreev, N. D. Maurina, Yu, A. Rusakova, *Dokl. Akad. Nauk S.S.R.*, *105*, 533 (1955).
54. J. A. Brown, A. R. Garabrant, J. F. Coburn, L. Avrami, Moderation of the Sensitivity of Certain Lead Azide/RDX Mixtures, Third ICRPG/AIAA Solid Propulsion Conference, Atlantic City, N.J., 1968.
55. J. A. Brown, A. R. Garabrant, J. F. Coburn, J. C. Munday, A Beerbower, Sensitivity and Rheology Studies on Lead Azide–RDX–Silica Mixtures, Esso Research and Engineering Co., Report GR-1SRS-69, Contract DAAA21-67-C-1108, Final Technical Report, Lindon, N.J., 1969.
56. L. Avrami, H. J. Jackson, Impact Sensitivity of Lead Azide in Various Solid and Liquid Media, Tech. Rept. 3721, Picatinny Arsenal, Dover, N.J., 1968.
57. L. Avrami, H. J. Jackson, The Behavior of Explosives and Explosive Devices in a Cryogenic Environment, Tech. Rept. 4351, Picatinny Arsenal, Dover, N.J., 1973.
58. L. Avrami, N. Palmer, Impact Sensitivity of Lead Azide in Various Liquids with Different Degrees of Confinement, Tech. Rept. 3965, Picatinny Arsenal, Dover, N.J., 1969.
59. G. G. Savel'ev, Yu. A. Zakharov, G. T. Shechkov, R. A. Vasuytkova, *Izv. Tomsk Politekh. Inst.*, *151*, 40 (1966).
60. R. W. Hutchinson, F. P. Stein, *J. Phys. Chem.*, *78*, 478 (1974).
61. K. Singh, *Indian J. Chem.*, *7*, 694 (1969).
62. Y. A. Zakharov, E. S. Kurochkin, G. G. Savel'ev, Y. N. Rufov, *Kinet. Katal.*, *7*, 377 (1966) (English translation).
63. Y. A. Zakharov, E. S. Kurochkin, *Russ. J. Inorg. Chem.*, *13*, 919 (1968) (English translation).
64. R. F. Walker, private communication.
65. Yu. N. Sukhushin, Yu. A. Zakharov, *Izv. Tomsk Politekh. Inst.*, *199*, 100 (1969).
66. V. P. Pichugina, S. M. Ryabykh, *Izv. Tomsk Politekh. Inst.*, *199*, 66 (1969).
67. R. W. Hutchinson, unpublished results, 1972.
68. F. P. Bowden, D. Tabor, *Br. J. Appl. Phys.*, *17*, 1521 (1966).
69. L. da Vinci, Il Codice Atlantico; Herausg, vonder Accademia dei Lincei (1508); quoted by G. Vogelpohl, *Oel Kohle*, *36*, 89, 129 (1940).
70. L. da Vinci, *The Notebooks of Leonardo da Vinci*, compiled by E. MacCurdy, Reynal and Hitchcock, 1938.

71. G. Amontons, De la Resistance Causée Dans les Machines, *Mem. Acad. R. Sci. Paris*, 206 (1699).
72. C. A. Coulomb, Théorée des Machine Simples, En Ayant Égard au Frottement de Leur Parties, et à la Roideur des Cordages, *Mem. Acad. R. Sci. Paris*, *10*, 161 (1785).
73. L. Euler, Sur le Frottement des Corps Solides, *Historie de l'Academie Roy. Sci et Belles Lettres*, Berlin, 1750, p. 122.
74. F. P. Bowden, D. Tabor, *The Friction and Lubrication of Solids*, Parts I and II, Oxford University Press, London, 1954, 1964.
75. E. Rabinowicz, *Friction and Wear of Materials*, Wiley, New York, 1966.
76. M. J. Furey, *Ind. Eng. Chem.*, *61*, 12 (1969).
77. J. A. Brown, A Study of Friction Fundamentals in Explosives, Final Tech. Rept. Contract DAAA21-69-C-0558, Berkeley Heights, N.J., December 1970.
78. C. D. Struber, A Microscopic Study of the Friction Process, Air Force Flight Dynamics Laboratory Tech. Rept. AFFDL-TR-69-7 February 1970.
79. P. G. Fox, J. Soria-Ruiz, *Proc. R. Soc.* (*London*), *A317*, 79 (1970).
80. K. Matsubara, *Kogyo Kayaku Kyokaishi* (*J. Ind. Expos. Soc. Jpn.*), *30*, 83 (1969).
81. H. Koenen, K. H. Ide, New Testing Methods for Explosive Substances, Proceedings of the 31st International Congress of Chemical Industry, Liége Belgium, 1958, Translation by Explosives Research and Development Establishment TIL/T.5194, Waltham Abbey, England, 1962.
82. H. Poeschl, H. Siedel, *Chem. Technol.*, *18*, 565 (1966).
83. F. P. Bowden, A. D. Yoffe, *Initiation and Growth of Explosion in Liquids and Solids*, Cambridge University Press, Cambridge, England, 1952.
84. S. Kinoshita, T. Arimura, *Kogyo Kayaku Kyokaishi* (*J. Ind. Explos. Soc. Jpn.*), *24*, 363 (1963).
85. M. Yamada, K. Kaishi, *J. Ordnance Soc.*, *25*, 379 (1932).
86. I. Yamamoto, *Kogyo Kayaku Kyokaishi* (*J. Ind. Explos. Soc. Jpn.*), *27*, 115 (1966).
87. L. Szulacsik, I. Lazar, *Mag. Kem. Lapja, Budapest*, *71*, 469 (1965).
88. K. K. Andreev, *Zh. Prikl. Khim.*, *35*, 1956 (1960).
89. A. P. Amosov, S. A. Bostandzhiyan, *Zh. A. Zinenko, Kokl. Akad. Nauk S.S.R.*, *209*, 1361 (1973).
90. A. P. Amosov, S. A. Bostandzhiyan, Y. S. Kozlov, *Fiz. Gor. Ivz.*, *8*, 362 (1972).
91. K. G. Hoge, Friction and Wear of Explosive Materials, UCRL-50134, Lawrence Livermore Laboratory, Livermore, Calif., 1966.

5

Electrostatic Sensitivity

M. S. Kirshenbaum

A. INTRODUCTION

Accidental detonations occur during the processing, handling, and transport of lead azide and have been attributed to electrostatic discharges. Although the initiation of azides has been extensively studied and various mechanisms offered to explain their behavior for a wide variety of situations, initiation by (gaseous) electrostatic discharge has received relatively little attention. There is little knowledge of the transfer of energy between a gaseous discharge and an explosive; for example, it is not known whether initiation is by electronic processes or whether the electrostatic energy first degrades to thermal energy.

Electrostatic sensitivity tests, however, provide an important measure of the hazards associated with processing and handling. The nature of the test and interpretation of the results are key factors. But it is only when the energy transfer mechanism is understood that reliable extrapolations and predictions for new situations can be made.

To assess the electrostatic hazards many approaches have been made to determine the threshold or minimum energy that will initiate detonation or fast decomposition. The conventional method used to determine a minimum energy is to subject the explosive to a single discharge from a known capacitor that has been charged to a high voltage. With successive samples the energy of the discharge is gradually decreased (by reducing the capacitance and/or the voltage across the capacitor) until no initiation occurs in a specified number of trials. Despite the apparent simplicity of the technique, widely varying minimum energy values (e.g., 4×10^{-10} to 2×10^{-2} J) have been reported for a single sub-

Table I. Initiation Energies for Electrostatic Initiation of Lead Azide

Azide	Apparatus and electrodes	Gap length (mm)	Discharge	Capacitance (pF)	Series resistance (ohms)	Relative humidity (%)	Energy (J)	Reference
RD1333								
	Approaching needle plane	0	Contact	300	0	Dry	$<6 \times 10^{-8}$	[37]
						75	3.8×10^{-5}	[37]
						40	3.0×10^{-6}	[2]
				1,000			$<3 \times 10^{-5}$	[22]
		0.48	Gaseous	3,000		12	$3\text{–}6 \times 10^{3}$	[24]
		0.25		50			$<5 \times 10^{-5}$	[24]
		0.19		2,000		Dry	3.6×10^{-3}	[11]
		0.19		1,200	100K[a]		3.5×10^{-4}	[11]
	Needle rubber	?		1,000		40	$<5 \times 10^{-4}$	[22]
	Fixed gap needle plane	0.25	Gaseous	300	0	Dry	$1\text{–}2 \times 10^{-3}$	[37]
		1.1		10,000		55	2×10^{-2}	[40]
		0.13		330		40	1.3×10^{-3}	[8]
		0.13			1.2M[b]		2×10^{-4}	[8]
	Parallel plate	0.19			0		3.3×10^{-4}	[8]
		0.19			1.2M		1.1×10^{-4}	[8]
	Vibrator needle	0.003	Contact	50	0	?	4×10^{-10}	[25]
Service	Approaching needle plane	0	Contact	100	0	40	2.0×10^{-6}	[2]
					20K		1.6×10^{-6}	[2]
		0.04	Gaseous	1,500	100K		2.2×10^{-4}	[5]
	Needle rubber			500			1.0×10^{-3}	[3]
				1,500			2.2×10^{-4}	[3]

Dextrinated								
	Approaching needle plane	?	Gaseous	300	0	Dry	$7\text{–}9.6 \times 10^{-3}$	[37]
		?		?			7.0×10^{-3}	[41]
		0	Contact	?	?	?	1.0×10^{-5}	[42]
		0.75	Gaseous	?	0	?	1.0×10^{-2}	[43]
		?		1,000		40	$3\text{–}4.5 \times 10^{-3}$	[21]
		?		20,000			1.0×10^{-2}	[3]
		?		1,000			$1\text{–}4.5 \times 10^{-3}$	[22]
	Plumb-bob plane	0	Contact	300			2.0×10^{-5}	[2]
	Needle rubber	?	Gaseous	10,000	100K		8.0×10^{-2}	[21]
		?		10,000			$0.5\text{–}1 \times 10^{-1}$	[22]
	Plumb-bob rubber	?		6,000			2.0×10^{-3}	[3]
	Fixed gap needle plane	1.3		10,000	100		$2\text{–}3 \times 10^{-2}$	[10]
		1.1		10,000	0	55	2.0×10^{-2}	[40]
	Parallel plate	0.19		550	5.6M	40	5.0×10^{-4}	[11]
PVA								
	Approaching needle plane	0	Contact	?	?	?	2.0×10^{-7}	[42]
		?	Gaseous	?	?	?	6.3×10^{-3}	[42]
	Needle (neg.) rubber	?		400	100K	40	2.2×10^{-4}	[7]
	Needle (pos.) rubber	?		200			1.0×10^{-4}	[7]
	Fixed gap needle plane	1.1		10,000	0	55	2.0×10^{-2}	[40]
		1.3		1,000	100	40	$3\text{–}4 \times 10^{-3}$	[10]

[a] K = kilohms
[b] M = megohms

stance such as lead azide with different apparatus and under different conditions (Table I).

Many of the differences have been explained by Moore, Sumner, Wyatt [1-7] and Kirshenbaum [8, 9]. They showed that the electrical circuit, spark-gap–explosive geometry, cathode properties, and nature of the explosive all play important roles. The circuit components and spark-gap geometry affect the rate of energy delivery, efficiency of energy transfer, and character of the spark. One of the problems is differentiation between the contributions to the observed results made by the spark and by the explosive.

B. ELECTROSTATIC SENSITIVITY APPARATUS

1. General

There is no standard electrostatic sensitivity test as all apparatus in current use were designed and fabricated by government or private laboratories for their internal purposes. However, the basic components of any apparatus include a fixed or variable high-voltage power supply, an electrical charging circuit, a triggering circuit, an electrode assembly, and an electrostatic voltmeter to measure the voltage. Although, in principle, the tests are similar, the parameters vary. In order to understand better the electrostatic data, it is helpful to review the two approaches most commonly used to measure the minimum energy: the fixed-gap method and the approaching-electrode method.

2. Fixed-Gap Apparatus

In the fixed-gap method two electrodes are placed at a desired fixed distance; although the distance is not standardized, the gap may be varied and measured as required even with a single apparatus. Normally, the electrodes have a needle-plane configuration, but other electrode arrangements are also used, e.g., sphere-plane and parallel-planes. The storage capacitor, charged to the desired voltage, is discharged between the electrodes by means of a fast, low-loss electronic switch, e.g., thyratron, mercury, vacuum. The voltage applied to the electrodes breaks down the gap and causes a spark to pass. Pictures of the needle-plane and the parallel-plane apparatus used by Kirshenbaum [8] are shown in Figures 1 and 2, respectively.

3. Approaching-Electrode Apparatus

In the approaching-electrode method, there is no switch between the electrodes and the charged storage capacitor. The gap is initially too wide for a

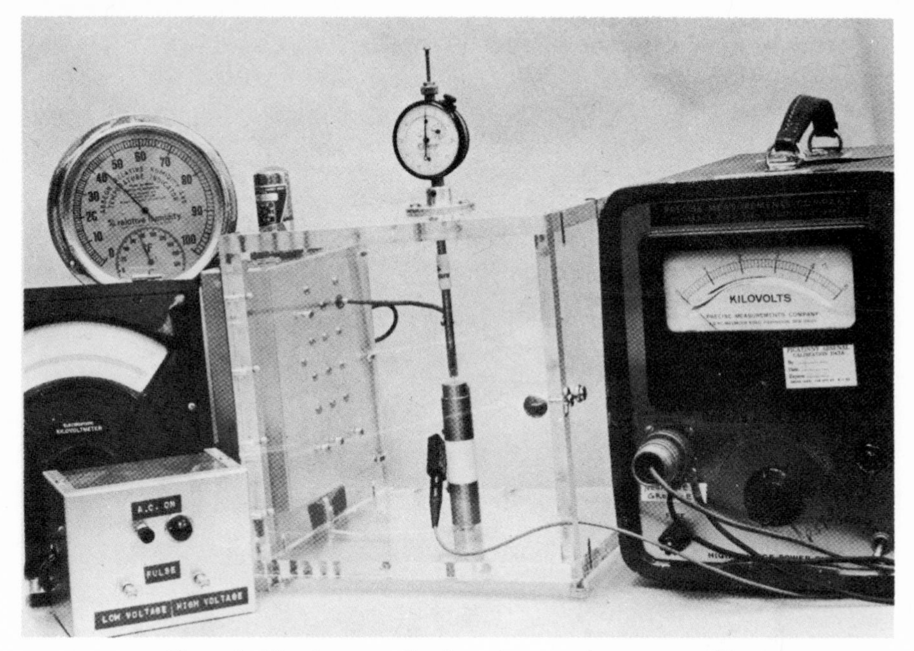

Figure 1. Fixed-gap, needle-plane electrostatic apparatus [8].

gaseous discharge to take place at the voltage applied. A spark is produced as one electrode (usually a needle) approaches the other (plane). The upper electrode is moved by gravity, a spring, or a motor to a preset distance or until it contacts the other electrode. The apparatus used by Wyatt [2] and Montesi [10] is shown in Figure 3. The apparatus used by Kirshenbaum [11] is shown in Figure 4.

The threshold voltage for gap breakdown is governed by Paschen's law [12], which states that in a given medium the sparking potential is a function of the product of the density of the medium and the distance between the electrodes. The critically sized gap varies, therefore, with the applied voltage, and there is a minimum voltage below which a spark will not pass between two electrodes, e.g., in air the voltage is about 300 V.

C. CURRENT–VOLTAGE CHARACTERISTICS OF GASEOUS DISCHARGE

When a charged capacitor is discharged through a spark gap (containing no added resistance) by an approaching electrode [1] or by a mercury-wetted relay

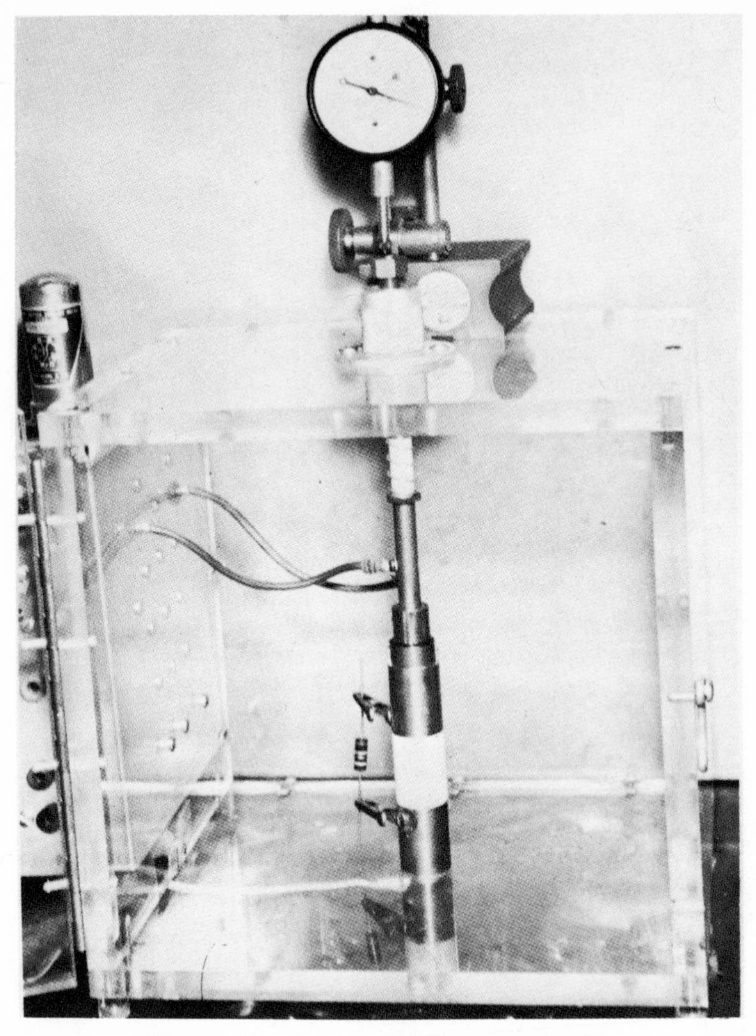

Figure 2. Fixed-gap, parallel-plate apparatus [8].

switch [8], the current and voltage vary in an oscillatory (underdamped) manner due to stray (inherent) inductance in the circuit. The current and voltage are slightly out of phase, the voltage leading the current. Addition of resistance to the circuit damps out the discharge and results in an overall shortening of the

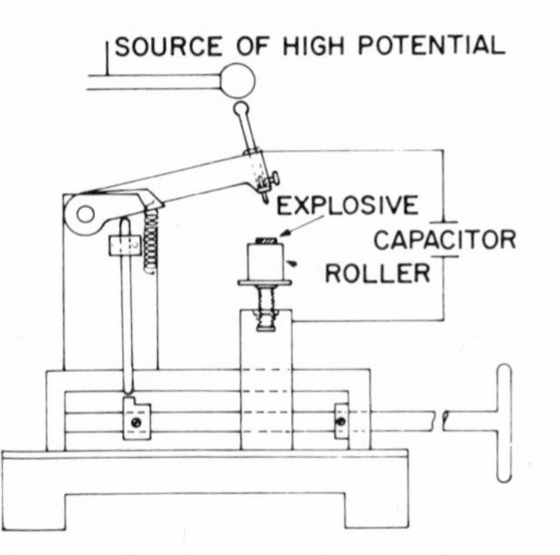

Figure 3. Approaching-electrode electrostatic apparatus [2].

discharge. The discharge time reaches a minimum when the added resistance results in a critically damped circuit. Further increases in resistance result in unidirectional (overdamped) discharges of longer duration. With large series resistances ($>10^5$ Ω), the discharge is no longer continuous but takes the form of bursts of sparks due to relaxation oscillations [8,9]. If a Krytron switch tube is substituted for the mercury switch (no added resistance), unidirectional instead of oscillatory discharges are obtained since the Kryton switch provides some rectification of the current [8].

For a better understanding of electrostatic sensitivity data, it is also important to recognize that capacitor discharge through a gas can occur as arcs and sparks (glow discharge). Both modes may occur in a single discharge, depending primarily on the resistance in the discharge circuit and to a lesser extent on the initial voltage [12,13].

An arc is a postbreakdown discharge in which thermionic emission from the cathode is responsible for sustaining the discharge. The most characteristic features of an arc are the low postbreakdown voltage drop across the spark gap, which is usually of the order of tens of volts (30-60 V), and the high current flow, larger than 0.3-0.5 ampere [13]. An arc may be formed for series resistances as large as 10-20 kΩ, for gaps as large as 1.3 mm (0.050 in.), and is not necessarily associated with touching electrodes. A spark discharge, on the other hand, is a postbreakdown regime in which the discharge is maintained by secondary emission of electrons from the cathode by ion bombardment [12]. The voltage drop across the spark gap is typically 300-400 V and the current is in the milliampere range [13].

Representative current and voltage waveforms are shown in Figure 5 for

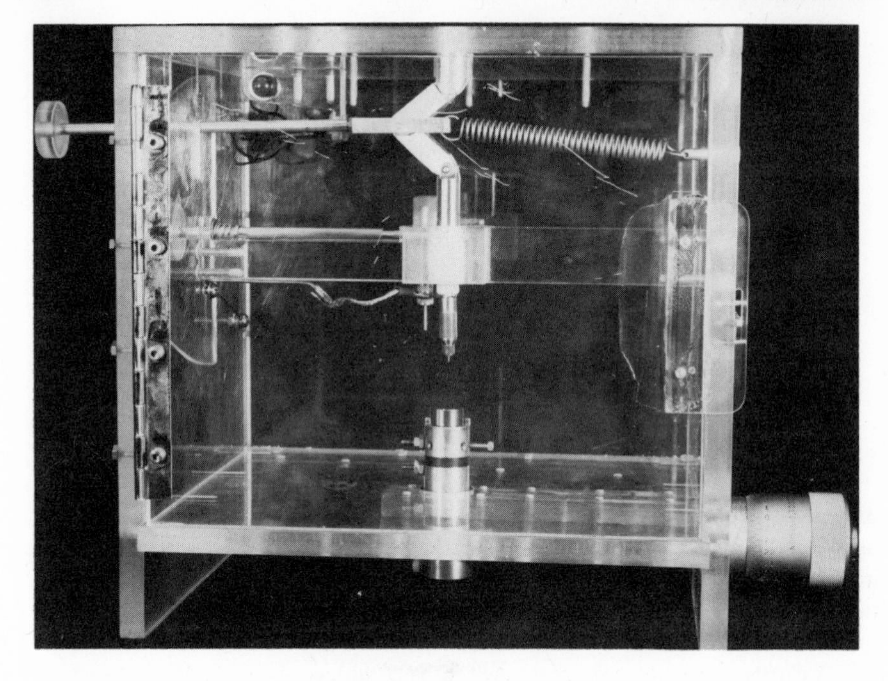

Figure 4. Approaching-electrode apparatus [11].

four values of series resistances, 1, 6.8, 10, and 100 kΩ. Figure 5A represents an arc discharge. It can be seen that the voltage across the gap, which initially increases to the threshold voltage of the gap to break down the gap (greater than 1500 V), rapidly decreases to the low postbreakdown voltage (approximately 35 V) (region a) and remains at this low voltage until the discharge ceases. The voltage across the gap then increases to the value of the voltage remaining on the storage capacitor (region b), since there is no longer a voltage drop across the series resistor. As the storage capacitor discharges, the current decreases in a conventional capacitor discharge pattern from several amperes to below a minimum sustaining value, at which time the discharge ceases.

With the 100-kΩ resistor (Figure 5D), a spark discharge is obtained. The spark discharge is recognized by the high postbreakdown voltage value of about 350 V (region a) and the low peak current of about 30 mA. When the discharge ceases, the gap voltage increases to the voltage remaining on the storage capacitor (region b).

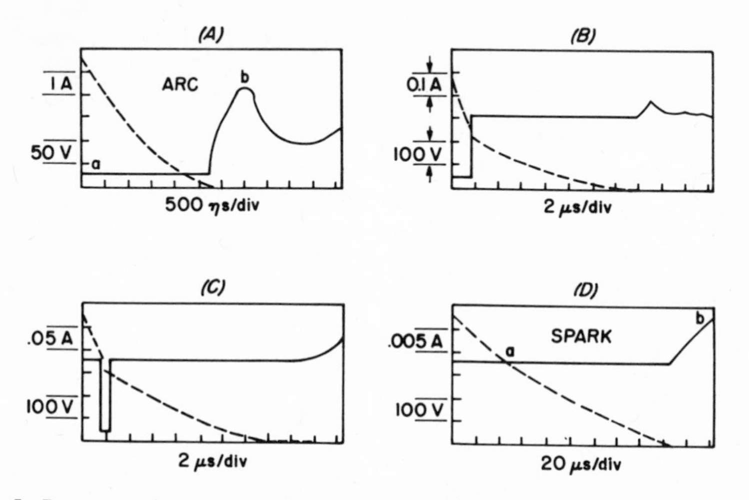

Figure 5. Representative current–voltage waveforms [9]. Dashed line: current vs. time; solid line: voltage vs. time. (A) $R = 1\text{k}\Omega$; (B) $R = 6.8\text{k}\Omega$; (C) $R = 10\text{k}\Omega$; (D) $R = 100\text{k}\Omega$.

A transition regime exists between the two regions (arc and spark) which is discontinuous. In Figures 5B and 5C, the waveforms show that the postbreakdown voltage varies between approximately 30 V (arc region) and 300 V (spark region) during short intervals throughout the discharge period. In the transition range the nature of the discharge can be different for repetitive tests [9].

An important feature of both sparks and arcs is the partition of energy delivered to the gap between the gas and the electrodes. In both forms of discharge, a large fraction of the total voltage drop occurs near the electrodes. In the spark (glow) regime, about 270 V of the total drop of 300 V occurs near the cathode [13]. A very much smaller drop occurs near the anode (approximately 10–15 V). In the region near the cathode the ions are accelerated in the large electric field to produce the secondary electrons necessary to sustain the discharge. In the arc nearly all of the total voltage drop also occurs near the cathode (approximately 15 V) [14]. Furthermore, in the arc the current is concentrated in a small filament so that the cathode is intensely heated in a small region or "spot." The intense heating is necessary to provide the emission of electrons to sustain the discharge. Thus, most of the energy delivered to a gap in either a spark or arc discharge is delivered to the cathode and possibly to the gas within a few microns of the cathode surface.

Energy can be coupled to an explosive by heat conduction, radiation, and by vapor jets of metal from the cathode [15]. Estimates of the coupling efficiencies for sparks and arcs are difficult to make without detailed information on the temperature of the cathode spots, discharge channel size, properties of the explosives, and, in the case of arcs, mechanisms for emission of electrons. If the boiling point of the cathode is lower than the temperature required for sufficient thermionic emission, metal vapor may be mixed with the gas near the cathode region [14]. Field emission has been proposed as a mechanism in some cases. Thus the observed sensitivities of explosives to arc discharges are almost certainly dependent upon the type of cathode material or cathode surface condition used. Experimental evidence that this is so is given in Section G.

D. EFFICIENCY OF ENERGY DELIVERED TO SPARK GAP

Most explosive minimum energy data are reported as the total energy stored in the capacitor (Calculated from $1/2\ CV^2$). A few experimentalists [1,8], however, calculated the actual energy dissipated in the spark gap. Moore and coworkers [1] calculated the efficiency of the energy delivered to a 0.13-mm (0.005-in.) needle-plane discharge gap. The data show that a fairly constant value of 10% of the original energy is delivered to the gap when resistances of 10^3-$10^7\ \Omega$ are placed in series with the gap. Kirshenbaum [8], using parallel-plane electrodes with a 0.19-mm (0.0075-in.) gap, obtained very similar results. The efficiency of the energy delivered to the spark gap decreases from 80-90% to 10-14% as the series resistance is increased from 0.15 to 1000 Ω. As the resistance is further increased to 1.2 MΩ, the efficiency remains practically constant at the 10-14% level.

The efficiency of the approaching needle apparatus was measured in the arc and spark modes [11]. It was shown that the marked difference in the postbreakdown voltage drop across the gap of the arc and of the spark had an important effect on the efficiency of energy delivered to the gap from the storage capacitor when the current is limited by a series resistance. This is because the differential of the charge (total current) flowing in an arc and a spark is only about 10% while the postbreakdown voltage differential is 400-500%. For an arc discharge (series resistance = 2 kΩ), the efficiency decreases from 6 to 3% as the storage capacitance is decreased from 2000 to 276 pF. Keeping the storage capacitance constant, the efficiency remains practically constant as the charging voltage is decreased from 4500 to 2000 V. For a spark discharge (series resistance = 100 kΩ), on the other hand, the efficiency remains practically constant as the storage capacitance is decreased from 2000 to 624 pF. For each capacitance, however, the efficiency decreases from 22 to 16% as the charging voltage is increased from 2000 to 4500 V.

E. CONTACT VS. GASEOUS DISCHARGE

If contact between the electrodes is permitted, contact discharge occurs. Contact discharge initially results in intense Joule heating of the touching electrodes at the point of contact. Upon separation the hot cathode emits sufficient electrons to maintain the discharge. This type of discharge is usually accompanied by ejection of hot metal and emission of relatively high-frequency radiation (into the ultraviolet). It can occur even at very low voltage.

Moore and coworkers [2,6] were the first to report the distinction between the contact and gaseous discharge in initiating explosives. They showed that the initiation probability of Service lead azide is almost independent of the capacity of the storage capacitor, if the circuit resistance is small in a fixed-gap apparatus (Figure 6). Markedly different results were obtained with the same explosive and electrode materials, however, when an approaching-electrode apparatus was used where contact between electrodes occurred. As shown in Figure 7, two distinct energy regions led to initiation of lead azide. In the low-energy region, the initiation energies were as low as 1.5-2.0 $\times$ 10^{-6} J. The peak in the initiation

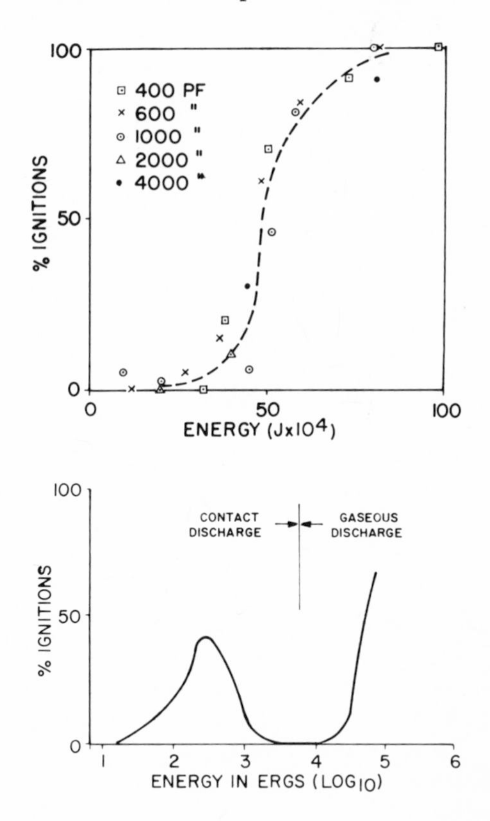

Figure 6. Dependence of ignition probability on energy of Service lead azide (fixed gap, 0.13 mm) [6].

Figure 7. Dependence of ignition probability on energy of Service lead azide (approaching electrode, C = 1000 pf) [6].

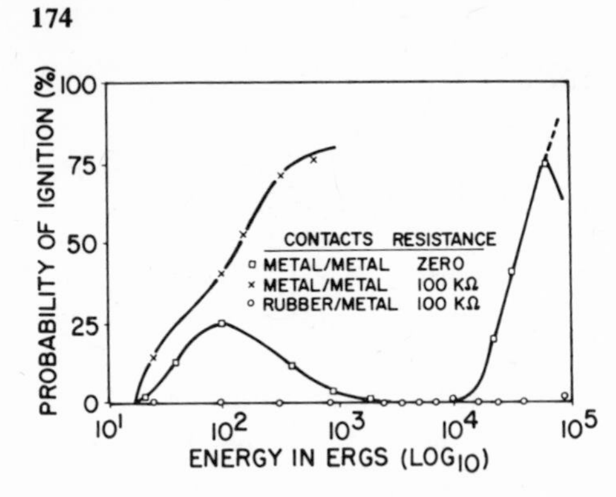

Figure 8. Ignition probabilities for Service lead azide (approaching electrode, $C = 500$ pf) [6].

probability at low energies occurred when the voltage across the storage capacitor was about 200–250 V. The peak thus occurred when gaseous discharge is not possible. Initiation at the lower voltages must, therefore, be due to contact discharge.

The decrease in probability of initiation above the low-energy peak is due to the onset of gaseous discharge. Initiation in the high-energy range is evidently due to gaseous discharge alone. The low initiation energy region was removed by precluding metal-to-metal contact with a sheet of conducting rubber in place of a steel-base electrode (Figure 8). The rubber electrode had an equivalent series resistance of 100 kΩ and, according to the authors, simulates the resistance of a human. (Different experimentalists [16–19] report values varying from 25 Ω to 10 MΩ for the resistance of a human, the majority reporting 500–5000 Ω. However, a single resistance does not adequately simulate a human or a hazardous situation. The cathode should ideally simulate the human skin, and it should be determined whether arcs or sparks occur.) The sensitiveness to both contact and gaseous discharge is considerably increased by increasing the duration of the discharge with a resistor in series with the gap. From Figure 9 it can be seen that

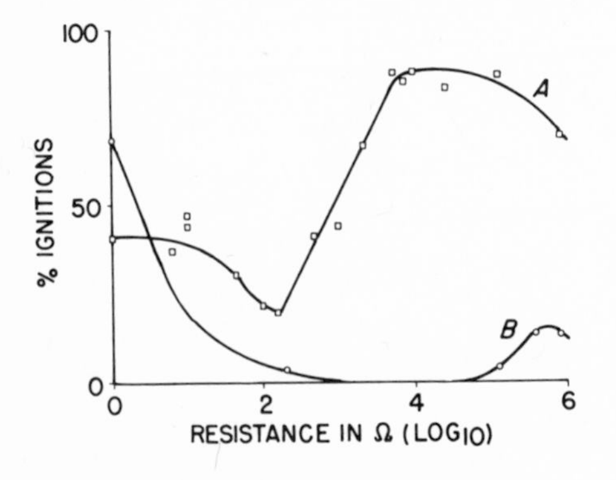

Figure 9. The effect of series resistance on ignition of Service lead azide in contact (A) and gaseous (B) discharge regions [6].

the initiation probability first decreases to a minimum at 10^2–10^3 Ω and then increases to a maximum at 10^4–10^6 Ω.

F. OPTIMUM GAP

The probability of initiation of Service lead azide was shown to vary with gap length for a fixed-gap, needle-plane apparatus having large resistances in series with the gap [2]. The optimum gap for maximum probability of initiation was between 0.075 and 0.175 mm (0.003 and 0.007 in.). Figure 10 shows a rise in the initiation probability to a maximum as the gap was increased from 0.0025 mm (0.0001 in.) to some value between 0.075 and 0.175 mm (0.003 and 0.007 in.). As the gap was increased further, the probability fell to a minimum between 0.75 and 1.25 mm (0.030 and 0.050 in.). With either 10-Ω or 0-Ω resistance in series [2], the above phenomenon was not observed.

G. EFFECTS OF CATHODE SURFACES

The sensitivities of explosives to arc discharges were shown to be dependent upon the type of cathode material or the cathode surface conditions [20]. Fixed-gap, needle-plane electrodes with a 27-Ω series resistance and a 2000-pF capacitor charged to 4500 V were used, but no ignitions of RD1333 lead azide were observed in more than 20 trials with the conventional cold-rolled steel cathode. When a steel cathode was oxidized with ceric ammonium nitrate and substituted for the polished cathode, the ignition probability was nearly unity. High ignition probabilities were also observed for epoxy coatings on steel and for waxed paper and filter paper placed on the cathode. The authors suggest that the oxide layer modifies the emissions of electrons from the cathode so that a higher efficiency energy delivery through spark formation occurs.

The high ignition probabilities observed for epoxy and paper are associated

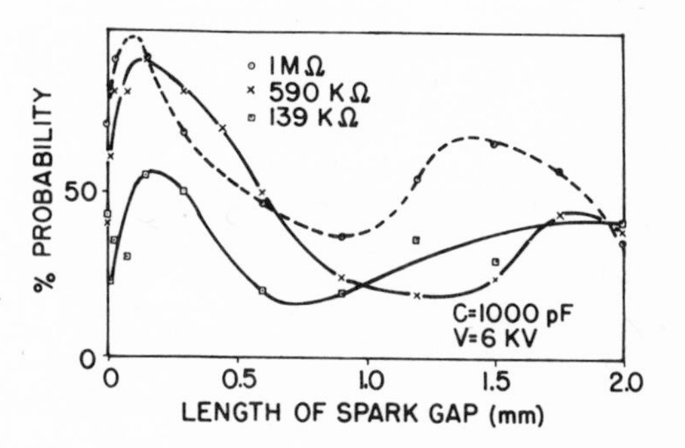

Figure 10. Effect of gap length on probability of ignition of Service lead azide [2].

with an unusual form of discharge. These materials are dielectrics, and the discharge does not proceed directly from the needle to the plane but passes through a weak spot in the material.

When a layer of Velostat (Custom Materials, Chelmsford, Massachusetts) was placed on the cathode, however, the ignition probability did not increase even though a spark formed. This was because the energy density was too small, since the spark was distributed over a volume (near the cathode) at least one order of magnitude larger than that for a spark with a high series resistance and the same delivered energy.

H. ENERGY RESPONSE CURVES

The 50% firing point is used in many studies to determine the sensitivity of the explosive. This is justified if the data for a complete firing curve are normally distributed. That this is approximately the case for RD1333 lead azide is shown in Figure 11, where the data for a complete firing curve [8] are plotted together with the straight line which represents a normal probability distribution.

I. NEEDLE-PLANE VS. PARALLEL-PLANE ELECTRODES

The minimum initiation energies obtained for RD1333 lead azide for a range of series resistances were lower using fixed-gap, parallel-plane electrodes than those using fixed-gap, needle-plane electrodes [8] (Table II). The energy values may be lower for the parallel-plane electrodes (or, equivalently, the initiation probability at a given energy is greater) because the configuration

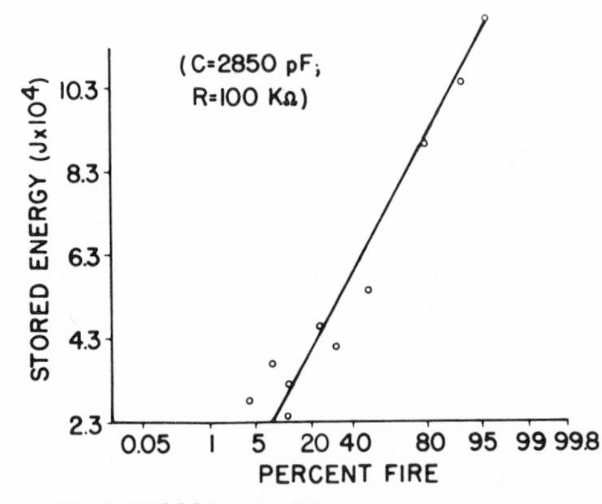

Figure 11. RD1333 lead azide energy–response curve [8].

Table II. Comparison of the Minimum Initiation Energy of Lead Azide with Needle-Plane and Parallel-Plate Electrodes [8][a]

Series resistance (Ω)	RC time constant (m/sec)	Energy in spark gap (J)	
		Needle plane	Parallel plate
1.5×10^{-1}	[b]	1.3×10^{-3}	3.3×10^{-4}
8.2×10^{1}	2.7×10^{-5}	$>5.9 \times 10^{-4}$	3.1×10^{-4}
1.0×10^{5}	3.3×10^{-2}	$>3.1 \times 10^{-4}$	3.1×10^{-4}
6.8×10^{5}	2.2×10^{-1}	2.2×10^{-4}	1.8×10^{-4}
1.2×10^{6}	4.0×10^{-1}	2.0×10^{-4}	1.1×10^{-4}

[a] RD1333 lead azide and a constant capacitance of 330 pF were used.
[b] Self-commutated by krytron switch.

confines the powder more than does a needle and plane. The spark results in a rapid movement of air which tends to blow the powder away from the electric discharge. The parallel-plane electrodes tend to confine the powder and permit more efficient "coupling" between the discharge and the powder. Another contributing factor is the larger volume of explosive sampled with the parallel-plane configuration than with the needle and plane, where only the explosive near the needle is sampled.

J. EFFECT OF ENERGY DELIVERY RATE ON INITIATION

Kirshenbaum [8] demonstrated that the initiation probability of RD1333 lead azide by gaseous discharge is a strong function of the properties of the discharge circuit, particularly as they affect the rate of the energy delivered to the spark gap. Two series of tests were conducted with a fixed-gap, parallel-plane apparatus.

1. Storage Capacitor Constant

In the first series of tests (Figure 12), the storage capacitor was kept constant at 2800 pF and the RC time constant of the spark electrical circuit was increased from approximately 0.1 μsec to 40 msec by increasing in increments the series resistance from approximately 1 Ω to 15 MΩ. The energy level at the 50% firing point was used as a measure of sensitivity. As can be seen in Figure 12, the energy decreases from 8×10^{-3} J to a minimum of 6.5×10^{-4} J as the RC time constant is increased from 0.1 μsec to some value between 0.1 and 1 msec and then increases to 1.4×10^{-3} J as the RC time constant is further increased to 43 msec.

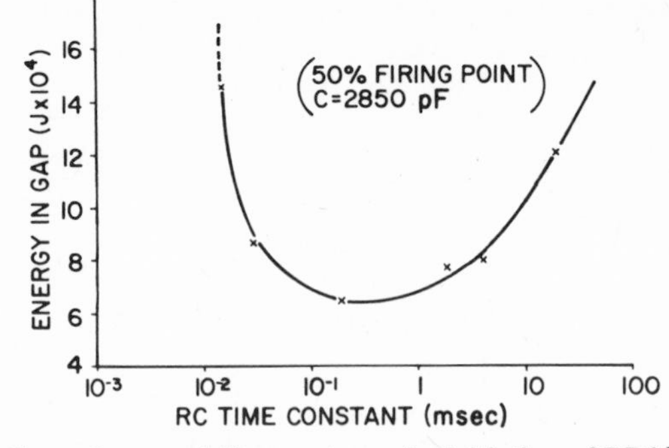

Figure 12. The effect of energy delivery rate on the initiation of RD1333 lead azide [8].

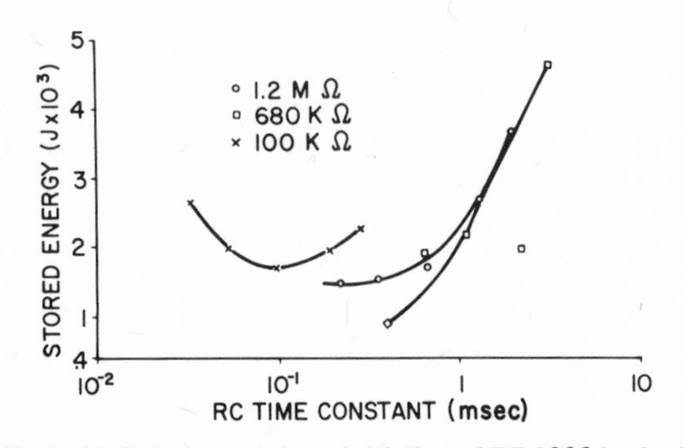

Figure 13. The effect of RC time constant on initiation of RD1333 lead azide [8]; minimum energy; parallel-plate electrodes.

2. Series Resistance Constant

In the second series of tests (Figure 13), the resistance was kept constant and the storage capacitance was increased from 300 pF to 3000 pF. In this set of tests the threshold energy values were used as a measure of sensitivity. It is again interesting to note that the time constants for all combinations of capacitance and resistance which resulted in a minimum of the threshold energy are approximately the same, 0.1–0.4 msec. This suggests that there is an energy delivery rate which results in the initiation of lead azide by the least amount of energy being dissipated in the spark gap.

K. INITIATION ENERGIES FOR VARIOUS AZIDES

1. Service Lead Azide

According to Wyatt [5], there is a minimum capacitance (400 pF) for the initiation of Service lead azide below which initiation does not take place, regardless of what potential is placed on it. The minimum capacitance was attributed to the occurrence of multiple incomplete low-energy discharges which occur consecutively as the electrodes approach. He also noted an optimum capacitance (1500 pF) at which the threshold energy is a minimum (2.2×10^{-4} J). The minimum and optimum capacitance were demonstrated with an approaching electrode apparatus using a rubber-base electrode. With a steel-base electrode and small series resistances the phenomena do not occur, and the initiation probability is almost independent of the capacitance. The contact initiation of Service lead azide [2,6] is discussed in Section E.

2. Polyvinyl Alcohol Lead Azide (PVA)

Wyatt [7] showed that the minimum energy value of PVA lead azide decreases from approximately 2.2×10^{-4} J to about 1×10^{-4} J when the polarity of the steel needle is changed from negative to positive (rubber-base electrode). The minimum capacitance for initiation is also lowered from about 400 pF to about 200 pF. This effect is not observed when the bottom (stationary) electrode is steel.

Montesi [10] reported a minimum energy value between 3.1×10^{-3} and 4×10^{-3} J for PVA lead azide. The value was obtained with a fixed-gap, needle-plane apparatus, using a 1.25-mm (0.050-in.) gap, 100-Ω series resistance, and a 1000-pF storage capacitor. The apparatus is used primarily to provide relative sensitivity data for explosives. The high minimum energy value obtained may have been due to the use of the 100-Ω series resistance and the large 1.25-mm gap.

3. Dextrinated Lead Azide

Moore [3] reported that dextrinated lead azide is difficult to initiate with energies less than 1×10^{-2} J (20,000-pF storage capacitor) when a metal–metal approaching-needle electrode apparatus is used. However, initiations were quite easily obtained when a plumb-bob electrode was substituted for the needle electrode. Minimum energy values as low as 2×10^{-5} J (300-pF storage capacitor) were obtained (contact discharge). When metal–metal contact is precluded by replacing the steel electrode with rubber, a minimum energy value of 2×10^{-3} J (6000-pF capacitor) was obtained. A minimum capacitance of 2000 pF was also noted.

Montesi [21] reported a minimum initiation energy between 3.1 and 4.5×10^{-3} J when he used Wyatt's approaching-needle, metal-base electrode apparatus and procedure. Substituting a rubber base increased the minimum initiation energy to greater than 8×10^{-2} J. In a more recent report, Montesi [22], using Wyatt's apparatus and the interim test procedure described in the United States Joint Service Safety and Performance Manual [23], reported minimum initiation energy values between 1.1 and 4.5×10^{-3} J and between 4.5×10^{-2} and 1×10^{-1} J, respectively, for needle metal-base and needle rubber-base electrodes. With a fixed-gap, needle-plane apparatus, Montesi [10] reported a minimum initiation energy value of $2\text{-}3 \times 10^{-2}$ J.

Two energy values are reported for dextrinated lead azide using a fixed-

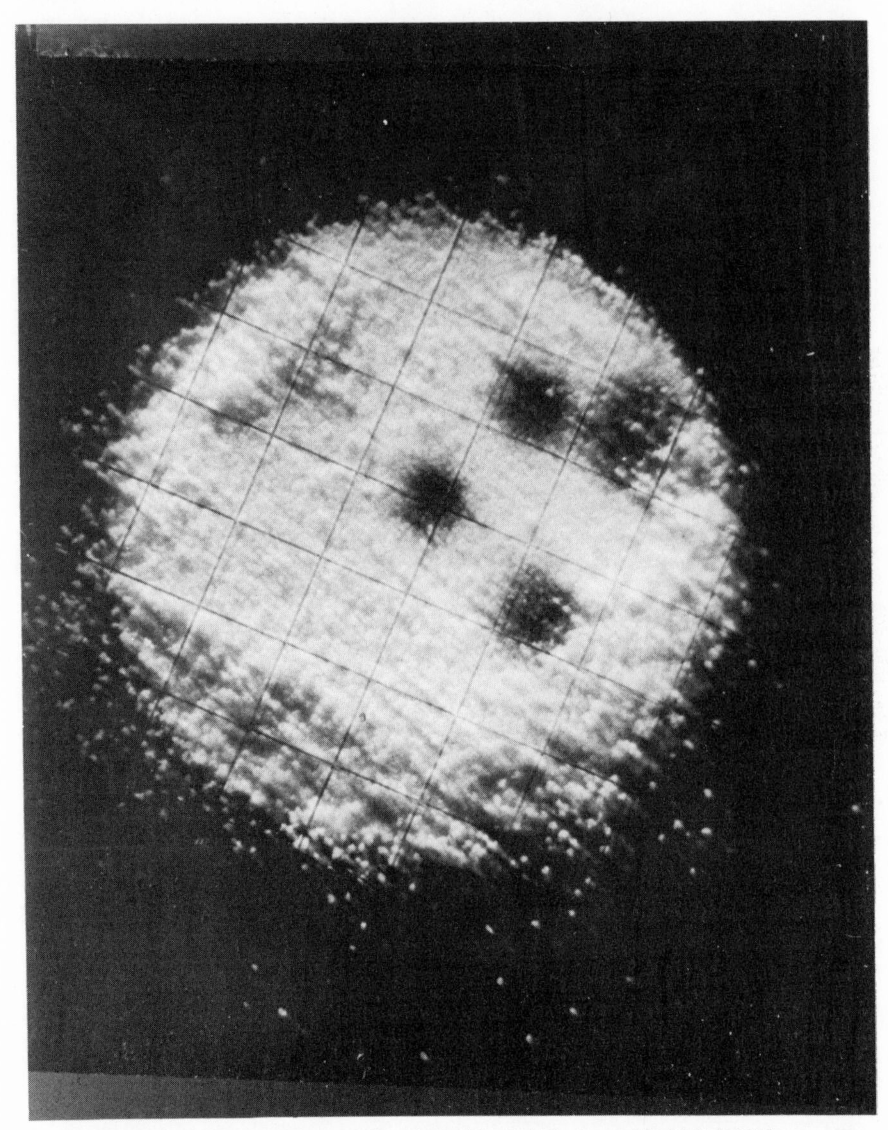

Figure 14. Partially burned dextrinated lead azide (magnified 15X) [11].

gap, parallel-plate apparatus: 5×10^{-4} J for complete detonation and 1.5×10^{-4} J for partial burning of the powder [11]. Figure 14 shows a picture of the partially burned powder. The four burn marks in the picture indicate more than one spark. (The energies reported in the literature do not always distinguish between complete or partial detonation or burning. Many values are reported as a "fire" if the explosive gives the slightest indication of an ignition.)

4. RD1333 Lead Azide

The effect of the energy delivery rate on the initiation of RD1333 with a parallel-plate apparatus is discussed in Section J. For an approaching-needle apparatus with a preset gap of 0.25 mm (0.010 in.), the minimum energy increased from 4.7×10^{-4} J for a spark discharge to 4.7×10^{-3} J for an oscillatory discharge and to 2.0×10^{-3} J for an arc discharge [11]. Lower initiation values are obtained for dry powder tested at 35% relative humidity than for powder maintained and tested at 60 ± 5%. The 50% initiation value increased from 2.8×10^{-3} J to 4.0×10^{-3} J for the humid condition. Lower energy values are also obtained when the spark occurs entirely in the explosive powder instead of starting above (out of) the powder (Table III). It is interesting to note that lower values were obtained for the fixed-gap needle-plane apparatus [8] than for the approaching-needle apparatus.

A minimum energy of less than 5×10^{-5} J was reported [24] for the initiation of RD1333 using an approaching-electrode apparatus having a preset gap of 0.25 mm (0.010 in.) and a 50-pF capacitor. Increasing the preset gap to 0.48 mm (0.019 in.) increased the minimum energy value to $3\text{–}6 \times 10^{-3}$ J (2000-pF capacitor).

Hanna and Polson [25] reported that freshly broken crystals of RD1333 can be detonated by as little as 4×10^{-10} J, using either a probe and plate or a

Table III. Comparison of the 50% Initiation Values of RD1333 Lead Azide Obtained for the Spark Starting Within or External to the Powder [11] (R = 100 kΩ, C = 1176 pF)

Base electrode material	Preset gap (mm)	Spark starting position	Energy ($J \times 10^{-3}$)
Steel	0.18	External	3.4
Steel	0.18	Within	2.8
Steel	0.63	External	3.6
Steel	0.63	Within	2.6
Stainless steel	0.18	External	4.2
Stainless steel	0.18	Within	3.3

mechanical vibrator apparatus. By the former method, the charged steel test probe (a 22-pF capacitor charged to 4 V) was manually manipulated over the azide crystals on a steel-base electrode. The movement of the probe caused the azide crystals to be broken into fine particles before detonation. With the latter apparatus, an electrified steel probe (a 50-pF capacitor charged to 4 V), vibrating and discharging 120 times/sec over the azide crystals (1 1/2 mg) in a small cavity in the steel-base electrode, caused fracture of the crystals before detonation. When no potential was placed on the capacitor, no detonations were obtained from the mechanical vibration. It was also demonstrated that ignition of the freshly broken crystals can be obtained without mechanical vibration. In this test, the azide crystals were first fragmented by an electrically shorted vibrator. Then, when the apparatus (capacitor charged) was operated by hand, the broken crystals detonated upon contact.

Wyatt [26] confirmed the observations of Hanna and Polson. Wyatt and coworkers used a charged, steel phonograph needle, vibrating and discharging 1-10 times/sec, in contact with azide powder in a 1-mm diam spherical cavity in the steel-base electrode. The needle also came in contact with the steel-base electrode. Wyatt noted that ignition usually occurred with an energy of 3.4×10^{-10} J (a 27 pF capacitor charged to 5 V) within the first 100 contacts with either crystalline lead azide, dextrinated lead azide, or silver azide. When no potential was placed on the capacitor, no ignitions were found after 3000 contacts.

Wyatt suggested two possible explanations for the very low initiation values: One is that freshly broken crystals are more reactive; the other is that it is largely a matter of particle size (breaking a crystal just provides smaller crystals). Preliminary tests indicate that it is the latter possibility. In one test, when no ignitions occurred after 100 contacts with the electrified vibrator, most of the azide crystals were removed, leaving about 0.2 mg of a finely ground explosive. Then, when the apparatus was operated by hand, the first contact (capacitor charged) caused an ignition. In another test, colloidal lead azide (prepared some time previously) showed that ignition could be obtained as easily as with the finely ground material. (Particle size of colloidal lead azide is between 1 and 10 μm).

In another series of tests carried out by Hanna and Polson [25], an RD1333 crystal, gripped between two steel needle electrodes, was subjected to an electric discharge. It was noted that a potential of 1000 V was required to detonate a 0.04-mm crystal, while 50 V were needed for fragments of crystals. The electric fields for the two tests were greater than 250 kV/cm and 130 kV/cm, respectively. These electric fields were much greater than the 30–40 kV/cm values reported by Downs and coworkers [27] as the threshold field for initiation of 0.2–0.8-mm-thick crystals (see Section M).

5. Lead Azide-Aluminum-RDX Mixtures

Montesi and Simmons [22] reported a minimum energy of less than 5×10^{-4} J for both RD1333 and for a 75:25 RD1333–aluminum mixture, using

Wyatt's apparatus with approaching-needle rubber-base electrodes. With a steel-base electrode, however, a value of less then 3×10^{-5} J was obtained for RD1333 and a value between 5×10^{-4} and 1.1×10^{-3} J for the mixture. (The 3×10^{-5} J value was due to contact initiation since the electrode potential difference was only 250 V, which is less than the threshold voltage for gap breakdown in air.)

Brown *et al.* [28] reported that the spark sensitivity of mixtures of RD1333 and RDX decreased monotonically with percent RDX. Mixtures wet with Freon TF, containing more than 60% lead azide, were more sensitive than the same mixture dry; below 60% lead azide, Freon desensitizes the mixture markedly. However, if the positive pointed electrode is raised so that it is in the Freon but not in the explosive, the explosive is markedly desensitized.

L. COMPARISON OF ELECTROSTATIC SENSITIVITY OF PRIMARY EXPLOSIVES

In many practical situations, an exact knowledge of the sensitivity of a particular explosive is not necessary. It is sufficient to know the relative sensitivity of the explosive being handled so that the appropriate precautions can be taken. The relative sensitivity at the 50% initiation point of RD1333 is compared to basic lead styphnate and tetracene in Table IV [9]. The energy required to initiate RD1333 is more than twice that needed for lead styphnate but less than half that required for tetracene under the particular test conditions.

The circumstances (equivalent electrical circuit) prevailing in a hazardous situation are not always known. Therefore, tables similar to Table IV, which rank explosives at only one energy delivery rate, can be misleading. Kirshenbaum [9] showed that the rank of the three explosives varies for different rates of energy delivery. The energy delivery rate was varied by changing the resistance in series with the gap. As can be seen in Figure 15, with no series resistance RD1333 is the most sensitive of the three explosives, followed closely by basic lead styphnate and then tetracene. With a 1-kΩ resistance in series with the spark gap lead styphnate is the most sensitive, followed by lead azide and then

Table IV. 50% Initiation Values of Primary Explosives (Spark Discharge) [11]

Explosive	Energy ($J \times 10^{-4}$) parallel plane, fixed gap	Approaching needle
Basic lead styphnate	2.6	6.7
RD1333 lead azide	5.5	27
Tetracene	>9.6	62

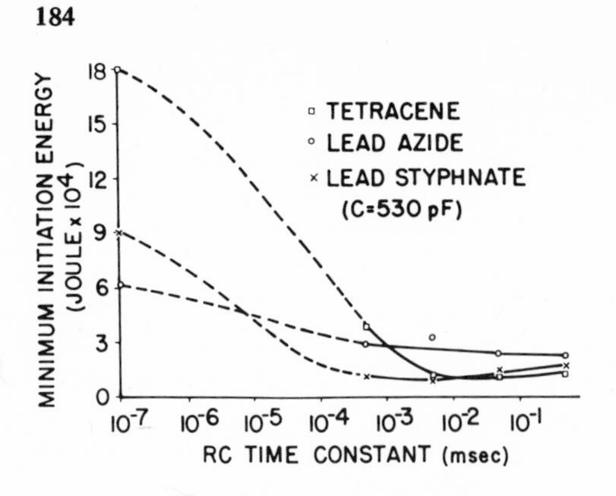

Figure 15. The effect of energy delivery rate on rank of primary explosives [9].

tetracene. However, with a 1-MΩ series resistance tetracene is the most sensitive, followed closely by lead styphnate and then lead azide. Small differences, however, might be reversed under other circumstances. No variation in rank was noted using the approaching-needle apparatus.

M. DIELECTRIC STRENGTH OF LEAD AZIDE

Until recently, little attention has been given to the dielectric breakdown strength of lead azide; values reported in the literature vary from 20 to 300 kV/cm. Past studies were primarily directed to the dependence of the dielectric strength on such factors as sample thickness, density, temperature, etc. Little attention was given to the type or shape of electrode, the electrode material, or the degree of contact between sample and electrodes. Thus the measured dielectric strength was determined by the conditions of the measurement, which were usually different for each investigation.

1. Single Crystal Measurements

To obtain a better understanding of the effect of an electric field on lead azide crystals, a study of the current–voltage characteristics was carried out by Downs and coworkers [27]. The dependence upon crystal thickness was also studied by varying the sample thickness from 0.19 to 0.76 mm. Gold electrodes were applied to both ends of a single crystal by vacuum evaporation. (The electrode diameter was 2.36 mm.) The crystal was subjected to a one-minute-on, one-minute-off application of voltage, increasing the voltage in 100-V increments until the sample detonated. The current was monitored continuously.

Typical current–voltage characteristics reflected the nonohmic nature of the Au/lead azide system. (Gold forms a nonohmic contact with lead). The voltage

Table V. Electric Field Effects on Lead Azide

Crystal thickness (cm)	Current at detonation (A)	Voltage at detonation (V)	Threshold field (kV/cm)
0.019	1.2×10^{-9}	800	42.0
0.022	7.0×10^{-9}	800	36.0
0.024	6.9×10^{-8}	1000	41.7
0.025	1.4×10^{-8}	1000	40.0
0.043	1.6×10^{-8}	1300	30.2
0.073	1.1×10^{-8}	2400	32.6
0.076	5.5×10^{-9}	2700	35.4

at which detonation occurred together with the current just prior to initiation are given in Table V. The average field for detonation (voltage/crystal thickness) was calculated for each crystal, and these are also shown in Table V. The fields for detonation ranged from 30.2 to 42.0 kV/cm. The threshold field values for the four thinnest samples (0.019–0.025 cm) were all higher than those for the thicker samples (0.043–0.076 cm).

2. Pressed Pellet Measurements

Pressed lead azide is a two-phase dielectric system consisting of azide crystals of various dimensions and air in the pores at ambient pressure. Its dielectric strength varies with the density, crystal dimensions, thickness, etc.

Leopold [29] determined the dielectric strength of a number of primary explosives for two different pellet thicknesses, 0.5 and 1.0 mm, pressed directly on the flat surface of a brass electrode. A polycarbonate resin charge holder was used to contain the explosive and to prevent arc-over around the periphery of the pellet. Another brass electrode had a 1.27-mm (0.05-in.) flat which was placed on the upper surface of the explosive pellet. The voltage field was raised in 30–60 sec from zero to the level where the explosive initiated. The dielectric strengths (average of four measurements) are summarized in Table VI along with the values for air as a comparison. The dielectric strength (per unit thickness) was higher for the thinner samples. The dielectric strength of RD1333 was slightly higher than PVA lead azide and dextrinated lead azide. Silver azide had the lowest dielectric strength of those tested.

The effect of density on the dielectric strength of dextrinated lead azide, normal lead styphnate, and silver azide was also investigated (Figure 16). The dielectric strengths of dextrinated lead azide and normal lead styphnate increased with density. The breakdown level of silver azide was relatively unaf-

Table VI. Dielectric Strength of Primary Explosives [29]

Material	0.5 mm pellet Voltage (V)	0.5 mm pellet Field strength (kV/cm)	1.0 mm pellet Voltage (V)	1.0 mm pellet Field strength (kV/cm)
Silver azide	570	11.4	990	9.9
Dextrinated lead azide	1300	26.0	2460	24.6
PVA lead azide	1580	31.6	2560	25.6
RD1333 lead azide	1580	31.6	2690	26.9
Normal lead styphnate	2200	44.0	3490	34.9
Basic lead styphnate	2540	50.8	4080	40.8
Air	2645	52.9	3830	38.3

fected. The effect of DLA particle size was such that milled material had an increase in the dielectric strength (Figure 17). However, the rate of increase with increasing density was considerably less than unmilled explosive.

The electric fields were calculated in terms of an average field strength (kV/cm), although the electric stress could have been more highly concentrated in certain regions due to shape of the electrodes. Leopold noted that arcing sometimes occurred around the circumference of the 1.27-mm electrode, indicating a certain amount of stress concentration.

Leopold also demonstrated that it is necessary to have contact between the explosive and the electrodes for initiation to occur. No initiations were observed with any of the test explosives when the explosive pellets were insulated from

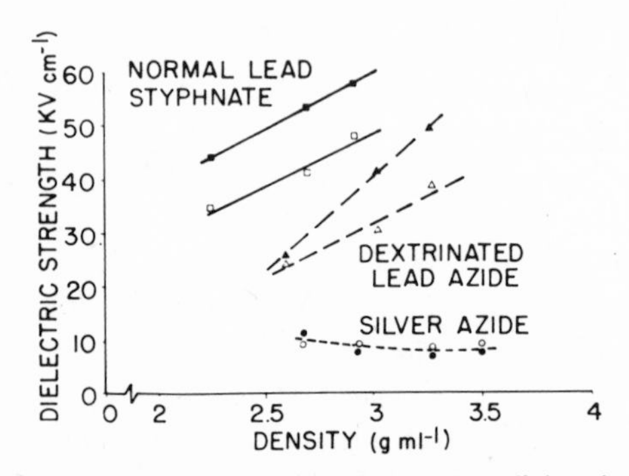

Figure 16. Effect of explosive density and thickness upon dielectric strength [29]: solid circles, 0.5 mm thickness; open circles, 1.0 mm thickness.

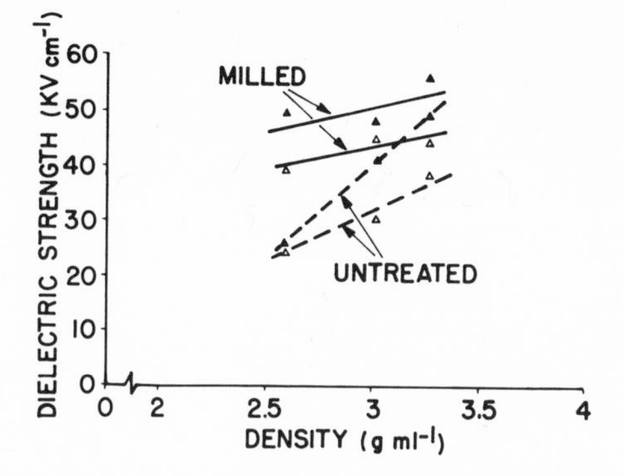

Figure 17. Effect of particle size on dielectric strength of dextrinated lead azide [29]: solid circles 0.5 mm thickness; 1.0 mm thickness.

one or both electrodes by a layer of 0.05-mm thick mylar film and exposed to an electric field strength of 70–80 kV/cm for 1 min.

Downs and coworkers [30] confirmed the observations of Leopold that electrical contacts are required to obtain field-assisted initiation. They used pellets of RD1333 lead azide (pressed to ~3.5 g/ml density, 1-3 $\times 10^{-2}$ cm thick) mounted between mylar insulators. The experiment was conducted in vacuum, 2 $\times 10^{-5}$ torr. The voltage was increased in steps of 500 V to a maximum of 5 kV, and was held constant for 60 sec at each step. No initiation occurred in the samples up to the highest values applied, 140 kV/cm. This is four times greater than the average field strength values that lead to initiation in contact samples [29]. The same result (no initiation) held for single-crystal samples with the highest field strength value attained being 102 kV/cm (although the value depends on the dielectric constant used). The highest value was maintained on the sample for a half an hour.

Pitts [31] reported a dielectric strength of 36 kV/cm for a 0.5-mm-thick lead azide pellet of 3.7 g/ml density. The explosive was confined in a block between two 4.2-mm-diam flat-bottom drill rods. A 500-pF capacitor discharge was used to determine the dielectric strength. No description was noted in the report of the type of azide investigated.

Zimmerschied and Davies determined the change in the dielectric strength of pellets as a function of lead azide type, density, thickness, and interstitial gas pressure [32]. Each explosive pellet was pressed or cemented into an acryllic disk to prevent electrical paths around the surface between the electrodes. Successively higher voltage pulses were applied between two Rogowski-type aluminum electrodes in contact with opposite sides of a lead azide disk until an arc or discharge and subsequent initiation of the lead azide occurred. The decay constant of the electric pulse was varied from 230 to 908 nsec with most experiments being performed with a time constant of 430-470 nsec. The desired pulse shape was formed by discharging a charged capacitor through a resistance. Experiments were also performed under d.c. conditions.

The measured dielectric strength of dextrinated lead azide varied from 30 to 160 kV/cm, depending on lead azide density, thickness, and gas pressure. Increasing the dielectric strength of the interstitial gas (helium or sulfur hexafluoride in place of air) increased the dielectric strength of the explosive pellet at the lower densities. Dextrinated lead azide showed a slightly higher dielectric strength than the other types, especially at lower pressures.

The effect of an electric field on dextrinated lead azide pellets as a function of density and thickness was also studied [11]. The explosive was pressed in a nylon sleeve between two 0.48-mm (0.187-in) -diam, flat-surface steel rods with rounded edges. The explosive was subjected to a one-minute-on, one-minute-off application of voltage, increasing the voltage in 100-V increments until the sample detonated. The current was monitored continuously.

Current began to flow through the azide pellet after the electric field reached a certain field strength. The prebreakdown current–voltage characteristics revealed three distinct regions. In the low-voltage region the current increased with increasing voltage slower than a resistive load (due to polarization effects). In the second region the current increased linearly at a faster rate which appeared to follow Ohm's law. At a certain critical potential the current started to increase more sharply (superlinearly). Most azide pellets detonated at a potential slightly higher than the critical potential. Several pellets detonated before the superlinear region was reached. The resistivity, calculated from the slope of the curve which followed Ohm's law, ranged from $1\text{-}2 \times 10^{12}$ Ω-cm for all the different density and thickness pellets.

The average electric field for detonation was a strong function of the density (Figure 18) and thickness (Figure 19), varying from 37 to 73 kV/cm. Detonation did not always occur immediately upon application of the voltage (electric field). Delay times varied from 0 to 60 sec. Thin, low-density pellets did not always detonate upon breakdown. This phenomenon occurred in several 0.04-cm-thick

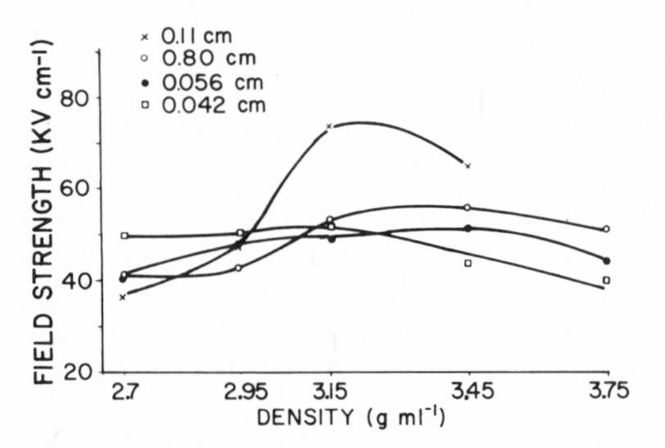

Figure 18. Average electric field strength for detonation of dextrinated lead azide as a function of density [11].

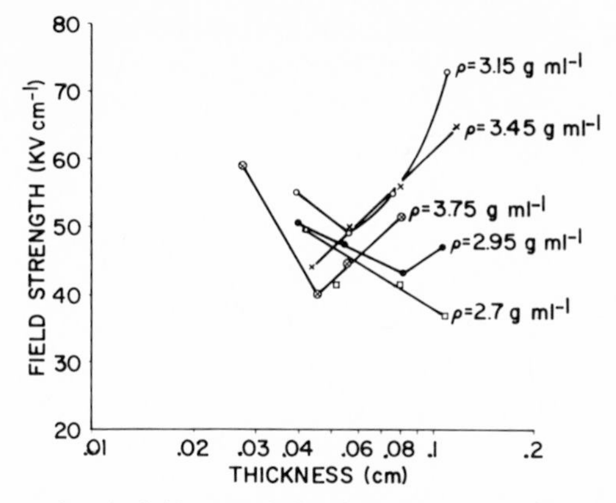

Figure 19. Average electric field strength for detonation of dextrinated lead azide as a function of thickness [11].

pellets, pressed to a density of 2.7 g/ml. It was also observed that as the voltage was being raised from zero to the next 100-V increment, several samples detonated at a lower potential than the previous voltage.

Extensive measurements have been made on lead azide in the U.S.S.R. in recent years, but no description is provided of the lead azide. Sten'gach [33] carried out a study on the sensitivity of lead azide to spark discharge in relation to density, size of crystals, interelectrode distance, temperature, and presence of inert impurities. The lead azide was pressed into a chamber containing a spark gap formed by two pointed electrodes. The mean density of the azide was varied from 1.1 to 2.7 g/ml and the interelectrode gap was varied from 0.020 to 0.50 mm (0.0008–0.020 in.). The 50% firing point was used as a measure of sensitivity.

Sten'gach reported that the sensitivity to spark discharge increased as the density of the azide was increased. The electrostatic sensitivity of samples pressed to a mean density of 2.7 g/ml decreased approximately linearly as the distance of the electrode gap was increased from 0.020 to 0.50 mm (0.0008–0.020 in.). Increasing the crystal size of the lead azide from 0.3–1 to 3–7 μm decreased the electrostatic sensitivity. Raising the temperature increased the sensitivity to spark discharge. Solid inert admixtures of up to 10% lead nitrate, barium titanate, or graphite had no substantial effect on the sensitivity. Additives like paraffin, however, which are capable of forming coatings on the particles, lowered the sensitivity. Increasing the moisture content of the lead azide decreased the sensitivity.

Sten'gach [34] also studied the dependence of the dielectric strength on density, crystal size, ambient air pressure, and electrode surface dimensions. Polished steel electrodes with a radius of curvature of 1.5 mm were used in most of the tests. The distance between electrodes was varied from 0.05 to 0.5 mm.

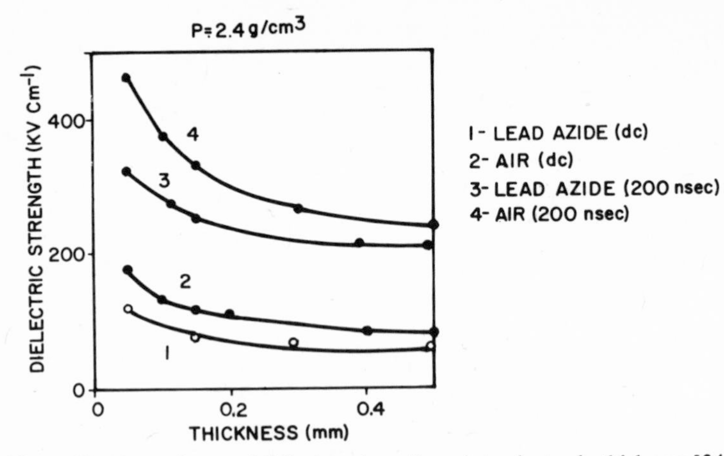

Figure 20. Dependence of dielectric strength on interelectrode thickness [34].

The breakdown voltages were determined for both d.c. voltages (applied across the electrodes at a steady rate of increase of about 100 V/sec) and pulsed voltages (square-wave voltage pulses of 2×10^{-7} sec duration).

The dielectric strength of pressed lead azide decreased with increasing interelectrode distance and was lower than that of air at one atmosphere for both d.c. and pulsed voltages (Figure 20). The dielectric strength of both air and azide were 2.5–3 times greater for pulsed voltages than for d.c. voltages. Moreover, the dielectric strength of 0.15-mm-thick pellets pressed to a density of 2.7 g/ml decreased with increasing electrode radius of curvature (Table VII). Increasing the crystal size of the azide from 0.3–1 to 4–7 mm (lead azide pellets of the same density) decreased the dielectric strength from 160 kV/cm to 110 kV/cm and the pulsed dielectric strength from 260 kV/cm to 170 kV/cm. The dependence of the dielectric strength on the volume fraction of lead azide is shown in Figure

Table VII. Lead Azide[a] Dielectric Strength as a Function of Electrode Curvature [34]

Radius of curvature (mm)	Dielectric strength (kV/cm)
0.3	400
0.5	390
1.2	290
1.5	250

[a]Lead azide: density = 2.7 g/ml, interelectrode separation = 0.15 mm.

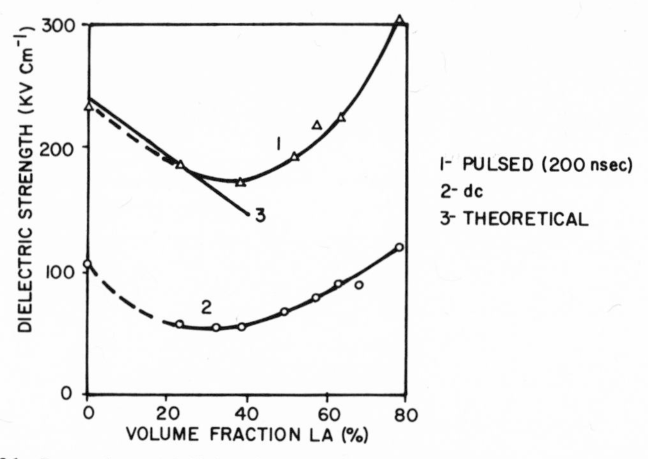

Figure 21. Dependence of dielectric strength on volume fraction of lead azide [34].

21. An increase of packing density results in both the constriction of pores and an increase in the electric field strength. The observed dielectric strength of a pressed pellet is therefore determined by the additive interaction of these two opposing effects. Sten'gach believes that when the volume fraction of lead azide was <33%, the dielectric strength decreased with increasing volume fraction as a result of an increase in the electric field strength in the air-filled pores in the pellet. For volume fractions increasing above 33%, the field strength in the pores continues to increase, but the decrease in air-channel cross-section and the subsequent increase in breakdown potential exerted an overriding influence and produced the increase in the dielectric strength of the azide. The dielectric strength of 2.4 g/ml density lead azide pellets increased linearly with gas pressure in the range 1–30 atmosphere (Figure 22). This also indicates that breakdown occurred in the gas phase, i.e., along the air channels.

In another study, Sukhushin and coworkers [35] studied the impulse breakdown strength of lead azide pellets of different thicknesses and densities. The effect of changing electrode polarity was also investigated. The specimen was a wafer pressed on a polished, hardened steel cylinder into a polyvinyl chloride sheath. The outer spherical electrode was of 10 mm diam and positioned against the outer surface of the specimen. Square-wave pulses, 1.5–40 μsec, were used to determine the breakdown voltages.

A minimum breakdown voltage and a minimum breakdown field strength of about 110 kV/cm occurred at a density of 3 g/ml (Figures 23 and 24). The effect of the electrode polarity was quantitatively similar, but the breakdown voltage for the positive spherical electrode was lower than for the corresponding flat electrode. The dielectric strength was practically independent of specimen thick-

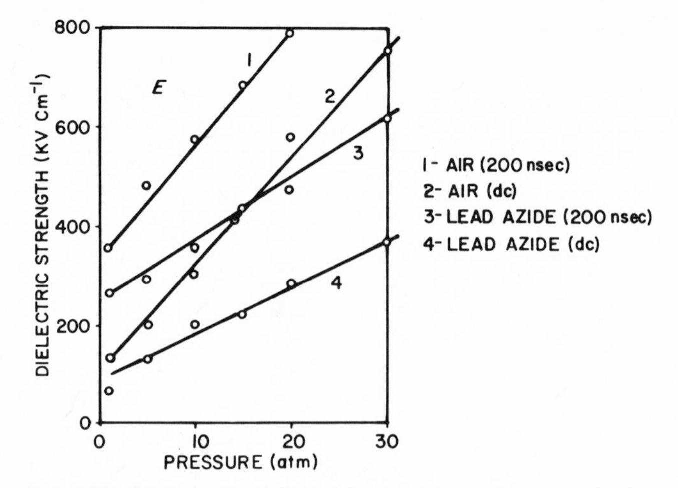

Figure 22. Dependence of dielectric strength on air pressure [34].

ness for densities ⩽3.5 g/ml (Figure 25). At greater densities, the dielectric strength decreased with increasing specimen thickness.

Saturation of pressed lead azide pellets with glycerin, cresol, or carbon tetrachloride produced relatively little change. However, saturation with transformer oil caused the breakdown voltage to increase monotonically with decreasing density. Apparently for specimens saturated with glycerin, cresol, and carbon tetrachloride, the process of discharge in air pores was partially or completely replaced by discharge in liquid. On the other hand, transformer oil has a high

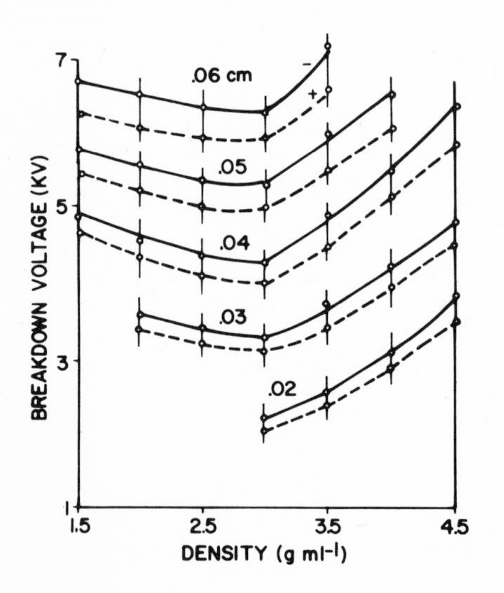

Figure 23. Dependence of breakdown voltage of lead azide on density and thickness for negative and positive flat electrodes [35].

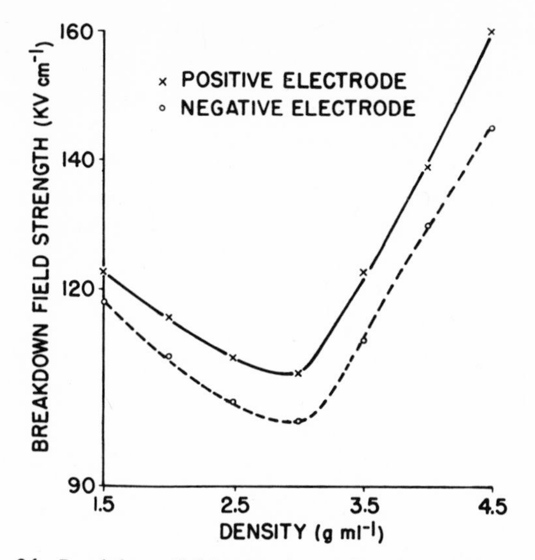

Figure 24. Breakdown field strength as a function of density [35].

dielectric strength, and breakdown occurred in the microcrystals of lead azide. The investigators suggested that for high porosities (density ⩽3 g/ml), breakdown is controlled by microdischarges in the pores. For low porosities (density >3 g/ml), the breakdown mechanism is controlled by the development of electron avalanches.

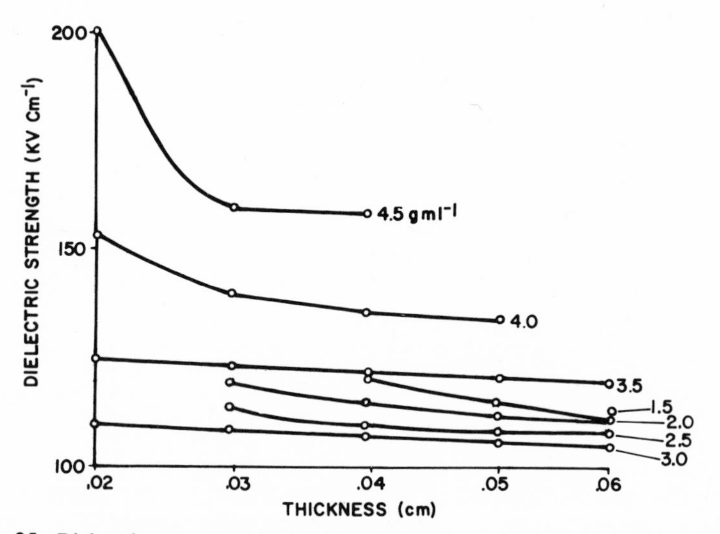

Figure 25. Dielectric strength of lead azide as a function of density and thickness [35].

The prebreakdown currents in pressed wafers of lead azide (0.05-0.35 mm thick) were studied by Gavrilin and Dimova [36]. Single square-wave pulses of up to 2.5 kV and 10^{-5} sec duration were used. The dependence of prebreakdown currents on wafer thickness at a fixed field strength showed that the current increases with increasing thickness. The current-voltage characteristics for a 0.15-mm-thick sample revealed that at a certain critical field (40 kV/cm in the given sample) the current started to increase sharply. The current reached its maximum value at a field strength of 50 kV/cm, and then began to decrease at higher field strengths.

N. SILVER AZIDE

Wyatt and coworkers [2,5] obtained a minimum initiation energy value of 2.8×10^{-6} J for silver azide powder when a plumb-bob approaching-electrode apparatus was used (contact discharge). Similar results were obtained with a blunt needle; only partial ignitions resulted with a regular needle (4.5×10^{-6} J). When a conducting rubber electrode (100-kΩ resistance) was substituted for the base electrode, a minimum energy value of 4×10^{-4} was obtained.

For sieved powder (passing through a 100 mesh, but retained on a 200 mesh screen) a minimum initiation energy of 7.5×10^{-3} J was reported [37]. In the subsieve range the minimum energy was 7×10^{-4} J. The values were obtained with an approaching-electrode apparatus with a preset gap of 0.50 mm (0.030 in.).

In an attempt to determine the mechanism of electrical initiation, Bowden and McLaren [38] conducted experiments with 1.0 × 0.5 × 0.5-mm silver azide crystals having an electrical resistance of about 10^{12} Ω. When subjected to a constant electrical field (450 V/cm), the current through the crystal increased with time, and after several minutes, when the current reached 150 μA, an explosion took place.

In another experiment, they determined the breakdown voltage of a crystal as a function of field strength at temperatures between 50 and −100°C (Table VIII). The breakdown voltage increased with decreasing crystal temperature. When the electrical field was applied for less than 10 sec, however, and removed

Table VIII. Breakdown Voltage of Silver Azide as a Function of Temperature [38]

Temperature (°C)	Breakdown voltage (V)
50.5	19
21.5	67
−45	120–130
−100	>400

Table IX. Time to Explosion in Silver Azide as a Function of Frequency [38][a]

Frequency (cps)	Time to explosion (min)
0	7
20	6
50	5
1,000	45
5,000, no explosion after 90 min	
10,000, no explosion after 90 min	

[a] Applied field: 900 V/cm.

before explosion occurred, the breakdown in resistance was found not to be permanent.

Bowden and McLaren also determined the time to explosion as a function of the frequency of the applied electrical field (900 V/cm). The results, summarized in Table IX, showed that the time to explosion was greatly increased if the frequency of the applied field was increased beyond 100 cycles/sec. They suggest that the breakdown is due to field emission from the cathode. If electrons from the cathode enter the crystal with sufficient energy to remove electrons from the ions of the crystal lattice, current will increase rapidly and decomposition will take place, followed by self-heating and explosion.

Leopold [29] found that pressed pellets of silver azide require higher field strengths for breakdown than the single crystals used by Bowden and McLaren [38] (Section M).

An investigation was carried out in the U.S.S.R. [39] to study the mechanism of initiation of silver and thallium azide by an electric discharge pulse. The azides were pressed at 1600 N/cm^2 into disks, 6 mm diam. by 0.07–0.4 mm thick, and were clamped between electrodes comprising a sphere and a flat surface. A square-wave pulse of 4.5 kV of 5 μsec duration was applied to the specimen. A minimum initiation energy was determined by monitoring the voltage and current through the specimen and the time to onset of detonation (inferred from the flash of light with the aid of a photoelectric multiplier). The energies were confirmed in other experiments where the electrical circuit had provisions for controlling the instant at which the applied voltage could be disconnected from the specimen during the breakdown stage. This allowed the amount of energy delivered to the specimen to be controlled. In the experiments, when a 2.2-kV pulse was applied to a 0.07-mm-thick silver azide disk, detonation occurred at 2×10^{-4} J but not at 1×10^{-4} J. The electric field strength was about 300 kV/cm, which is about 30 times greater than the dielectric breakdown value reported by Leopold [29] for a 0.5-mm pellet; it is also higher than the dielectric strength of lead azide (Section M). The authors also found that the minimum energy decreased linearly with increasing ambient temperature. They concluded that the initiation mechanism is thermal.

REFERENCES

1. P. W. J. Moore, J. F. Sumner, R. M. H. Wyatt, The Electrostatic Spark Sensitiveness of Initiators: Part I–Introduction and Study of Spark Characteristics, Explosives Research and Development Establishment Report 4/T/56, Waltham Abbey, Essex, 1956.
2. P. W. J. Moore, J. F. Sumner, R. M. H. Wyatt, The Electrostatic Spark Sensitiveness of Initiators: Part II–Ignition by Contact and Gaseous Electrical Discharges, Explosives Research and Development Establishment Report 5/R/56, Waltham Abbey, Essex, 1956.
3. P. W. J. Moore, The Electrostatic Spark Sensitiveness of Initiators: Part III–Modification of the Test to Measure the Electrostatic Hazard Under Normal Handling Conditions, Explosives Research and Development Establishment Report 22/R/56, Waltham Abbey, Essex, 1956.
4. D. B. Sciafe, The Electrostatic Spark Sensitiveness of Initiators: Part IV–Initiation of Explosion by Spark Radiation, Explosives Research and Development Establishment Report 9/R/59, Waltham Abbey, Essex, 1959.
5. R. M. H. Wyatt, The Electrostatic Spark Sensitiveness of Initiators: Part V–Further Study of Ignition with Metallic and Antistatic Rubber Electrodes, Explosives Research and Development Establishment Report 24/R/59, Waltham Abbey, Essex, 1959.
6. R. M. H. Wyatt, P. W. J. Moore, R. J. Adams, J. F. Sumner, *Proc. R. Soc. London*, *A246*, 189 (1958).
7. R. M. H. Wyatt, The Electrostatic Spark Sensitivity of Bulk Explosives and Metal/Oxidant Mixtures, NAVORD Report 6632, Naval Surface Weapons Center, Silver Spring, Maryland, 1959.
8. M. S. Kirshenbaum, Response of Lead Azide to Spark Discharges Via a New Parallel-Plate Electrostatic Sensitivity Apparatus, Techn. Rept. 4559, Picatinny Arsenal, Dover, N.J., 1973.
9. M. S. Kirshenbaum, Electrostatic Sensitivity of Explosives As a Function of Circuit Parameters, Picatinny Arsenal, Proc. of the DAE-AF-F/G-7304 Technical Meeting, April 29,–May 1, 1974, Naval Surface Weapons Center, Silver Spring, Maryland.
10. L. J. Montesi, The Development of a Fixed Gap Electrostatic Spark Discharge Apparatus for Characterizing Explosives, Proc. Sixth Symp. on Electroexplosive Devices, The Franklin Institute, Philadelphia, Pa., 1969.
11. M. S. Kirshenbaum, Response of Primary Explosives to Gaseous Discharges in an Improved Approaching-Electrode Electrostatic Sensitivity Apparatus, Techn. Rept. 4955, Picatinny Arsenal, Dover, N.J., 1976. (Other reports in preparation.)
12. L. B. Loeb, *Fundamental Processes of Electrical Discharges in Gases*, Wiley, New York, 1939.
13. J. M. Meek, J. D. Craggs, *Electrical Breakdown of Gases*, Oxford University Press, London, 1953.
14. J. M. Somerville, *The Electric Arc*, Wiley, New York, 1960.
15. J. R. Haynes, *Phys. Rev.*, *73*, 891 (1948).
16. T. J. Tucker, Spark Initiation Requirements of a Secondary Explosive, Sandia Laboratories Report No. SC-R-68-1759, Albuquerque, New Mexico, 1968.

17. I. G. Gwillim, J. Nicholson, The Electrostatic Ignitability of Dust Clouds of Powdered Explosives, Atomic Weapons Research Establishment, Explosives Research Note No. 25/55, Waltham Abbey, Essex, 1955.
18. W. E. Perkins, A Survey of the Methods of Testing the Electrostatic Sensitivity of Solids, Frankford Arsenal Memo. M69-29-1, Philadelphia, Pa., 1969.
19. J. T. Petrick, Discharge of an Electrostatically Charged Human, The Naval Weapons Laboratory, Dahlgren, Va., Proc. of the Sixth Sym. on Electroexplosive Devices, The Franklin Institute, Philadelphia, Pa., 1969.
20. C. R. Westgate, M. S. Kirshenbaum, B. D. Pollock, Electrical and Photographic Characterization of Low Intensity Capacitor Spark Discharges, Techn. Rept. 4737, Picatinny Arsenal, Dover, N.J., 1974.
21. L. J. Montesi, The Electrostatic Spark Sensitivity of Various Organic Explosive and Metal Oxidant Mixtures, NOLTR 65-124, Naval Surface Weapons Center, White Oak, Silver Spring, Maryland, 1966.
22. L. J. Montesi, H. J. Simmons, Sr., Interim Qualification Test Results of the New Explosive Compositions Used in the APAM Rotor Assembly, NOLTR 73-47, Naval Surface Weapons Center, White Oak, Silver Spring, Maryland, 1973.
23. Joint Service Safety and Performance Manual for Qualification of Explosives for Military Use, Prepared by Joint Technical Coordinating Group for Air Launched Non-Nuclear Ordnance, Working Party for Explosives, Picatinny Arsenal, Dover, N.J., 1971.
24. D. Hopper, F. Arentowicz, Unpublished results, Picatinny Arsenal, Dover, N.J., 1968.
25. H. A. Hanna, J. R. Polson, Investigation of Static Electrical Phenomena in Lead Azide Handling, Mason and Hanger-Silas Mason Co., Inc., Tech. Rept 98-A., Burlington, Iowa, 1967.
26. R. M. H. Wyatt, Explosive Research and Development Establishment, Waltham Abbey, Essex; private communication to Professor L. A. Rosenthal, College of Engineering, Rutgers University, New Brunswick, N.J., 1974.
27. D. S. Downs, W. Garrett, D. A. Wiegand, T. Gora, M. Blais, A. C. Forsyth, H. D. Fair, Jr., Photo and Electric Field Effects in Energetic Materials, Tech. Rept. 4711, Picatinny Arsenal, Dover, N.J., 1974.
28. J. A. Brown, A. R. Garabrant, J. F. Coburn, J. C. Munday, A. Beerbower, Sensitivity and Rheology Studies on Lead Azide RDX-Silica Mixtures, Esso Research and Engineering Company, Government Research Laboratory Report No. GR-1SRS-69, Linden, N.J., 1969.
29. H. S. Leopold, Exposure of Primary Explosives to Applied Electric Fields, NOLTR 73-125, Naval Surface Weapons Center, White Oak, Silver Spring, Maryland, 1973.
30. D. S. Downs, T. Gora, P. Mark, Unpublished results, Picatinny Arsenal, Dover, N.J., 1976.
31. L. D. Pitts, Designing Electro-Explosive Devices for Electrostatic Insensitivity, Proc. Fifth Sym. on Electroexplosive Devices, The Franklin Institute, Philadelphia, Pa., 1967.
32. A. Zimmerschied, F. Davies, Investigation of Lead Azide Dielectric Strength, The Boeing Company Report No. T2-4096-1, Seattle, Washington, 1974.
33. V. V. Sten'gach, *Fiz. Goreniya Vzryva Nauka Sib. Otdel. Novosib.*, *6/1*, 113 (1972).
34. V. V. Sten'gach, *Zh. Prikl. Mekh. Tekh. Fiz.*, *1972*, 128 (1972).
35. Yu. N. Sukhushin, Yu. A. Zakharov, G. A. Rappaport, *Izv. Tomsk Politekh. Inst.*, *251*, 219 (1970).
36. A. I. Gavrilin, N. I. Dimova, *Izv. Tomsk Politekh. Inst.*, *162*, 194 (1967).
37. T. B. Johnson, Remington Arms Company Inc., Static Sensitivity of Lead Azide, Lake City Army Ammunition Plant, Independence, Missouri, 1966.
38. F. P. Bowden, A. C. McLaren, *Proc. R. Soc. London*, *A246*, 197 (1958).
39. M. A. Mel'nikov, A. I. Gavrilin, N. I. Dimova, A. L. Kalashnikov, *Russ. J. Phys. Chem.*, *44*, 1314 (1970).

40. W. L. Shimmin, J. H. Huntington, L. Avrami, Radiation and Shock Initiation of Lead Azide at Elevated Temperatures, Air Force Weapons Laboratory Report TR 71-163 and Physics International Final Report PIFR-308, Kirtland Air Force Base, New Mexico, 1972.
41. F. W. Brown, D. J. Kusler, F. C. Gibson, Sensitivity of Explosives to Initiation by Electric Discharges, U.S. Bureau of Mines Report of Investigations 5002, Bruceton, Pa., 1953.
42. L. D. Pitts, Demythologizing Electrostatics, Franklin Institute, Proc. of the Sixth Symp. of Electro-Explosive Devices, Philadelphia, Pa., 1969.
43. Ordnance Explosive Train Designer's Handbook, NOLR 1111, Naval Surface Weapons Center, Silver Spring, Maryland, 1952.

6

Sensitivity to Heat and Nuclear Radiation

L. Avrami and J. Haberman

A. INTRODUCTION

This chapter is concerned with data on the sensitivity of explosive azides to heat, nuclear, and other intense radiation. These stimuli, when directed at solid explosives, may have very similar consequences, even though they may have very different origins. The parallelism is attributable to the conversion and degradation processes which occur when powerful stimuli impinge on condensed matter.

The role of heat in the initiation of fast reactions in the azides is fundamental to many theories of initiation. Most interpretations of the sensitivity of solid explosives to mechanical, electrical, and other stimuli treat the possibility that part of the delivered energy is converted to heat prior to initiation (see for example, Chapter 8, Volume 1). The heat evolved during the early states of decomposition sustains further reaction. Detailed studies of the slow, non-self-sustaining thermal decomposition of the azides are described in Chapter 6, Volume 1, where it is shown that the rates are structure-sensitive and strongly dependent on the temperature. At sufficiently high temperatures, usually at temperatures in the region of the melting point, thermal decomposition becomes self-sustaining, deflagrations or detonation regimes ensue, and samples are consumed in periods of a millisecond to a microsecond.

Because of the extensive discussion in Volume 1, the sensitivity of azides to heat is only briefly touched upon in this chapter. The effects of high-energy (UV) radiation of low intensity are also discussed at length in Chapter 7 of Volume 1.

While such radiation may be a component of the radiation resulting from electron excitations, radioactive decay, or nuclear fission processes, the consequences for azides are usually mild and non-self-sustaining. When intense beams and the full spectrum of species associated with nuclear and electromagnetic radiation are considered, more spectacular effects can be observed, and it is these which take up most of the discussion here.

In 1922 Henderson [1] postulated that α-particles might be used to detonate explosives, and there has been a continuing interest in whether self-sustaining reactions can be initiated by nuclear or optical radiation of selected energies. The molecular or electronic aspects of the subject are described in Volume 1; this chapter, therefore, is devoted primarly to macroscopic effects.

B. SENSITIVITY TO HEAT

The factors which affect the initiation of fast decomposition of azides have not been investigated to the extent reported for slow decomposition. However, such information as is available suggests that fast decomposition is affected by some, if not all, of the factors that affect the kinetics of slow decomposition. Among the kinetic parameters is a short, temperature-dependent induction period or ignition delay, which is commonly related to the induction period of slow decomposition and is the parameter most often used to define a thermal sensitivity of explosives. Normally, thermal sensitivity is expressed in terms of the temperature at which detonation ensues after a specified delay. The delay may occur while the sample is being heated to give a programmed, linear temperature rise or may follow the insertion of a sample into a constant-temperature bath. The delay selected depends on the precision with which time can be measured with available instrumentation. Nevertheless, of all the measures of sensitivity, thermal sensitivity is the property which can be measured with the highest precision and reproducibility.

A variation of these methods, in which the volume of gas evolved in a vacuum at a fixed temperature is measured, is also in common use. This approach is particularly useful when treating long-term, relatively mild thermal effects and the compatibility of explosives with other materials. Still other techniques for determining thermal stability are differential thermal and thermogravimetric analysis. These are coming into more common use, but are not yet standardized with respect to the explosive azides.

1. Explosion Temperature Tests

As indicated, the thermal sensitivity of solid explosives is studied by two basic methods. The first heats a sample to give a constant rate of temperature rise, the explosion temperature being taken as the temperature at the time of explosion. The explosion temperature depends on the amount of sample and the rate of heating. A typical apparatus is the Mettler thermoanalyzer, which allows the use of small quantities of azide, and the differentiation between decomposition and detonation is feasible [2,3]. In this approach differential thermal and thermogravimetric analyses are performed simultaneously.

In the second method, the explosive, in a container such as an aluminum blasting cap, is plunged into a hot bath maintained at a constant temperature; the induction time or ignition delay, τ, is measured and is related to the temperature (T) by an Arrhenius-type equation [4]:

$$\log \tau = a + b/T$$

For reproducible results, the temperature of the bath must be accurately controlled, and the time to explosion measured with an electronic timer set to start at the instant the container is lowered into the bath and to stop at the impulse generated by a microphone picking up the sound of the explosion. A modification of the apparatus [5] inserts the explosive, in a cap, into a Wood's metal bath at constant temperature, and the time to explosion is recorded. In another method a hot plate and electronic timing are used. A sample of 3–4 mg is dropped on the hot plate, and a light beam directed across the hot plate to a photocell is interrupted starting the timing. When explosion occurs the light emitted by the explosion stops the timer [6]. Normally, many trials are conducted to obtain a statistically significant determination of the ignition delay, and the above equation for τ is usually developed by the method of least squares from the data generated [4].

As contained in a blasting cap, the explosive is open to the environment, and thus the sensitivity pertains to ambient conditions. To study the effect of other chemical atmospheres on sensitivity, a Pyrex flame-seal may be cemented to the blasting cap [7]. This allows the removal of air and moisture by evacuation and their replacement by a selected atmosphere.

2. Thermal Sensitivities in Air

Routinely, the explosion temperature test is carried out with the azides exposed to air at normal pressures. No significant differences were found in testing in air, argon, or vacuum; the only significant factor appeared to be the age of the lead azide, as there was an increasing sensitivity with aging until a maximum was reached [7]. Figure 1 shows time-to-explosion curves as a function of the storage time.

In routine testing the temperature corresponding to a given ignition delay is generally reported. Table I shows such temperatures for various azides and ignition delays. Table II lists the explosion temperatures of a number of azides for a 5-sec ignition delay, calculated from the data of Wöhler and Martin [15]. Also listed are the minimum temperatures that result in explosion, determined

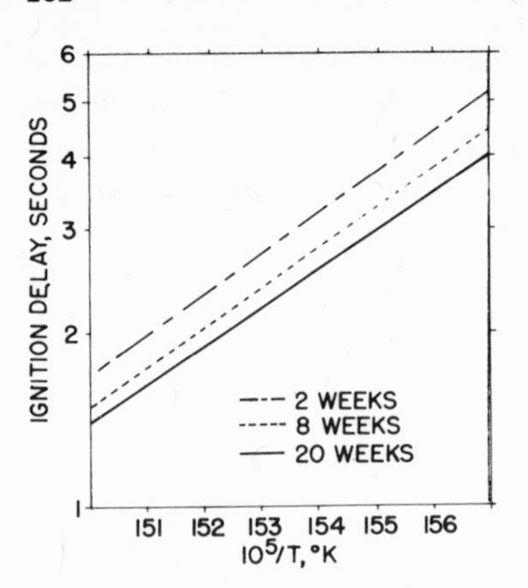

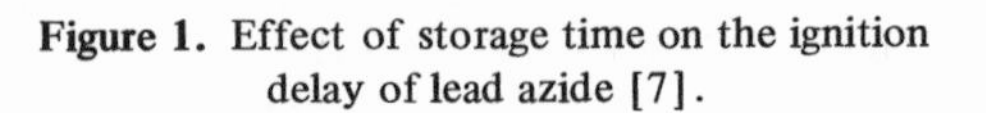
Figure 1. Effect of storage time on the ignition delay of lead azide [7].

by finding the minimum temperatures at which 0.005, 0.01, and 0.02 g of the substances would explode and plotting the minimum temperature of explosion against the mass of explosive.

Table I. Explosion Temperatures of Metal Azides

Azide	Explosion temp. (°C)	Ignition delay (sec)	Sample wt (mg)	Reference
Lead azide				
Dextrinated	340	5	10	[8]
	326	5	20	[13]
Service	350	5	10	[8]
	358	5	20	[13]
Colloidal	344	5	10	[8]
PVA	340	5	10	[8]
RD1333	345	5	10	[8]
Unspecified	315–330	300	50	[9]
	340	5	10	[11]
	340	1	3	[13]
	315	>300	3	[13]
Silver azide	310–320	300	50	[9]
	>250	?	?	[10]
	290	5	10	[11]
	340	1	~2	[12]
Thallium azide	>334	?	?	[10]
	500	>1	~2	[12]
Calcium azide	158	?	~2	[12]
Barium azide	152	?	~2	[12]
	210	25	~2	[12]
Cuprous azide	205	~2	?	[14]

Table II. Minimum Explosion Temperatures of Metal Azides

Azide	Explosion temp. (°C) at 5 sec[a]	Sample weight (mg)	Minimum temp. (°C)[b]
Cobalt	210	5	148
Barium	213	20	152
Calcium	261	20	158
Strontium	245	20	169
Cuprous	197	10	174
Nickelous	279	10	200
Manganous	245	15	203
Lithium	317	20	245
Mercurous	320	20	281
Zinc	316	20	289
Calcium	336	20	291
Silver	337	20	297
Lead	342	20	327

[a] Data from reference [15].
[b] Calculated from data of reference [15].

3. Effect of Impurities

From a practical viewpoint, it is probable that an azide in storage will absorb ions and organic species from the environment or by handling; it is desirable to understand the influence of these impurities on the sensitivity of the stored material. Several investigations of the effects have been made with lead azide.

To test the theory that metal deposited on the surface of lead azide would act as an electron trap, Reitzner *et al.* [6] deposited silver on the surface of lead azide. A sensitization was found for both slow and explosive decomposition. Lead nuclei are produced on the surface of lead azide by the action of ultraviolet light with the concomitant production of nitrogen. It has been postulated that the lead nuclei behave at elevated temperatures as electron sinks during the induction period and account for the observed shortened ignition delays [4,7].

Singh [16] showed that bismuth incorporated in the lead azide lattice sensitized the explosive to thermal initiation; Figure 2 shows the explosion–temperature curves obtained.

The thermal sensitivity of lead azide is affected by adsorbate–adsorbent interactions. Adsorbates with terminal polar groups such as OH, NH_2, and COOH are strongly attracted by the electrostatic field of the surface. This appears to be a necessary but not sufficient condition for reaction with the surface. The time of adsorption should be long enough, and the activation energy for reaction low enough, to ensure participation in surface processes during the induction period leading to explosive decomposition [7].

Lead azide exposed to strongly adsorbed normal propylamine and di-

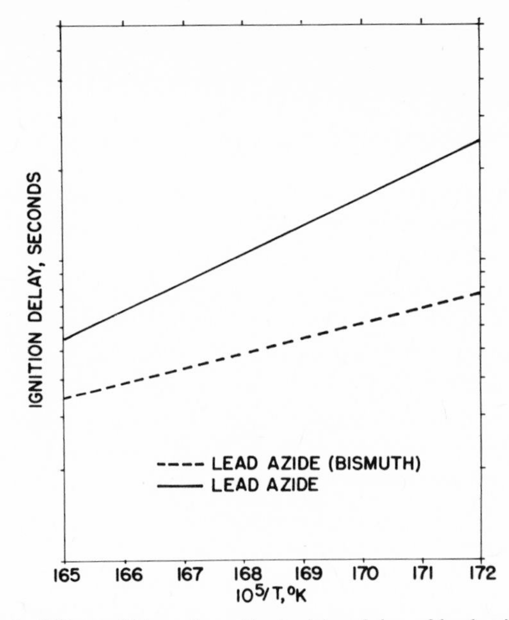

Figure 2. Effect of bismuth on the ignition delay of lead azide [17].

propylamine was significantly sensitized (Figure 3). These amines are Lewis bases, and the adsorbed states may form a charge-transfer complex with the lead azide surface.

The postulated, ground-state, molecular-charge-transfer complex,

$$PB(N_3)_2 + 2RNH_2 \longrightarrow Pb(N_3^- H_2 N^+ R)_2$$

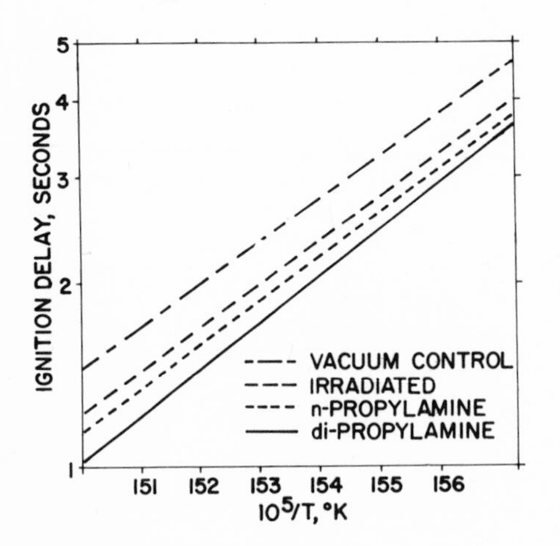

Figure 3. Effect of electron-donor vapors on the ignition delay of lead azide [7].

is based on the property [17] of amines to complex with metal salts in the solid state. During the induction period, the complex presumably disproportionates, releasing nascent lead:

$$Pb(N_3^- H_2 N^+ R)_2 \longrightarrow Pb + 2(RN^+ H_2 N_3^-)$$

The process of lead nucleation is thus favored by the conditions, which in turn increase the rate of free-radical formation and the onset of autocatalysis, resulting in the observation of shorter times to explosion.

If there are adsorbates which can effectively shorten the induction time by promoting the formation of lead nuclei, then it should be possible to increase the induction time by removing lead nuclei. When lead azide was exposed to formic and acetic acids during the induction period, a significant prolongation of the time to explosion was observed (Figure 4). The prolongation only occurred when the acids were present during the induction period. When propionic acid was present, no prolongation of the time to explosion was observed. During the induction period protons from the acids evidently scavenge the lead nuclei already present at the surface and, being produced during the induction period, effectively delaying the explosion. The effectiveness of the acids was in the same order as their dissociation constants as measured in aqueous solution.

The time to explosion was significantly prolonged when lead azide was preheated in the presence of water and Freon before testing [18]. In this case the retardation may be due to the scavenging of lead nuclei over a period of time at an elevated temperature by the water and Freon. The presence of water for shorter periods of time at ambient temperature did not influence the induction time. It is of interest to note that the induction period of lead azide was lengthened during thermal decomposition in the presence of adsorbed water [19], but

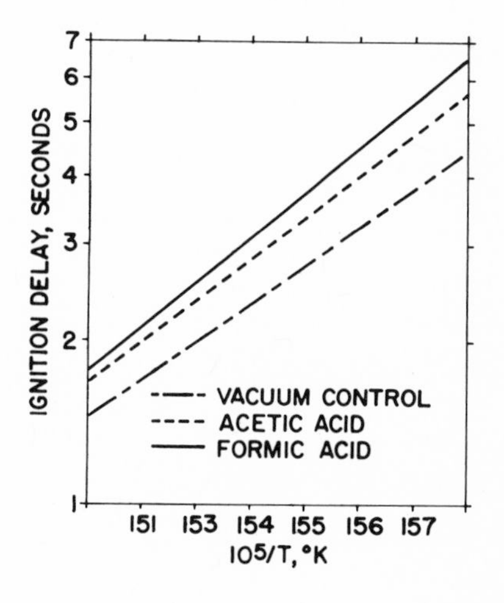

Figure 4. Effect of proton-donor vapors on the ignition delay of lead azide [7].

Table III. Sensitization of Service Lead Azide by Preheating [13]

Bath temp. (°C)	Preheat time (sec)[a]	Ignition delay (sec)
320	0	47
	15	34.8
	25	24.8
	30	24.8
	35	22.8
	43	22.2
	45	0[b]
330	0	20.3
	5	18.3
	8	16.4
	10	14.6
	15	12.3
	18	11.7
	19	10.3
	20	0[b]

[a] At bath temperature.
[b] Exploded on withdrawal.

no effect was noted when adsorbed water was present during fast explosive decomposition.

It has been shown that lead and other azides can be sensitized by heating at a certain temperature and that subsequent heating at that temperature shortens the induction period by approximately the length of the previous exposure [4,13,20] (Table III). It has been postulated [20] that the metal decomposition product catalyzes the thermal decomposition, and that during the induction period, the metal is formed and is retained as metal in the crystal as an impurity. In air there is the possibility, however, that with time the metal will react with ambient species, and this will inevitably complicate the picture.

4. Effect of Sample Mass

The mass of azide used determines the minimum temperature of explosion, as first reported in 1917 [15]. Below a certain mass or crystal size (~5 μm) explosion is not observed, as has been pointed out many times (see for example references [4,21]). The most recent work of Chaudhri and Field [21] is illustrated by Figure 5, which gives the explosion temperature of lead azide vs. crystal thickness (mass). The usual explanation for the phenomenon is that for small samples self-heating does not keep pace with heat dissipation.

The time to thermal explosion may be viewed as a balance between the heating by decomposition, which increases exponentially with temperature, and

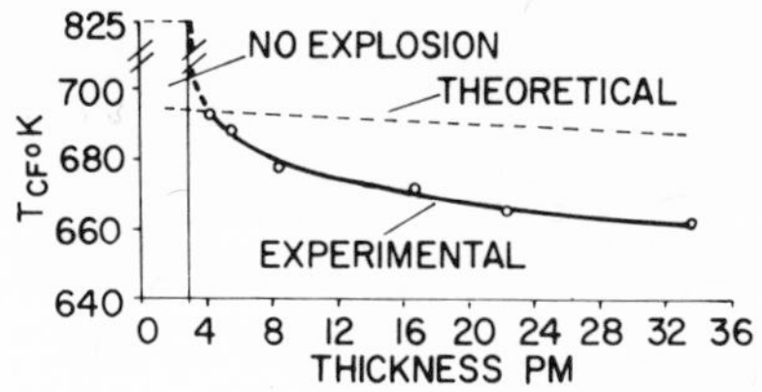

Figure 5. Effect of crystal thickness on the critical explosion temperature of lead azide [21].

the rate of heat removal, which is linear according to Newton's law of cooling. At the lower temperatures, a steady state is attained; at higher temperatures, the rate of heat production increases preferentially. A point is reached at which enough heat is produced to effect an explosion [22].

5. Vacuum Stability Tests

The rationale for the vacuum stability test [23] is that the azide either alone or in contact with some material, at an elevated temperature, may in time, cause a reaction leading to the deterioration of the explosive and a consequent evolution of gas. Generally, 1 g of the explosive and of the material are intimately mixed and placed into a suitable vessel connected to a mercury manometer. After evacuation the mixture is heated for a fixed 40 hr or as specified. The quantity of evolved gas is noted. Reactivity is measured by the difference between the volume of gas evolved by the mixture and by a control (the pure explosive) in 40 hr. The reactivity is defined qualitatively as follows:

Excess gas (ml)	*Degree of reactivity**	*Compatibility*
0.0–3.0	Negligible	Good
3.0–5.0	Moderate	Fair
>5.0	Excessive	Poor

Generally, the temperature of the test is 100°C; however, higher temperatures may be used. The test is used to establish compatibility of the explosive with various components of explosive devices. Table IV shows vacuum stability data for various lead and silver azides. Table V shows some typical compatibility data.

Lead azide reacts with copper to form a very sensitive cuprous azide. Polished copper strips in contact with lead azide for 9–20 days at 90% relative humidity and 50°C react. In contrast, magnesium-aluminum alloys and magnesium strips show a negligible reaction under the same conditions even after 18 months exposure [24].

*Negative values can also be obtained.

Table IV. Effect of Heat on Azides *in Vacuo* [27]

Azide	Temp. (°C)	Weight (g)	Time (hr)	Gas evolved (ml, STP)
Lead				
PVA	100	1.0	40	0.37
	150	0.2	40	0.32
DLA	100	1.0	40	0.34
	150	0.2	40	0.46
RD1333	120	0.2	40	0.40
Silver	150	1.0	40	0.49
	150	1.0	40	0.34
	150	1.0	40	0.40

Table V. Typical Compatibility of Lead Azide with Various Materials[a]

Material	Compatibility
Lead azide (RD1333)	
Tungsten	Good
Molybdenum	Good
Tungsten disulfide	Fair
Boron nitrate	Good
Talc	Good
Molybdenum disulfide	Good
Aluminum silicate hydrate	Good
Lead azide dextrinated	
Zinc	Good
Sodium bicarbonate	Good
Orange shellac	Good

[a]Other representative data will be found in reference [24].

6. Differential Thermal Analysis (DTA)

The technique of DTA measures the thermal sensitivity of metal azides in terms of the difference in temperature between an inert reference sample and the explosive while both are heated at an identical, controlled rate (~10°C/min). The record obtained is the DTA curve (Figure 6), an exotherm indicating self-heating of the explosive.

The temperature at the start of an exotherm is a measure of the thermal sensitivity of the sample, and the lower the temperature at which this occurs the more sensitive is the material to thermal energy. The difference (ΔT) be-

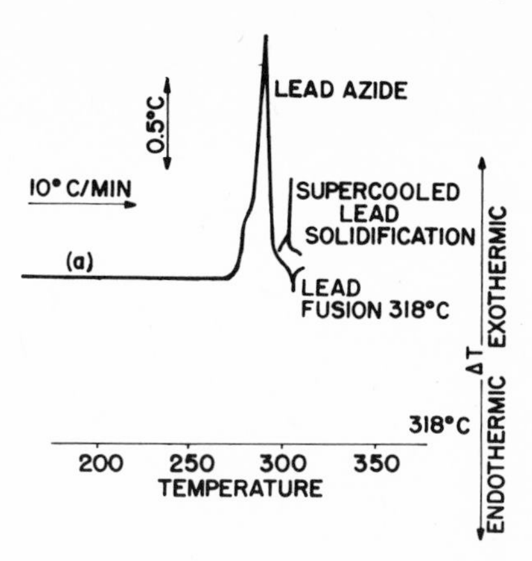

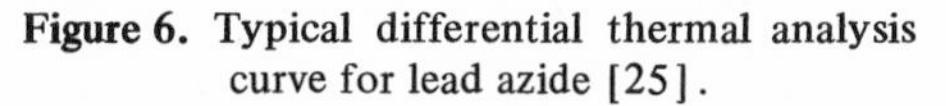
Figure 6. Typical differential thermal analysis curve for lead azide [25].

tween the start of the exotherm and its peak is a measure of the rate of self-heating, with a smaller ΔT indicating a faster reactivity, and hence greater sensitivity.

Table VI gives data obtained for RD1333 lead azide. A statistical analysis using the 11 data points in the table gives a mean of ΔT of 15°C with a standard deviation of 2°. The 95% confidence interval is ±1.3°. The importance of making more than one measurement is made clear because the 95% confidence interval for a single measurement using the same mean and standard deviation is ±4°. If we compare data for lead azides (Table VII) for heating rates of 5° and 10°/min, dextrinated material is significantly more sensitive than the other products tested. The explosion temperature is also significantly lower. A higher heating

Table VI. Temperatures of DTA Exotherms for RD1333 Lead Azide [25][a]

$Temp._{(onset)}$ (°C)	$Temp._{(peak)}$ (°C)	Lead MP (°C)	Sample weight (mg)
287	305	317	2.4
289	305	316	2.6
288	305	320	1.1
293	305	320	1.0
294	306	321	0.8
290	304	320	0.5
288	304	320	0.4
290	304	316	0.4
287	302	316	0.5
290	303	320	0.3
287	302	320	0.3

[a]Heating rate: 10°C/min in He.

Table VII. Differential Thermal Analyses and Explosion Temperatures of Lead and Silver Azides [26]

Azide	Exotherm, onset (°C)	Exotherm peak (°C)	Δ°C	Explosion temperature, (°C) (5 sec delay)
Heating rate, 5°/min				
Lead				
PVA	285	303[a]	18	349
DLA	290	296	6	297
RD1333	288	306	18	335
	297	308	11	335
Silver	295	~300	5	
Heating rate, 10°/min				
Lead				
PVA	296	312	16	
DLA	298	306	8	
RD1333	296	312	16	
	301	317	16	
High purity	287	302	15	
Heating rate, 20°/min				
Lead				
PVA	298	316	18	
DLA	302	314	12	
RD1333	308	323	15	
	320	330	10	

[a]Samples detonated.

rate tends to blur such differences. Again it must be emphasized, as with other thermal sensitivity tests, procedures must be rigidly standardized if apparent sensitivity differences are to be attributed to real differences.

It has been noted that silver azide explodes at a lower temperature when tested in copper blasting caps than when tested in aluminum blasting caps. It has been observed that the thermocouple catalyzes a strong exothermic reaction just above the azide melting point, whereas when the thermocouple is encapsulated the exothermic reaction is milder [27].

7. Thermogravimetric Analysis (TGA)

In TGA the loss or gain in weight of a sample is recorded as a function of increasing temperature. Data are shown as weight change vs. temperature. Weight loss by gaseous products gives a quantitative measure of the amount and rate of decomposition at each temperature. To differentiate between the sensitivities of different kinds of lead azide, the weight of sample should be the same and the heating rate may be used to differentiate among the sensitivities.

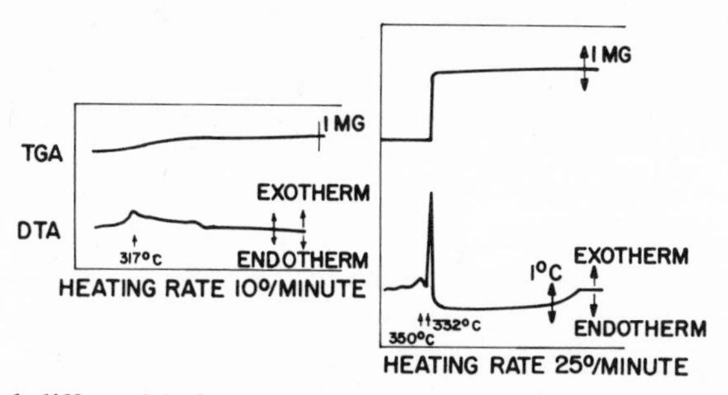

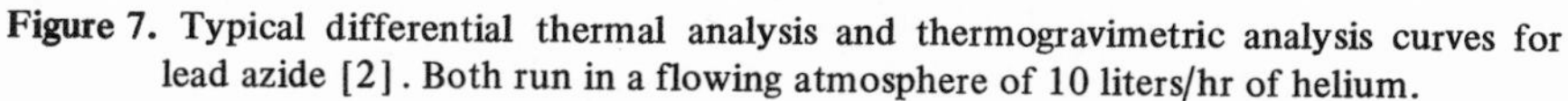
Figure 7. Typical differential thermal analysis and thermogravimetric analysis curves for lead azide [2]. Both run in a flowing atmosphere of 10 liters/hr of helium.

The lower the heating rate at which a given mass of PbN_6 explodes, the higher its sensitivity. This approach requires further work before a routine test can be established. The advantage of TGA (and DTA) is that small amounts of sample may be used; however, TGA does not appear to be as useful as DTA since at low heating rates a small change in weight is spread out over a large temperature interval and at high heating rates the sample detonates with damage to the apparatus (Figure 7).

C. SENSITIVITY TO NUCLEAR RADIATION

Early investigators found that nitrogen iodide could be initiated with α-particles [1,28,29] and that TNT, mercury fulminate, silver acetylide, nitrocellulose, picrates, and azides could be ignited with high-speed electrons or intense ion beams of hydrogen, argon, and mercury [30,31]. In contrast to the conclusions of Bowden and Yoffe [32], some investigators questioned whether the results implied that explosions were caused by the activation of a few neighboring molecules, rather than by hot spots of 10^{-3}-10^{-5} cm diam.

In succeeding years, various types of radiations were used, such as α-particles, neutrons, fission products, reactor radiation (fast and slow neutrons plus gammas), γ-rays, X-rays, and electrons. Initiations of even the more sensitive azides were rarely observed, but the radiations induced, for example, weight loss, gas evolution, increased mechanical and thermal sensitivity, and decreased explosive performance.

1. α-Particle Irradiation

Henderson [1] reported that nitrogen iodide detonated after 20 sec when placed 1 cm from a radium source. Increasing the strength of the source decreased the time to detonation. Poole [28,29] concluded that a single α-particle in a flux of 10^7–10^8 was responsible for the explosion, but that for more stable explosives more α-particles would be required. Muraour [33], in fact, was unable to detonate nitrogen iodide, picric acid, lead picrate, TNT, nitroglycerine, RDX, silver acetylide, and lead azide with α-rays from polonium.

Garner and Moon [34] reported effects of radiation other than color changes; the effects produced by the emission from radium on barium azide were dependent on temperature and led to the acceleration in the thermal decomposition. On the other hand the thermal decomposition of mercury fulminate was not affected in the same environment.

Haissinsky and Walen [35] subjected nitrogen iodide to 5-MeV α-particles from a polonium source. The dryness of the sample and the intensity of the source decreased the time to explosion. The investigators stated that "the detonation of nitrogen iodide could be explained by a local heating of a grain of the powder." However, they thought this was an exceptional result because lead azide, silver azide, and diazo-*m*-nitroaniline perchlorate did not detonate from 1 mCi of Po within 20 min. In lead azide a yellowing of the material occurred.

McLaren [36] questioned the theory of Poole and Henderson that a single α-particle caused the detonation. For dry nitrogen iodide (Table VIII) the time to explosion decreased as the intensity increased, but the total dose of α-particles required was not constant. Ammonia and water vapor in the ambient prevented initiation. The range of α-particles in nitrogen iodide is greater than 1 μm, but McLaren suggested that the mechanism of initiation was the removal of ammonia (which served as a stabilizer) from the surface of nitrogen iodide by the α-particle, causing an explosion at room temperature that resembled the thermal ignition of nitrogen iodide.

Bowden and Singh [37,38] irradiated lead, cadmium, silver, and lithium azides and silver acetylide with α-particles but failed to cause explosion in any

Table VIII. Explosion of Iodine Azide by α-Particles [36]

α-particles striking azide (sec^{-1})	Time to explosion (sec)	Total dose
150	>3600	>540,000
400	2200	888,000
539	2100	1,131,900
550	1080	594,000
770	420	323,400
810	1080	874,800
970	180	174,600
1150	10	11,500

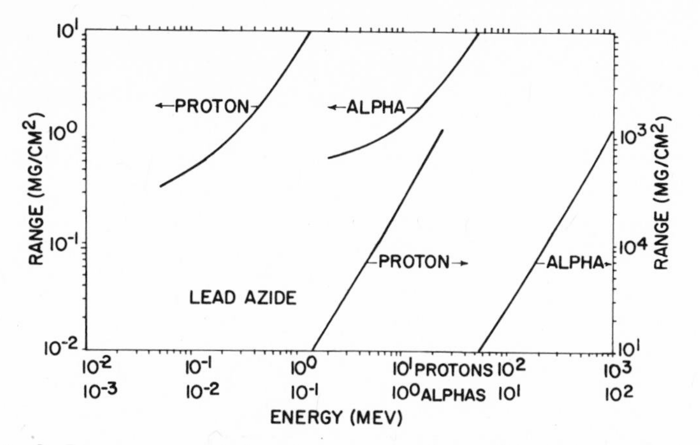

Figure 8. Range–energy relations for protons and α-particles in lead azide [41].

of the substances. Bowden [39] later reported that the passage of a single α-particle through silver azide will liberate a large amount of silver.

Ling [40] calculated the ranges in several explosives of α-particles and fission fragments of different energies. An error was subsequently found in the α-particle calculations [41], but the results showed that both types of particle have greater ranges in copper chlorotetrazole, potassium dinitrobenzofuroxan (KDNBF), and nitrogen iodide than in lead azide, lead styphnate, TNT, RDX, and PETN. The results were interpreted to mean that the greater range increased the probability of reaching sensitive regions of crystals, thus giving some insight into the relative sensitivities. Assuming complete thermalization of energy transferred in nuclear collisions, a fission fragment was more efficient in causing ignition than an α-particle of the same incident energy.

Cerny *et al.* [41] calculated range–energy relations for protons and α-particles in various explosives, including lead azide (Figure 8). The α-particle ranges were obtained from established proton ranges.

2. Neutron Irradiation

a. Slow or Thermal Neutrons

Bowden and Singh [37,38] utilized an Sb–Be source with a slow neutron flux of about 10^6 n/cm^2/sec, and a cyclotron for fluxes up to 3×10^8 n/cm^2/sec. Explosives were irradiated for 1 hr so that the maximum total slow neutron dose was 1.08×10^{12} n/cm^2. In most cases a large number of high-velocity recoil atoms are formed during the irradiation of the explosives with slow neutrons (Table IX). In no case did an explosive (even nitrogen iodide) detonate as a result of the slow neutron irradiation.

Table IX. Slow Neutron Irradiation of Primary Explosives [38]

Explosive	Density (g/ml)	Flux ($n/cm^2/sec$)	Total atoms (per ml)	Metal nuclei reacting (per sec)	Nitrogen nuclei reacting (per sec)	Nuclear reaction
Cadmium azide	0.729	3×10^8	1.56×10^{22}	3×10^8	6×10^6	$^{14}N(n, p)^{14}C$, 0.6 MeV protons, 40 keV ^{14}C recoil $^{113}Cd(n, \gamma)$, ^{114}Cd, 5.1 MeV γ cascade.
Lithium azide	0.570	7×10^7	4.01×10^{22}	2×10^7	2×10^6	Nitrogen reaction: $^6Li(n, \alpha)^3H$, 2.1 MeV α and 2.7 MeV ^3H
Silver acetylide	0.381	3×10^7	3.8×10^{21}	^{107}Ag 7×10^5		$^{107}Ag(n, \gamma)^{108}Ag$, 4.5 MeV γ cascade
				^{109}Ag 2×10^6		$^{109}Ag(n, \gamma)^{110}Ag$, 10 MeV γ cascade Principal decay processes: ^{108}Ag, 2.12 MeV β, 0.6 MeV γ; $^{110*}Ag$, 2.8 MeV β
Silver azide	0.745	1×10^8	1.21×10^{22}	^{107}Ag 4×10^6 ^{109}Ag 1×10^7	1×10^6	Silver and nitrogen reactions
Lead azide	1.875	2×10^7	2.71×10^{22}	17	8×10^5	Lead reactions unimportant, nitrogen reaction
Nitrogen iodide	1.40	6×10^6	1.64×10^{22}	^{127}I 5×10^5	4×10^4	$^{127}I(n, \gamma)^{128}I$, 8.5 MeV γ Principal decay process: ^{128}I, 2.1 MeV β

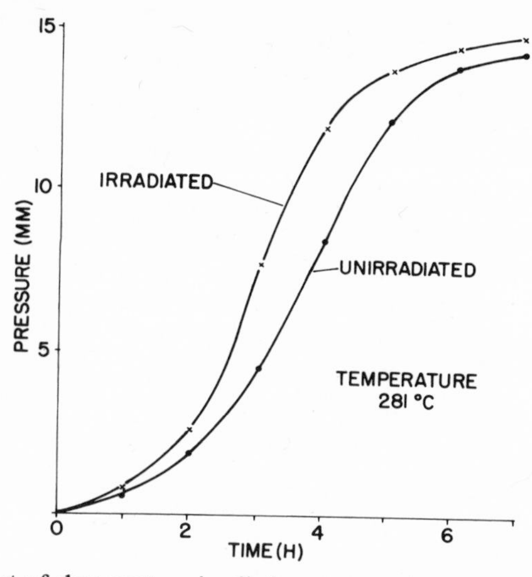

Figure 9. The effect of slow neutron irradiation on the subsequent thermal decomposition of lead azide [37].

Slow neutron irradiation caused color changes and darkening in most of the explosive crystals, indicating the formation of imperfections, which subsequently affected the thermal decomposition of the explosives. The rates of thermal decomposition of irradiated lead azide are shown in Figure 9.

Irradiation increased the rate of thermal decomposition of lead azide, but the effect was not as pronounced as with lithium azide, for which the induction period was reduced to about one half and the rate increased considerably. Cadmium azide produced pressure–time curves similar to lead azide. Irradiated silver azide was unaffected, but the experiment was conducted at 315°C, which caused the silver azide to be molten.

b. Fission-Product Neutrons

The detonation of nitrogen iodide by nuclear fission products was reported by Feenberg [42]. Small samples of nitrogen iodide mixed with uranium oxide (about 0.5 g) were exposed to a 200-mg Ra–Be neutron source surrounded by 6 cm of paraffin. The average exposure required to produce detonation was 40 min, but individual initiations took from 1 min to several hours. Similar results were obtained by Fabry *et al.* [43].

More detailed experiments were conducted by Bowden and Singh [38, 39]. Crystals of lead and cadmium azides and nitrogen iodide (coated with uranium oxide) were irradiated with slow neutrons; only nitrogen iodide exploded. Lead and cadmium azides were irradiated with fission fragments at elevated temperatures up to 290°C, but neither exploded (Table X).

The damage caused in an explosive by fission fragments was investigated by Bowden and Montagu-Pollock and their coworkers [44-49]. Disorder was produced in the lattice, together with holes and tunnels having track widths of 100-120 Å. It was concluded that even the intersection of two tracks did not produce initiation in azides, while the intersection of three or more tracks within 10^{-11} sec was an unlikely event.

An analysis of the hot-spot model was attempted by Cerny and Kaufman [50], who irradiated several explosives with π^- mesons (pions). The decay of mesonic atoms formed by the capture of π^- mesons can result in the emission of 12-17 charged particles from a single lattice site. It was estimated that a temperature of 10^4°C would be produced over a 10-Å radius for 10^{-11} sec. The calculations indicated that the temperature would decrease rapidly, but that the radius of a hot-spot site would increase to meet the criterion of Bowden.

However, when lead azide, lead styphnate, mercury fulminate, RDX, TNT, and PETN were subjected to bombardment with a negative pion beam, no explosions or decompositions were observed for any of the explosives. The analysis had predicted initiation only for RDX. Also it had indicated that nuclear fission events would produce higher energy densities and greater temperature increases than were actually observed.

Subsequently, Mallay *et al.* [51] irradiated RDX, HMX, PETN, and nitroglycerin with fission fragments from the spontaneous fission of californium-252 at 160, 215, 125, and 180°C, respectively. The californium-252 was mixed through the explosive pellet or liquid. No explosions were obtained nor were any signs of accelerated thermal decomposition evident, even though the explosives were exposed to 200-2000 fission fragments. An analysis of the presumed heating pre-

Table X. Fission-Fragment Irradiation of Explosive Azides [39]

Sample[a]	Flux (m/cm^2 sec)	^{235}U atoms per ml coating	Fissions per ml sample per hr	Result
Nitrogen iodide	2×10^6	6×10^{16}	2×10^5	Explosion
	4×10^6	6×10^{16}	4×10^5	Explosion
Lead azide				
150°C	3×10^7	7×10^{17}	4×10^7	No explosion
225°C	4×10^7	7×10^{17}	5×10^7	No explosion
280°C	3×10^7	7×10^{17}	4×10^7	No explosion
290°C	2×10^7	7×10^{17}	4×10^7	No explosion
Cadmium azide	2×10^7	3×10^{18}	2×10^8	No explosion
280°C	2×10^7	3×10^{18}	1×10^8	No explosion

[a]Coated with 1 μm ^{235}U.

dicted initiation of each explosive. However, because of uncertainity about appropriate values of the parameters used, the investigators were unable to conclude with assurance that the hot-spot model fails to describe microscale events in explosives.

c. Reactor Radiations

Steady-State Fluxes. A nuclear reactor is a source of fast and thermal neutrons accompanied by a gamma-ray field; therefore, azides exposed in a reactor are subjected to the combined effect of these radiations.*

Muraour and Ertaud [52] exposed lead azide, mercury fulminate, diazo-*m*-nitroaniline perchlorate, lead trinitroresorcinate, and tetracene to total thermal neutron fluxes up to 10^{13} n/cm^2. The γ component was not given. No explosions occurred. The only effect noticed was a slight color change from white to light brown in tetracene. The lead azide was exposed to a total thermal neutron dose of 3×10^{14} n/cm^2.

Although Muraour and Ertaud confirmed the results of Bowden and Singh [37,38], a different environment was used. The former utilized reactor radiation at higher dose rates and doses compared to the slow thermal neutron radiation of the latter. For example, the thermal neutron dose rate for lead azide was 4.2×10^9 compared to 2×10^7 n/cm^2/sec and the total dose was 3×10^{14} compared to 7.2×10^{10} n/cm^2.

Alpha-lead azide crystals, wrapped in a thin aluminum foil, were subjected to fast and thermal neutrons in a heavy-water reactor [53]. With a thermal flux rate of about 10^{14} n/cm^2/sec, the crystals were irradiated for 8, 17, and 170 hr. The crystals decomposed to a brown powder, which was identified as lead carbonate by X-ray diffraction and infrared absorption. From a mass spectrographic analysis of the isotopes of carbon and oxygen in the decomposition products, it was determined that the carbonate was formed from the atmosphere by the breaking of surface bonds by the neutrons. It was subsequently reported [54] that the total dose required for conversion to lead carbonate is approximately 7.5×10^{16} n/cm^2.

A total neutron exposure of 2.2×10^{18} n/cm^2 enhanced the decomposition of lead styphnate monohydrate by a factor of three [55] (Figure 10), the rate increasing monotonically with the amount of exposure. A sample decomposed 8 days after irradiation produced a decomposition curve almost identical with one stored 46 days after irradiation.

Groocock [56] irradiated α-lead azide for 5 hr in the Harwell BEPO pile. The doses were not given, but the decomposition curves were similar to those for the

*The gamma (γ) radiation exposure units used to express radiation doses are as follows: 1 R (roentgen) (of dry air) = 87.7 ergs (absorbed/g(C) = 87.7×10^{-7} J/g(C); 1.14 R = 100 ergs/g(C) = 10^{-5} J/g(C) = 1 rad.

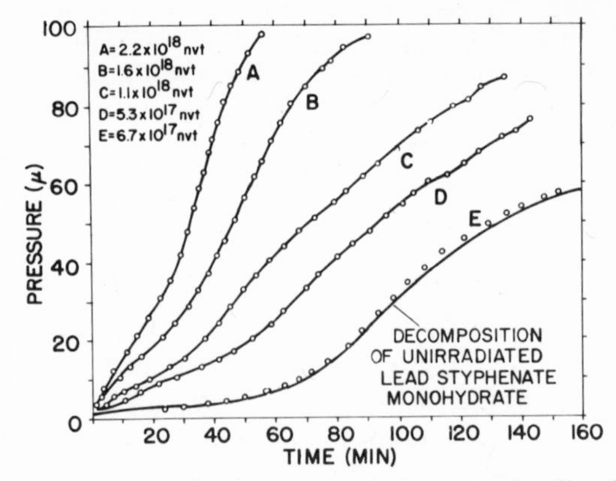

Figure 10. The thermal decomposition curves of reactor-irradiated lead styphnate monohydrate (222.5°C) [55].

same material subjected to X-rays for a total dose of 6.35×10^6 R (see Section C.4 below). Assuming the thermal neutron rate to be 6×10^{11} n/cm^2/sec, the approximate total neutron dose was 1.08×10^{16} n/cm^2. The 5-hr exposure lowered the explosion temperature 15–20°C.

Abel *et al.* [57,58] studied the effect of reactor and gamma-ray irradiation on the impact sensitivity of colloidal lead azide. Reactor irradiations ranged from 3.3×10^{17} to 1.57×10^{18} nvt (n/cm^2) (fast plus slow neutrons) with an accompanying reactor gamma dose rate of 2×10^6 R/hr. The results (Figure 11) indicated an increase in the impact sensitivity of colloidal lead azide as a function of neutron dose. The studies also revealed an incompatibility of colloidal lead azide with Teflon and aluminum during long-term reactor exposures.

Jach [59, 60] studied the effect of reactor irradiation on the thermal decomposition of colloidal α-lead azide (average particle size 7 μm), irradiated for 35 hr in a flux (fast plus slow neutrons) of approximately 7–8 $\times 10^{12}$ n/cm^2/sec and a

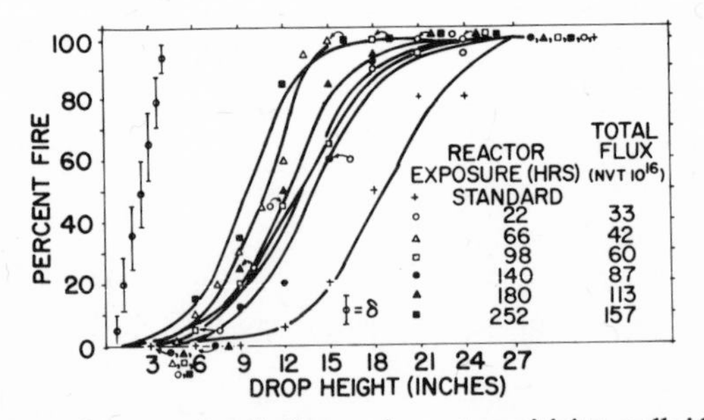

Figure 11. The effect of reactor irradiation on impact sensitivity, colloidal lead azide [57]; Picatinny Arsenal impact test, 2-kg weight, 20 trails for height.

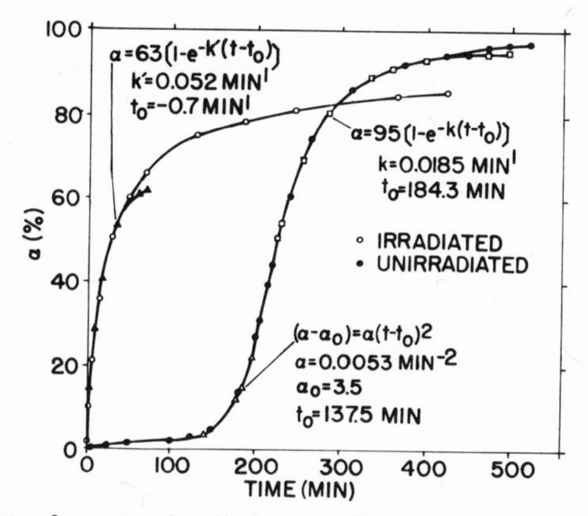

Figure 12. The effect of reactor irradiation on the thermal decomposition of colloidal α-lead azide [59]. Fraction of total decomposition α against time. Solid circles: experimental points for unirradiated material at 240.9°C; open circles; experimental points for irradiated material at 238.5°C; other points are the attempted fits indicated by arrows.

gamma component of 2×10^6 R/hr. Figure 12 shows that reactor irradiation significantly altered the decomposition kinetics. The induction period was greatly reduced, and the rate of decomposition increased. Derived rate constants were

$$k(\text{unirradiated}) = 10^{12.0\pm1.0} \exp(-1.52 \pm 0.10\, MJ/RT)$$

$$k(\text{irradiated}) = 10^{7.9\pm1.0} \exp(-1.07 \pm 0.10\, MJ/RT)$$

Jach suggested that the decrease in activation energy was associated with changed electronic excitation energies, while the decrease in frequency factor arose from structural changes accompanying the irradiation.

Urizar *et al.* [61] subjected organic explosives (including TNT, RDX, HMX, tetryl, and PETN) to total neutron fluxes as high as 3×10^{16} n/cm^2 and 2×10^8 R gamma radiation with no explosions.

Primary explosives were among 32 substances which Avrami and Voreck [62] subjected to total doses ranging up to 5.0×10^{17} n/cm^2 fast neutrons, 3.1×10^{18} n/cm^2 thermal neutrons, and 4.3×10^9 R gamma. The primary explosives were 1.8-g samples of KDNBF (potassium dinitrobenzofuroxan), silver tetracene, and lead styphnate. The exposures and effects on some of the tests are listed in Table XI. Included are the results for tetryl, a booster explosive.

Lead styphnate detonated within 2 min, and a runaway reaction occurred with tetryl after 2 hr. The weight loss of the primary and booster explosives due to the reactor irradiation is shown in Figure 13.

In utilizing nuclear reactors (or any radioactive source) for the irradiation of explosives, precautions have to be taken to prevent damage to the reactor in

Table XI. Effects of Reactor Irradiation on Some Explosives [62][a]

Irradiation time (min)	Total neutron dose		Total gamma (R)	Weight loss (%)	10% loss on TGA (°C)	150° avg. gas evolution (ml/g/hr)	DTA peak exotherm at 20°C/min (°C)	5-sec explosion temperature (°C)
	Fast (n/cm^2)	Thermal (n/cm^2)						
KDNBF								
0					210	2.55	212[b]	194
60	5.04×10^{16}	3.10×10^{17}	4.0×10^{8}	0.46	188	10.95	198[b]	185
120	1.01×10^{17}	6.19×10^{17}	8.0×10^{8}	1.52	188	11.0	188[b]	170
180	1.51×10^{17}	9.29×10^{17}	1.2×10^{9}	4.84[c]	175	11.0	182[b]	169
Silver tetracene								
0						0.26	380[b]	375
38	3.07×10^{16}	2.1×10^{17}	2.3×10^{8}	7.52		4.02	360	247
126	1.00×10^{17}	6.3×10^{17}	8.0×10^{8}	23.6		5.92	315	214
Tetryl								
0								
25	2.2×10^{16}	1.2×10^{17}	1.9×10^{8}	2.14				
125	10.0×10^{16}		8.8×10^{8}			Sample ignited		
Lead styphnate								
0					285	1.02[c]	287[b]	299
~2	1.7×10^{15}	1.1×10^{16}	1.4×10^{7}			Sample detonated		
~1 msec	7.3×10^{13}	2.2×10^{14}	3.6×10^{6}	0.26		2.53[c]	270[b]	268
~1 msec	3.5×10^{14}	1.6×10^{15}	1.5×10^{7}			Sample detonated		

[a] Average reactor exposure rates: fast neutrons $\phi > 0.18$ MeV 1.4×10^{13} n/cm^2 sec; thermal neutron $\phi > 0.17$ eV 8.6×10^{13} n/cm^2 sec; gamma 4.0×10^{8} R/hr (3.38×10^{10} ergs/g(C)/hr gamma).

[b] Sample detonated.

[c] 200°C.

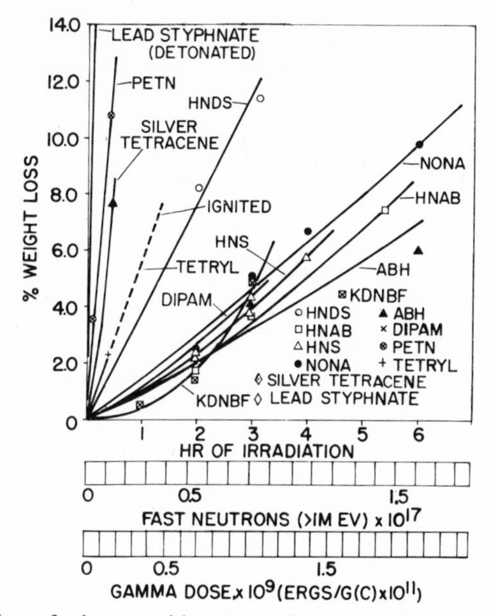

Figure 13. Weight loss of primary and booster explosives as a function of reactor irradiation [62].

the event of an explosion. This led to the development of explosion-proof irradiation capsules [57, 62–67].

Pulsed Fluxes. Urizar *et al.* [61] subjected explosives to a pulse of about 7×10^{12} n/cm^2. The peak flux was estimated by 10^{17} n/cm^2/sec with a pulse half-width of about 90 μsec and included a small gamma dose. None of the several explosives detonated, and negligible damage was observed.

Nine explosives and propellants were subjected to a transient burst of about 1 msec duration which resulted from fission of about 10^{21} atoms of ^{235}U [62]. The materials were tested with 3.6-g samples of TATB, DATB, TACOT, HNS, lead styphnate, black powder, and three composite propellants. The lead styphnate sample detonated, but since the samples were hurled 500 ft by the nuclear excursion there was some question whether this resulted from impact rather than irradiation. The melting point was lowered from 216 to 208°C and the 5-sec explosion temperature (Table XI) was lowered from 299 to 268°C. None of the other materials tested showed evidence of gross radiation damage.

Bulk samples of RD1333 were exposed to pulsed nuclear irradiation giving a total dose of 2.0×10^{14} n/cm^2, during a pulse of 40–50 μsec (full width at half-maximum) [68]. Vacuum stability, explosion temperature, and detonation velocity tests did not show any changes due to the irradiation.

Mixtures of RD13333 lead azide and fine-particle boron detonated [69] in

Table XII. Pulsed Reactor Irradiation of Lead Azide/Boron Mixtures [69]

Ratio lead azide/boron	Density (g/ml)	Neutron environment	Result
100 : 1	2.0	1.5×10^{15} nvt, $E > 10$ keV 0.5×10^{15} nvt, thermal	Detonation
	2.5		Detonation
	3.0		Detonation
	3.5		Detonation
50 : 1	2.0	3.5×10^{14} nvt, $E > 10$ keV 1.2×10^{14} nvt, thermal	No detonation
	2.5		No detonation
	3.0		No detonation
	3.5		No detonation
	2.0	3.0×10^{14} nvt, $E > 10$ keV 1.0×10^{14} nvt, thermal	No detonation
	2.5		No detonation
	2.0	9.0×10^{14} nvt, $E > 10$ keV 3.0×10^{14} nvt, thermal	Detonation
	2.5		Detonation
10 : 1	3.0	3.0×10^{14} nvt, $E > 10$ keV 1.0×10^{14} nvt, thermal	No detonation
	3.5		No detonation
	3.0	9.0×10^{14} nvt, $E > 10$ keV 3.0×10^{14} nvt, thermal	Detonation
	3.5		Detonation

the neutron environment of a water-moderated annular core pulsed reactor (ACPR) (Table XII). The absorption of thermal neutrons by boron produces α-particles according to the following interaction:

$$^{10}_{5}\mathrm{B} + ^{1}_{0}n \longrightarrow ^{7}_{3}\mathrm{Li} + ^{4}_{2}\mathrm{He} + 2.78\ \mathrm{MeV}$$

The Li ion and α-particle are stopped in the lead azide-boron mixture, converting their kinetic energy of 2.78 MeV to thermal energy.

The pulse width during the tests was 4.7 msec (FWHM) so that the maximum thermal neutron dose rate was about 3.6×10^{17} n/cm^2/sec.

If the mean temperature rise of the mixtures must be 350°C for detonation to occur, the authors calculated, assuming the boron to be uniformly distributed:

Thermal neutron fluences (nvt)	PbN_6/B ratios
1.95×10^{13}	0
5×10^{13}	3.24
10^{14}	8.56
5×10^{14}	51
10^{15}	104

Detonation of samples containing 1% B at 0.5×10^{15} nvt thermal would not be expected on this model, but the remaining results are consistent with the model's predictions. The results suggest that at the higher fluences detonation resulted from hot spots ($T > 350°C$) within the explosive.

3. Gamma Irradiation

Steady-state γ-irradiation of any explosive has not been known to initiate a detonation. The effect of such irradiation appears to result in slow decomposition with, in the case of primary explosives, a deterioration in the functional properties of the samples.

One of the first investigations of the effect of gamma rays on explosives was conducted by the Los Alamos Scientific Laboratory and the Oak Ridge National Laboratory in 1948 [70]. RDX, tetryl, TNT, and Composition B were placed for 10 days in activated uranium slugs for a total gamma dose of 8.6×10^6 R. The low intensity of radiation produced no visible changes, the gas evolution was slight, and the melting-point changes were negligible.

Subsequently, using 0.41 MeV $^{198}_{79}$Au γ-rays, a group of explosives was irradiated at three temperatures (70°C, ambient, and -40°C) to determine the effect on thermal stability [71-73]. The volumes of gas produced were measured during and after irradiation. The amount of gas evolved as a function of gamma dose for each of the explosives irradiated is shown in Figure 14. Figure 15 shows the effect of temperature at the time of irradiation for dextrinated lead azide. The total gamma dose that dextrinated lead azide received was 5.77×10^7 R.

A continued evolution of gas from lead azide after the radiation source had been removed suggested that either gas trapped in the sample during irradiation or that decomposition continued after irradiation [74]. When samples was irradiated at 71°C and cooled, no such gasing occurred. In all cases more gas was evolved than could have resulted from heat alone.

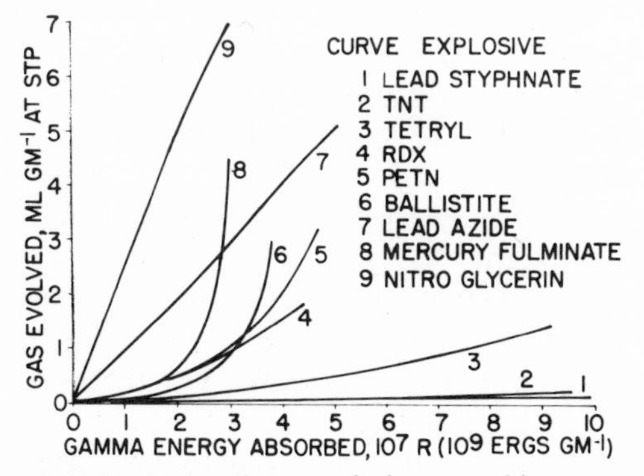

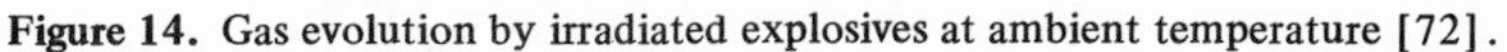
Figure 14. Gas evolution by irradiated explosives at ambient temperature [72].

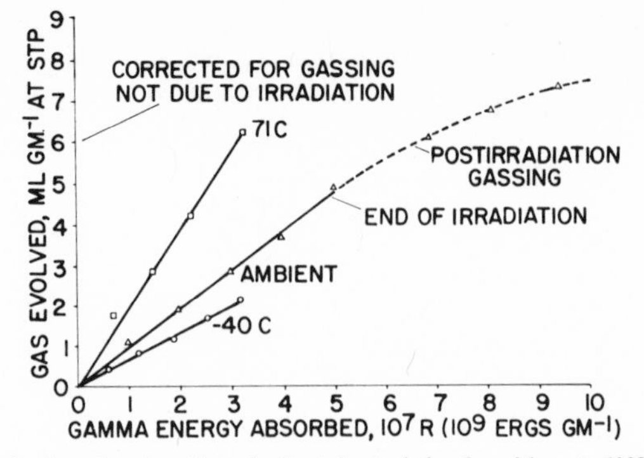

Figure 15. Gas evolution by irradiated dextrinated lead azide at different temperatures [72].

With detonators which included lead azide in the explosive train, the work of Warren *et al.* [72] also revealed that long-term gamma irradiation affected the functioning in stab tests.

Bowden and Singh [37] irradiated lead and cadmium azides with α-particles and γ-radiation at 290°C when mixed with uranium oxide without initiations occurring.

A dose of 10^6 R ^{60}Co gamma rays reduced the diamagnetic susceptibility of α-lead azide more than threefold [53]. Colloidal lead azide, irradiated with a ^{60}Co gamma source with exposures of 2.5, 7.5, and 22.5 $\times$ 10^7 R, had its impact sensitivity increased from 12 to 6 in. with the Picatinny impact test [57,58].

Rosenwasser *et al.* [75] noted that sodium azide turns brownish-yellow when subjected to radiation. Subsequently, when mechanically deformed crystals of sodium and potassium azide were irradiated with 10^7 gamma radiation, Dreyfus and Levy [76] observed the formation of pyramidal etch pits. These were also evident in ammonium perchlorate crystals [77].

Thermal decompositions of barium and strontium azides, preirradiated with ~1 MeV gamma rays, were conducted by Prout and Moore [78,79]. With dehydrated barium azide a total gamma dose of 20 Mrad (2.24 $\times$ 10^6 R) eliminated the induction period and increased the acceleration of the decomposition. A somewhat greater effect was evidenced with strontium azide. Avrami *et al.* [80] subjected barium azide to ^{60}Co gamma radiation to exposure levels up to 1 $\times 10^9$ R (Table XIII). Differential thermal analyses (Figure 16) showed a steady decomposition of the sample, and after 1 $\times$ 10^9 R exposure (10^4 hr at room temperature), infrared analysis indicated that the residue was in the form of barium carbonate.

Lead azide and thallous azide were subjected to long exposures of ^{60}Co

Table XIII. Effect of Gamma Irradiation on Barium Azide [80]

	Gamma radiation dose (R)[a]			
	Control	1.4×10^7	1.2×10^8	1.0×10^9
Melting point (°C)	201.5	205.5	Exploded at 260°C	No reaction at 290°C
Weight loss		Negl.	4.5%	
5-sec explosion temp. (°C)	312		249	
Activation energy/mole	110.2		77.8	
PA impact, 50% reaction (in.)	11.22 ± 1.30	7.03 ± 5.18	11.42 ± 1.67	36+
Vacuum stability, 100°C (ml/g/40 hr)	0.39		11+	0.59

[a]Gamma rate—8.0×10^5 R/hr.

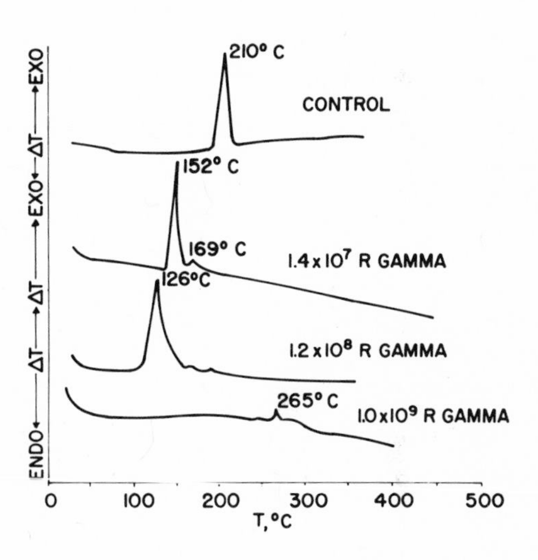

Figure 16. Differential thermal analysis thermograms for BaN_6 as a function of gamma dose [80].

γ-irradiation by Avrami *et al.* [81]. Powdered samples lost more weight than pellets (Table XIV). In the Picatinny impact test no reactions were obtained at the maximum height of 36 in. for dextrinated lead azide after the longer exposures, and infrared spectra of the residues revealed lead carbonate as a decomposition product. Ammonium nitrate was also formed on the top of capsules containing lead or thallous azides which had been subjected to a gamma dose of over 10^9 R (Figure 17).

Thallous azide was exposed to 5.4×10^9 R and, according to infrared analysis, was reduced to a form of thallous oxide.

Table XIV. Effect of Gamma Irradiation on Various Lead Azides [81]

Lead azide	Gamma radiation dose (R)[a]		
	Control	1.0×10^8	1.16×10^9
Dextrinated			
Color			
Powder	White	Dark gray	Yellow
Pellet	White	Dark brown	Light brown
Weight loss (%)			
Powder		–7.5	–14.5
Pellet		–0.9	–9.2
PA impact, 10% reactions (in.)	7	6	>36 (powder) >36 (pellet)
Vacuum stability (ml/g/40 hr at 100°C)	0.47	1.35	0.57
RD1333			
Color			
Powder	White	Dark gray	Brown
Pellet	White	Dark brown	Light brown
Weight loss (%)			
Powder		–6.4	
Pellet			
PA impact, 10% reactions (in.)	6	6	11 (powder) 16 (pellet)
Vacuum stability (ml/g/40 hr at 100°C)	0.38	7.18	2.06
PVA			
Color			
Powder	White	Dark gray	Brown
Pellet	White	Dark brown	Brown
Weight loss [%)			
Powder		–7.8	–14.0
Pellet		–.15	
PA impact, 10% reactions (in.)	8	4	11 (powder) 21 (pellet)
Vacuum stability (ml/g/40 hr at 100°C)	0.28	5.58	1.04

[a]Dose Rate–^{60}Co Gammas, 3.2×10^5 R/hr.

4. X-Irradiation

Irradiation of explosives with X-rays dates from the 1930s with the experiments of Muraour and Ertaud [52] on nitrogen iodide, and Günther *et al.* [82, 83] on barium azide.

In 1948 [70], TNT, tetryl, lead azide, black powder, and three propellants were exposed to 1 MeV X-rays for 1 hr at a dose rate of 12 R/sec. With a total

Figure 17. Formation of ammonium nitrate on top of irradiation explosion-proof capsule containing lead azide subjected to 1.16×10^9 R gamma dose [81].

dose of 4×10^4 R, no rise in temperature was observed and no significant changes in sensitivity were detected.

Decomposition studies were pursued by Heal [84,85], who used 44.5-kV X-rays with a beam strength of 45 mA on sodium azide and established that no nitride ions were formed during the decomposition because the nitride ion was rapidly converted to the colorless amide ion [86].

Bowden and Singh [38], in their efforts to initiate primary explosives, used 220-kV X-rays with a beam strength of 15 mA and an intensity of 700 R/min. Crystals of lead and silver azide were irradiated for 2 hr and no explosions were obtained. In the wavelength range between 0.06 and 0.4 Å, the absorption of energy takes place through photoelectric effect and Compton recoil. With crystals 2 mm thick only a small percentage of the energy is absorbed, and the only effects noticed were color changes and metal nuclei produced on the surfaces.

Alpha- and beta-lead azide crystals showed considerable deterioration in an intense X-ray beam, marked red coloration, and a breakup of crystal morphology; however, they did not detonate [53]. Appreciable reduction of the optical absorption band at 270 nm was attributed to F and V centers.

The effect of high-energy X-rays was studied by Groocock and Phillips [56, 87]. Small samples (2 mg) of Service lead azide were irradiated in air, and the effect on the time to explosion was observed. Below a dose of 10^6 R no effect was found, but with 6.35×10^6 R the time to explosion at a given temperature was only one fifth that for the unirradiated material. The minimum temperature for explosion was reduced from 310 to 295°C. Below 10^4 R the thermal decomposition and explosion temperatures of α-lead azide were unchanged [56]. With the higher doses the decomposition rates were changed and the explosion time reduced (Figure 18).

For the same energy deposition, reactor irradiation was found to be slightly less effective than high-energy X-rays in altering the reaction kinetics of subsequent thermal decomposition. Groocock also observed that the friction sensitivity of Service lead azide, irradiated with X-rays at total doses of 10^4 and 10^7 R, did not show any effects or changes when tested with the sliding-block friction test.

Radiation-induced changes in α-lead azide caused by X-rays were noted by Todd and Parry [88, 89]. The hardness was changed by exposure to soft X-rays. With an X-ray dose rate of 1.4×10^5 R/min in air, the decomposing crystal expanded prefentially along the *b* axis of its crystal structure. More than 97% destruction of the azide was achieved with a total dose of 3.4×10^8 R. It was also shown that the stable endproduct of X-ray decomposition in air was basic lead carbonate of the formula $2PbCO_3 \cdot Pb(OH)_3$. After a dose of 6.7×10^7 R, there was evidence of residual lead azide together with an unidentified phase. Higher doses produced a further unidentified phase before stable basic lead carbonate was finally formed.

Most of the studies conducted on azides irradiated with X-rays have involved the changes caused in the subsequent thermal decomposition of the samples. Erofeev and Sviridov [90] studied the effects of moisture and aging on barium azide. An increased decomposition rate was noted with strontium azide preirradiated with X-rays and decomposed at 126°C [91]. Zakharov and coworkers

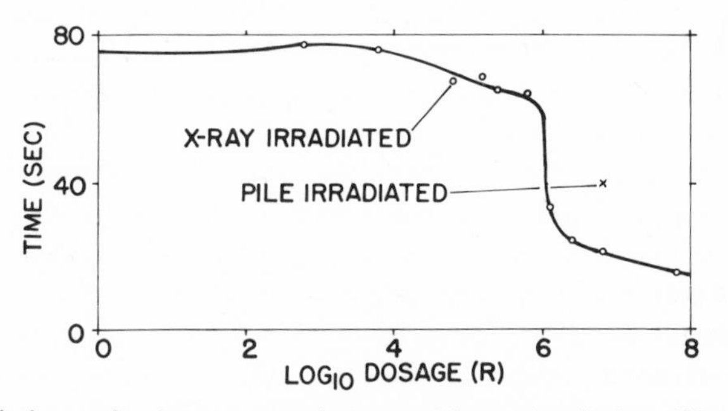

Figure 18. Variation of time to explosion with preirradiation X-ray dosage for decompositions at 298°C [56].

[92–96] studied the decomposition of lead and silver azides by X-rays and fast protons, and the degree of decomposition was determined from the amount of nitrogen retained in the samples. Decomposition was found to be proportional to the absorbed dose. The addition of Ag^+ in lead azide accelerated radiolysis, while Cu^{2+} decelerated it. With silver azide, Pb^{2+} substantially increased the rate of formation of radiolytic nitrogen. The effects of heterophase, semiconductor additives on the radiation stability of lead azide were also investigated.

As with gamma radiation, steady-state X-ray irradiations have not produced any initiations in any explosives. Pulsed, intense irradiations may have more severe consequences, but investigations have not yet been reported.

5. Electron Irradiation

a. Steady-State Irradiation

The electron bombardment of explosives has been undertaken by various investigators in an effort to initiate or decompose the material under study. One of the early investigations was undertaken by Kallmann and Schränkler [30], who bombarded TNT, mercury fulminate, nitrocellulose, and to some extent, picrates and azides with 10-kV, 1-mA electrons *in vacuo* but were unable to produce explosions. However, when heavy ions of argon and mercury were used, initiations were achieved with several substances with each of the ions. Muraour [31] subjected lead azide and silver acetylide to 90 kV at 3 mA for 3 min and only achieved explosion with silver acetylide. Both explosives blackened upon electron irradiation. Muraour believed that the explosion was either a thermal effect or that, by chance, a sufficiently large number of molecules decomposed at one point to bring about complete decomposition.

Bowden and Singh [37, 38] achieved explosion of lead and silver azides when crystals were irradiated with an electron beam of 75 kV and 200 μA. Explosion was partly due to heating of the crystals by the electron beam. To substantiate this, crystals of potassium chlorate with a melting point of 334°C readily melted in the beam, showing a temperature rise close to the explosion temperature of the azides. Sawkill [97] investigated with an electron microscope the effect of an electron beam on lead and silver azides. If explosion did not take place, color changes and nucleation occurred; cracks developed within the crystals which broke up into blocks about 10^{-5} cm across and were believed to be associated with a substructure in the crystals. In silver azide the progression to silver was pronounced but did not follow the thermal decomposition route.

The decomposition of various azides by electron bombardment has been studied. Müller and Brous [98, 99] studied sodium azide, Groocock and Tompkins [100] investigated barium and sodium azides, and Groocock [101] investigated α-lead azide. In each case only gas evolution was studied and detected [102].

Steady-state electron bombardment of high explosives has been conducted,

Table XV. Effects of 60-MeV Electron Irradiation on Explosives [105]

Explosive	Weight (mg)	Dose (R)	Dose rate (R/sec)	Weight loss	Color change	Test	Result
Lead azide	25.0	1.14×10^{9}	1.6×10^{6}	None	Dark brown	Heat[a]	Explosion
	4.0	3.65×10^{10}	2.9×10^{6}	None		Heat	No explosion
	11.0	2.96×10^{10}	2.7×10^{6}	None		Heat	No explosion
	12.8	1.14×10^{10}	1.1×10^{6}			Impact[b]	No explosion
	19.3	1.37×10^{10}	2.3×10^{6}	9%		Impact	No explosion
Lead styphnate	25.0	1.14×10^{9}	1.1×10^{6}		Dark rust brown	Heat	Explosion
	25.0	1.14×10^{9}	1.6×10^{6}			Heat	Explosion
	3.3	3.65×10^{10}	2.9×10^{6}			Heat	No explosion
	7.9	2.96×10^{10}	2.7×10^{6}			Heat	Explosion
	10.3	22.7×10^{10}	2.9×10^{6}			Impact	Explosion
TNT	8.0	1.14×10^{9}	1.7×10^{6}	None	Caramel	Impact	Explosion
HMX	7.7	1.14×10^{9}	1.7×10^{6}			Impact	No explosion
PETN	14.0	1.14×10^{9}	0.6×10^{6}	None	Gray	Heat	Sample evaporated

[a] Put on hot plate with temperature above explosion temperature.
[b] Samples tested at 100 cm with NOL apparatus.

one of the earliest investigations being on nitroglycerin [103]. Berberet [104] subjected RDX, HMX, TNT, TNB, DATB, tetryl, and some fluoroderivatives to 3- and 40-MeV electrons up to 2 hr for total doses of about 7.3×10^8 R with no detonations.

Using 60-MeV electrons, Farber [105] subjected lead azide, lead styphnate, TNT, HMX, and PETN to total doses up to 3.65×10^{10} R (3.2×10^{10} rad). The ionizing radiation desensitized lead azide to impact and heat (Table XV).

b. Electron Pulses

Much interest has been focused recently on the initiation of explosives by intense beams of fast electrons with pulse durations in the nanosecond range. For the most part the critical energy deposition doses observed are much lower than those required for direct thermal initiation.

Phung [106] applied the hot-spot theory to pulsed electron initiation and considered the spatial distribution of energy arising from the absorption of electrons or photons. In this model Phung divided the sample into cells of about 5×10^{-7} cm radius. With increasing dose, the spatial distribution becomes more random and the probability that packets of energy will be absorbed in the same cell increases. Using an appropriate heat-conduction equation, the average temperature (or dose) and corresponding hot-spot temperature required for initiation were determined. Critical doses were calculated for PbN_6, BaN_6, lead styphnate, PETN, RDX, HMX, and nitroglycerin, (Table XVI). With lead azide the calculations overestimated the dose required for initiation, and the ignition temperature was much below that observed experimentally. For secondary explosives, the model predicted values in good agreement with experience. Phung considered the results generally to suggest a thermal mechanism for the initiation of explosives by high-energy radiations.

Table XVI. Critical Doses and Temperatures to Initiate Explosives by Pulsed High-Energy Electrons [106]

Explosives	Critical dose (J/g)		Ignition temperature (°C)	
	Calculated	Experimental[a]	Calculated	Experimental
PbN_6	158.2	46	57	327–360
PETN	89.6	>167	222	205–225
RDX	1.61	>167	300	229–260
HMX	156.1	>167	310	335
Nitroglycerine	90.4		255	215–218
Pb styphnate	92.1	normal 42		
		basic 75		
BaN_6	94.2	80		

[a]Estimated 50% initiation dose.

Pulsed electron irradiations are usually conducted with accelerators charged from 0.1 to 6.0 MeV, and pulse durations range from 3 to 60 nsec. The explosives are pressed pellets with thicknesses of about 0.2 of the electron range. If the samples are too thick, the electrons will be trapped in the material causing a nonuniform dose. Further, the resulting space charge causes dielectric breakdown, and initiation is caused by electric discharge rather than by absorbed radiation.

Tests on dextrinated and RD1333 lead azide (Table XVII) showed that ambient pressure, sample thickness, and the type of lead azide are important factors in determining the sensitivity to pulsed electron beams [107].

The question arises as to what mechanism can explain the observed pressure, thickness, and material dependence. A purely thermal initiation mechanism or a compressive shock initiation (see Chapter 7, Volume 2), resulting from nearly instantaneous energy deposition, can account for some but not all of the observations.

If it is assumed that all the energy deposited in the lead azide is used to heat the material instantaneously and that heat losses do not occur during the energy-deposition time, the maximum temperature rise is given by

$$T_{\text{max}} = T_{\text{ambient}} + \frac{\text{Energy deposited (J/g)}}{C_p\,(\text{J/g}\ ^\circ\text{C})}$$

For an initiation dose of 151 J/g, an initial sample temperature of 23°C, and a heat capacity of 0.46–0.57 J/g °C, the maximum temperature for the lead azide is 288–349°C. The 5-sec explosion temperature reported for lead azide is about 340°C. Thus, a thermal initiation mechanism may account for the highest initiation dose responses, but the mechanism does not account for most of the data.

The response of the lead azide to the compressive shock waves resulting from the nearly instantaneous energy deposition may also account for some of the

Table XVII. Initiation of Lead Azide by Pulsed Electrons [107]

Type of lead azide	Sample: density (g/ml)	Sample: Thickness (in.)	Sample: Gas pressure	Responses produced: Number initiated	Responses produced: Number surviving	Average dose to lead azide (J/g): Highest surviving sample	Average dose to lead azide (J/g): Lowest initiated sample
Dextrinated	2.78	0.030	1 atm	1	8	124.6	150.7
	2.80	0.060	1	6	9	92.1	83.7
	2.82	0.090	1	4	11	100.5	92.1
	2.80	0.060	2 μm	9	5	46.0	41.9
	2.82	0.090	2	11	2	6.7	10.0
RD1333	2.95	0.060	1 atm	3	9	92.1	96.3
	2.95	0.060	2 μm	3	0	0	23.0

observations, but this mechanism does not account for the observed initiations at the low doses. A calculation of the stress wave history of a sample exposed to the electron beam was made using the electron-beam pulse width, the energy-deposition profile, and the equation-of-state parameters for the materials irradiated [108]. A compressive shock initiation mechanism could account for initiations of samples receiving doses in excess of 84 J/g. Electric-field-induced initiation [109] is an alternative mechanism which may account qualitatively for all the observed results. During energy deposition from the electron beam, charge deposition occurs throughout the sample and produces electric fields internally. Charge-deposition curves roughly similar to the energy-deposition profiles for the sample geometries can be calculated in principle. However, the absence of radiation-induced conductivity data for the materials prevents calculation of the critical electric field(s) at the observed initiation-dose thresholds. The initiation data can be explained qualitatively by the proposed electric-field initiation mechanism. The mechanism is applicable to photon as well as electron irradiations, since photon irradiation can produce nonzero charge distributions (and electric fields) in dielectric materials.

The electric-field-induced initiation concept is related to pulsed radiation. If a determination of a critical electric field is made in the absence of radiation, it is probably not relevant to this area of radiation-induced effect since the critical field for an ionized material is likely to be very different from that for a nonionized material (see Chapter 9, Volume 1).

6. Other Types of Nuclear Irradiation

Other radiations used to induce changes in different chemical and physical properties of explosive azides have included protons, deuterons, ultraviolet and visible light, and lasers.

In Figure 8 the range–energy relations for α-particles and protons were plotted. The ranges for protons were calculated by Cerny *et al.* [41] as listed in Table XVIII and, as stated previously, the α-particle ranges were obtained from the established proton ranges.

The experimental results of Ryabykh *et al.* [93] on single crystals of lead azide, pure and containing Ag^+ and Cu^{2+}, indicate that proton irradiation produces a more rapid radiolysis than X-ray irradiation. With an energy of 4.8 MeV the protons were entirely retained by the largest crystals, permitting the range of protons to be measured directly (~0.7 mm). During the experiments it was found that the sensitivity of PbN_6 crystals to radiation increases with aging. The yield of nitrogen from crystals six months old was more than twice that from fresh crystals. As with X-irradiation the addition of Ag^+ accelerated radiolysis, while the reverse was evident with Cu^{2+}.

Crystalline azides, bromates, and nitrates of alkali metals were irradiated by

Table XVIII. Ranges of Protons in Various Explosives [21]

Proton energy (MeV)	Range (mg/cm²)						
	RDX or HMX	TNT	PETN	Tetryl	Lead styphnate	Mercury fulminate	Lead azide
0.020	0.069	0.070	0.073	0.072			
0.040	0.10	0.10	0.10	0.10			
0.050					0.22	0.35	0.34
0.060	0.12	0.13	0.13	0.13			
0.080	0.15	0.15	0.15	0.15	0.29	0.45	0.44
0.100	0.17	0.18	0.18	0.18	0.33	0.52	0.51
0.150	0.24	0.24	0.25	0.25	0.44	0.69	0.67
0.211	0.31	0.31	0.32	0.32	0.56	0.86	0.85
0.250	0.40	0.40	0.41	0.41	0.69	1.06	1.05
0.300	0.49	0.49	0.50	0.50	0.85	1.27	1.27
0.400	0.71	0.71	0.72	0.72	1.19	1.75	1.76
0.500	0.96	0.96	0.98	0.98	1.58	2.31	2.33
0.600	1.25	1.25	1.27	1.27	2.03	2.94	2.97
0.700	1.57	1.57	1.59	1.60	2.52	3.63	3.68
0.800	1.92	1.92	1.95	1.96	3.06	4.39	4.45
0.900	2.30	2.31	2.33	2.35	3.64	5.20	5.28
1.00	2.71	2.72	2.75	2.76	4.27	6.06	6.17
1.50	5.15	5.19	5.22	5.23	7.95	11.1	11.4
2.00	8.22	8.32	8.36	8.37	12.5	17.5	17.8
3.00	16.3	16.4	16.6	16.6	24.1	33.3	34.1
5.00	39.5	39.7	40.0	40.1	57.1	76.7	78.9
7.00	71.5	72.0	72.6	72.8	102	134	138
10.00	135	136	137	137	190	244	251
13.00	216	217	219	220	300	382	392
16.00	315	316	318	319	433	545	560
19.00	430	431	433	435	587	733	753
22.00	561	562	564	567	761	944	969
25.00	707	708	711	715	957	1177	1209

4.7-MeV protons and also by ^{60}Co gammas by Oblivantsev *et al.* [110] and Boldyrev [111], and their radiation stabilities were compared to their thermal stabilities. Also determined were the range and loss of energy of 4.7-MeV protons in alkali azides.

Berberet [74] attempted to produce ignitions with 12- and 24-MeV deuterons in samples of Composition B and HMX/Exon instrumented with microthermocouples. Ignitions were achieved only at elevated temperatures. The thermal behavior indicated by the thermocouples suggested that stored energy as well as the elevated temperature played a role in the ignitions. There was evidence of a dependence of the ignition on the total exposure rather than on temperature.

The action of UV radiation on lead azide was reported by Garner and Gomm

[112], who found it was not possible to detonate lead azide crystals by UV light from a mercury vapor lamp. However, the crystals turned black from the irradiation and were more sensitive to heat. Muraour [33] tried unsuccessfully to initiate nitrogen iodide as well as other explosives with UV rays as well as with α- and X-rays.

UV irradiation of other azides was conducted by Müller and Brous on sodium azide [99], by Garner and Maggs on barium and strontium azides [113], by Mott on metal azides [114], and by Boldyrev and Skorik on silver and barium azides [115]. Sodium, strontium, and barium azides are decomposed by UV light at room temperature, and their thermal decomposition is accelerated by preirradiation. Boldyrev *et al.* found that irradiating silver azide with UV light or X-rays at the instant of decomposition had no effect on the rate of its thermal decomposition.

Dodd [116] found that UV radiation can partially decompose lead azide and that wavelengths shorter than 3200 Å are required.

The effect of UV radiation on the sensitivity of lead azide was investigated by Abel and Levy [117], who observed that changes occurred in the impact and thermal sensitivity of colloidal material (Figure 19).

More recent experiments involving UV and X-ray irradiations were performed by Wiegand [118], who determined the changes in the optical properties of lead and thallous azides, and are discussed in Chapter 7, Volume 1.

The initiation of azides and other sensitive primary explosives by means of a high-intensity, short-duration light pulse has been studied extensively during the past 25 years. Eggert and coworkers [119–124], Bowden and Yoffe [32], and Roth [125] found critical initiation energies that depend on the explosive, its

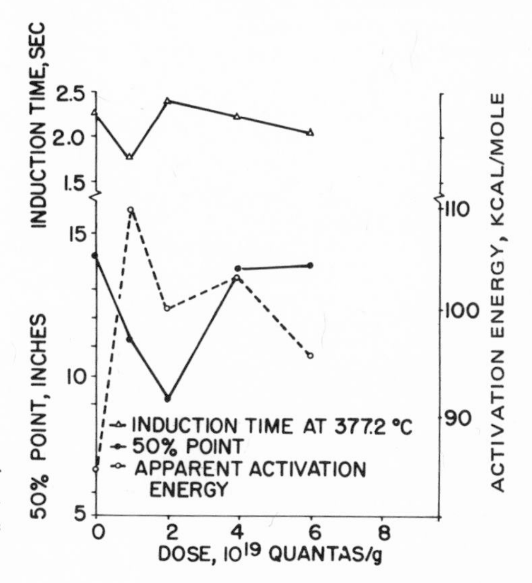

Figure 19. Effect of ultraviolet irradiation on induction time, 50% point, and activation energy of colloidal lead azide [117].

initial temperature, the flash duration, and less clearly on the absorption coefficient of the sample.

Materials studied by Eggert and his colleagues are listed in Table XIX. Eggert [123] and Courtney-Pratt and Rogers [126] suggested that the initiation of explosion by light is thermal. The azide is excited electronically by the absorption of light in a thin surface layer of the crystal, and the energy is degraded to heat in 1-50 μsec. The subsequent explosion then occurs by a normal thermal mechanism.

However, Deb *et al.* [127] indicated, at least for TlN_3, AgN_3, and CuN_3, that the initiation of decomposition is photochemical, with the growth to explosion being thermal. They suggested that the initial step is the formation of an

Table XIX. Minimum Ignition Energies of Explosives by Light Flashes [119, 123]

Compound	Formula	Color	Thermal ignition temperature[a] (°C)	Electrical energy of flash, half-life 0.8 msec (J)	Light intensity (J/cm²)
Nitrogen iodide	$NI_3 \cdot NH_3$	Brown-black	50	19	0.16
Silver nitride	Ag_3N	Black	100	24	0.20
Cuprous acetylide	Cu_2C_2	Brown-black	120	75	0.63
Silver acetylide	Ag_2C_2	White	165	95	0.79
Silver acetylide	$Ag_2C_2 \cdot AgNO_3$	Yellow-white	225	230	1.9
Mercurous acetylide	Hg_2C_2	Light grey	280	>350	>2.8
Mercuric acetylide	HgC_2	White	260	180	1.5
Silver azide	AgN_3	White	250	310	2.6
Lead azide	$Pb(N_3)_2$	Yellow-white	350	240	2.0
Mercurous azide	$Hg_2(N_3)_2$	White		>300	
Mercuric azide	$Hg(N_3)_2$	White	270	310	2.6
Silver fulminate	AgONC	White	170	250	2.1
Mercuric fulminate	$Hg(ONC)_2$	Light grey	190	200	1.65
Benzene diazonium nitrate	$C_6H_5N_3O_3$	Yellow	90	110	0.92
Benzene diazonium perchlorate	$C_6H_5N_2ClO_4$	White	155	110	0.92
p-Diazo-diphenyl-aminoperchlorate	$C_{12}H_{10}N_3ClO_4$	Yellow	170	95	0.79
Ammonium perchromate	$(NH_4)_3CrO_8$	Red-brown	90	135	1.1
Silver oxalate	$Ag_2C_2O_4$	White	140	>300	

[a] Small quantities placed in a heated aluminum block and exploded within 30 sec.

azide radical and an electron in the conduction band

$$N_3^- \underset{}{\overset{h\nu}{\rightleftharpoons}} N_3 + e$$

followed by the exothermic steps

$$2N_3 \longrightarrow 3N_2 + 0.879 \text{ MJ}$$

and

$$M_n + M^+ + e \longrightarrow M_{n+1}$$

where M represents the metal forming the azide.

The question whether initiation by light is primarily a thermal or a photochemical process has been pursued on an analytical and experimental basis. Blanchard [128] assumed initiation by monochromatic light to be thermal, without photochemical effects. Deriving differential equations relating temperature, time, and distance below the surface explosive crystal to specific heat (C), density (ρ), thermal conductivity (K), heat of reaction, and activation energy (E_a) and applying these to PbN_6 and AgN_3, a relationship was obtained between the intensity of illumination and ignition delay that agreed with experimental data. The variation of the specific heat with temperature was not taken into consideration. Experimental data were not available for the absorption coefficients for monochromatic light. The simplified equation of heat flow used was as follows:

$$C_p \rho \frac{\partial T}{\partial z} = K \frac{\partial^2 T_2}{\partial z} + \rho Q A e^{-E_a/RT} + \alpha E_0 e^{-\alpha z}$$

where α is the absorption coefficient, z the thickness coordinate of the crystal, and E_0 the incident light flux per second.

Roth [125] made measurements of the ignition delay for different absorbed energies by initiating PVA lead azide pellets with light from an argon flash bomb. The results indicated that the product of the rate of energy absorption and the ignition delay is constant and that ignition is thermal in origin; for the experimental conditions used the photochemical reactions were unimportant. Light absorption was believed to raise the azide surface above a critical temperature, leading to thermal explosion and detonation. The ignition delays were of the order of 1 μsec and the critical temperature ~900°C.

Courtney-Pratt and Rogers [126] studied the ignition of silver azide crystals by high-intensity, ultraviolet light flashes and determined the critical initiation energy to be 0.33 J/cm^2. The ignition delay for silver azide was estimated to be 15-20 μsec [32].

Bowden and coworkers concentrated their efforts on the monovalent in-

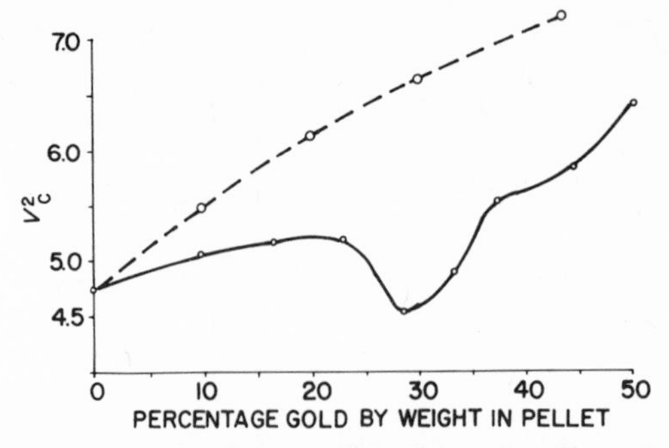

Figure 20. The sensitization of silver azide to light due to the effect of gold particles [127]. The dotted line corresponds to ignition for equal intensities; the solid line shows the experimental curve. V_c is the charging voltage in kV; capacity, 8 μF.

organic azides and derived an order of sensitivity to light initiation: $KN_3 < TlN_3 < AgN_3 < CuN_3$. Lead azide and cuprous azide were exploded by a high-intensity light flash, but required a lower flash energy for ignition than silver azide.

Azides can be sensitized to light by the introduction of foreign anions [127, 128] and by solid impurities. Metal particles such as gold (Figure 20) sensitize silver azide, but HgI_2 desensitizes it (Table XX).

The effect of crystal size on the critical light energy and ignition delay was studied by Rogers and reported by Bowden and Yoffe [32] for silver azide and by Roth [125] for lead azide. Each found that crystal size had no effect. Roth further reported that the density range 1.6-3.2 g/ml and the particle size range 0.5-10 μm had no effect on ignition delay. Formvar and 1-4 min of sunshine also did not affect the ignition delay.

Table XX. Light-Flash Ignition Energies for Azides and Mixtures [127]

Explosive	Critical flash ignition energy (J)
PbN_6	28.8
PbN_6 + 0.1% graphite	32.0
PbN_6 + 0.35% graphite	34.0
PbN_6 + 1.65% graphite	52.0
AgN_3	48.0
AgN_3 + 10% PbI_2	62.0
AgN_3 + 10% HgI_2	72.0
TlN_3	92.2
TlN_3 + 7% Tl_2CN_2	<200

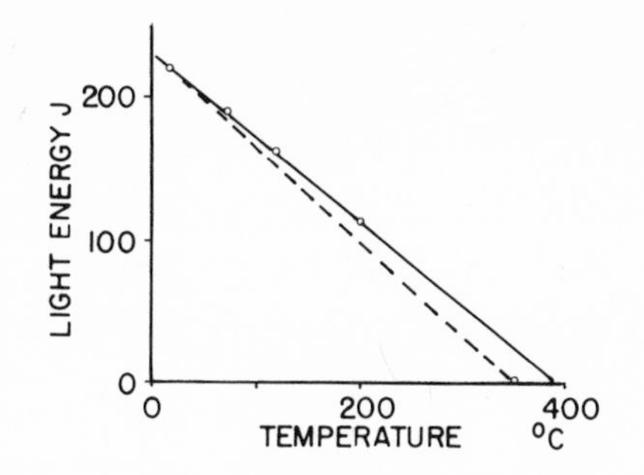

Figure 21. The relationship between light energy for ignition and temperature for lead azide [119]. The solid line gives the experimental curve; the dotted line connects the value for the ignition energy at 20°C and the point on the temperature axis corresponding to the thermal ignition temperature.

The effect of temperature on the light energy required was studied by Berchtold and Eggert [119] with lead azide (Figure 21) and silver acetylide, and by Meerkämper [124] with nitrogen iodide. Berchtold and Eggert regarded their data as further evidence for the thermal initiation mechanism. However, Bowden pointed out that the heating rate was not considered, i.e., for thermal ignition a delay of 30 sec was required, while in photoinitiation the delay may be of the order of microseconds.

Another compound, silver nitride, is a black material and consequently absorbs light over the whole visible spectral range [119]. There was no evidence that silver nitride is more sensitive to light of a particular wavelength, the minimum critical energy for ignition being identical for visible, ultraviolet, or infrared radiation. Berchtold and Eggert found that the energy required was dependent on the duration of the incident flash:

$$E = 8\sqrt{t}$$

where E is the energy of the incident light (J/cm^2) for initiation and t the duration of the flash in seconds.

With nitrogen iodide, Meerkämper [124] found that the incident energy to be inversely proportional to the ignition delay:

$$Et = \text{constant} = 1.35 \times 10^{-6} \text{ J sec/cm}^2$$

Lead styphnate and mercury fulminate can be initiated by light but show two requirements for propagation. They either burn with a velocity of a few meters per second or detonate. The minimum critical ignition energy for lead styphnate monohydrate crystals covered with a quartz plate is 29 J [32]. If a water filter was used, the value was 31 J, showing that infrared radiation plays little part in the initiation, the temperature rise being produced by absorption of the blue or ultraviolet light emitted by the flash.

The minimum ignition energy for lead styphnate is also a linear function of ambient temperature (Figure 22) [32]. As in Figure 21, extrapolation of the data to zero light energy gives a value (480°C in this case) which is higher than

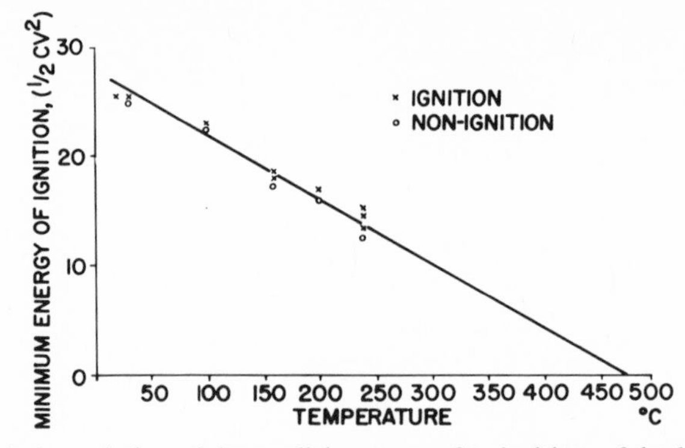

Figure 22. Variation of the minimum light energy for ignition of lead styphnate with ambient temperature [32].

the "dark" ignition temperature. However, the 5-sec explosion temperature is 250°C, so the value of 480°C for a time <10 μsec is not unreasonable.

Solid covalent organic azides and nitrates have not been exploded by an intense light, but molten explosives such as PETN explode with sufficient light energies. PETN ignites just above the melting point if the energy of the flash is 480 J and the time duration is 20 μsec. Nitroglycerin also decomposes during irradiation with ultraviolet light, and at 100°C it can be made to explode with a flash of 900 J. Khlevnoi and Kalmykova [129] subjected nitroglycerin powder with 1% lampblack to a light flux as a function of ignition time, 167 J/cm^2 for 30 sec. Mikheev *et al.* [130] obtained similar results with ballistite propellant and 1% lampblack.

The initiation of condensed explosives by a laser beam has been reported by several investigators [131–134]. Brish *et al.* [131] used a neodymium laser in the Q-switched mode to initiate detonations in lead azide and PETN. The explosives were pressed in transparent plastic shells to a density of about 1 g/ml. The laser pulse had a power up to 10 MW, a duration of 0.1 μsec (energy 0.5 J), and a beam diameter of 15 mm. The power density on the surface of the lead azide reached 0.08 MW/mm^2 (energy 0.1 J), but the ignition delay varied from 10^{-8} to 10^{-4} sec, while the energy required remained constant [134]. Above 10^{-4} sec the energy increased with ignition delay [135]. This effect was not observed for PETN for which a much higher density was required.

Vollrath [132] and L'Ast [102] reported the behavior of primary explosives with a pulsed-neodymium-glass (10,600 Å) laser of 100 MW, emitting 30-nsec flashes. With a maximum energy of 3 J, diazodinitrophenol burned and potassium dinitrobenzofuroxan deflagrated or detonated. Lead and cadmium azides and lead styphnate gave consistent results:

	Average energy densities	
	No reaction (mJ/cm^2)	100% Initiation (mJ/cm^2)
Lead azide	1.5	10
Cadmium azide	30	110
Lead styphnate	100	1200

Experiments were also conducted using a ruby laser of 50 MW emitting 25-nsec flashes at 6943 Å. Vollrath reported that lead and cadmium azides are less sensitive at 6943 Å than at 10,600 Å, while lead styphnate is more sensitive.

Kessler [133] conducted experiments with Q-switched and unswitched ruby lasers and found that initiation was dependent on the rate of energy application and that explosives activated by Q-switching had significantly shorter ignition delays. At a distance of 3 ft lead styphnate is initiated with 0.25 J with an unfocused laser. With lead azide, focusing was required and ignition occurred at 0.005 J in a focused area of 0.007 cm^2. Using a collimated laser beam and 0.3 J, red safety matches were ignited at 80 ft and blue safety matches at 120 ft, showing the increased absorption of the red laser radiation by the blue surface. Dextrinated lead azide was initiated at 250 ft.

Brish *et al.* [134] considered possible mechanisms of laser initiation, including shock caused by the light, pressure, electrical breakdown, and photochemical ion and thermal processes. An analysis of the experimental results for lead azide and PETN indicated that only the thermal theory of initiation could explain the data. The light energy is converted into shockwave energy which initiates the explosives.

Other investigators [135-140] have studied the initiation of secondary explosives by laser radiation. With direct initiation of secondary explosives, for example, by incorporating a fiber optic cable interfacing the main charge, primary explosives become unnecessary, thus providing increased safety and reliability in explosive trains.

D. SUMMARY AND CONCLUSIONS

The sensitivity of azides to heat is one of their properties which can be most precisely determined. The more practically useful substances, such as lead and silver azides, do not detonate until temperatures close to or at their melting points are attained. Among technologically important solid explosives such as TNT, tetryl, and RDX, the relatively high melting points of lead and silver azides (<300°C) and the good vacuum stability in standard tests are perhaps not representative of their overall sensitivity. Once a threshold temperature has been attained in the azides, the transition from slow decomposition to detonation is extremely rapid and requires only a few milligrams of sample. Nevertheless, the comparatively good thermal stability of lead and silver azides is a property which makes them attractive for use not only as primary explosives but as constituents of explosive trains.

While not a matter peculiar to azides, the understanding of the thermal behavior of these substances will continue to benefit from more precise and standardized tests. In particular, it would be beneficial to develop measurement techniques to quantify detailed reaction kinetics at temperatures close to the threshold for initiation. Even more informative would be investigations which combine the effects of heat with high pressures and differing confinements in this temperature region.

Of the many explosives subjected to nuclear radiation, the number that can be made to detonate by such stimuli has increased as higher intensities of particles have become available for experimentation. The azides do not show any unusual sensitivities in this respect, with the exception of nitrogen iodide, which explodes under the action of low-intensity α-particles and fission fragments. However, because the substance explodes spontaneously at room temperature when dried thoroughly, it is probably anomalous, even among azides, with respect to radiation sensitivity. Heavy ions of mercury and hydrogen ignite not only lead azide and mercury fulminate but also TNT and nitrocellulose. Reactor environments will initiate lead styphnate, tetryl, and lead azide mixed with boron. Protons and deuterons can ignite 60 : 40 RDX/TNT and HMX but only at elevated temperatures.

Electron beams can initiate lead azide, among other primary explosives, although there are cases of holes being burned in azide particles in electron microscope beams without ignition occurring. The mechanism of ignition is usually interpreted to be thermal. However, intense pulsed beams of electrons present a more complex situation in which electric fields, induced at sample interfaces, have been postulated to cause initiations below the thermal ignition threshold. Good correlations between the behavior of azides (and other explosives) under steady and pulsed irradiations are, however, not yet generally possible.

The detailed, slow photolysis of azides by UV- or X-irradiation is described in Chapter 7, Volume 1. Direct ignition or initiation of azides by light flashes has normally been interpreted in terms of a thermal mechanism, perhaps preceded by a photochemical step. More sophisticated approaches involving the careful selection of the wavelength and the application of an electric field are described in Chapter 9, Volume 1. These indicate much lower thresholds for ignition than those predicated on a thermal mechanism.

The ignition of azides and other explosives is possible with laser beams of nonoptimized wavelength. A thermal mechanism of ignition is again the usual interpretation of data, but the precise role of laser-induced shock waves in these experiments has not been determined. Continuation of these investigations is desirable because of the potential of lasers for direct initiation of secondary explosives or ignition of propellants. The trend toward the development of smaller, battery-operated lasers with tunable outputs will enhance the practical usefulness of this approach to azide replacement. In the meantime, data on the effects of radiation on azides are essential aids to defining the shelf-life and vulnerability of explosive trains under more exacting conditions.

REFERENCES

1. G. H. Henderson, *Nature*, *109*, 749 (1922).
2. V. R. Pai Verneker, J. N. Maycock, *Anal. Chem.*, *40*, 1325 (1968).
3. J. N. Maycock, *Thermochim. Acta*, *1*, 389 (1970).
4. A. R. Ubbelohde, *Phil. Trans. R. Soc. London*, *A241*, 1981 (1949).
5. H. Henkin, R. McGill, *Ind. Eng. Chem.*, *44*, 1391 (1952).
6. B. Reitzner, J. V. R. Kaufman, E. F. Bartell, *J. Phys. Chem.*, *66*, 421 (1962).
7. J. Haberman, T. C. Castorina, *Thermochim. Acta*, *5*, 153 (1972).
8. B. T. Fedoroff, *Encyclopedia of Explosives and Related Items*, Vol. 1, 1960, p. A559, Picatinny Arsenal, Dover, N.J.
9. F. A. Baum, K. P. Stanyukovich, B. I. Shekhter, *Physics of an Explosion*, translation of *Fizmatgiz*, Moscow, by Research Information Service, New York, 1959.
10. B. L. Evans, A. D. Yoffe, *Proc. R. Soc. London*, *A238*, 568 (1957).
11. W. R. Tomlinson, Jr., O. E. Sheffield, Properties of Explosives of Military Interest, Tech. Rept. No. 1740, Revision 1, Picatinny Arsenal, Dover, N.J., 1958.
12. A. D. Yoffe, *Proc. R. Soc. London*, *208*, 188 (1951).
13. A. S. Hawkes, C. A. Winkler, *Can. J. Res.*, *25B*, 548 (1947).
14. B. L. Evans, A. D. Yoffe, *Proc. R. Soc. London*, *A250*, 346 (1959).
15. L. Wöhler, F. Martin, *Z. Angew. Chem.*, *30*, 33 (1917).
16. K. Singh, *Ind. J. Chem.*, *7*, 694 (1969).
17. R. S. Mulliken, *J. Phys. Chem.*, *56*, 801 (1952).
18. T. C. Castorina, J. Haberman, E. W. Dalrymple, A. Smetana, A Modified Picatinny Arsenal Explosion Temperature Test for Determining Thermal Sensitivity of Explosives Under Controlled Vapor Exposures, Tech. Rept. 3960, Picatinny Arsenal, Dover, N.J., 1968.
19. B. Reitzner, *J. Phys. Chem.*, *65*, 948 (1961).
20. M. M. Jones, H. Jackson, *Explosivstoffe*, *9*, 2 (1959).
21. M. M. Chaudri, J. E. Field, *J. Solid State Chem.*, *12*, 72 (1975).
22. W. H. Andersen, *I. E. C. Proc. Des. Dev.* *4*, 286 (1965).
23. A. J. Clear, Standard Laboratory Procedures for Determining Sensitivity, Brisance, and Stability of Explosives, Tech. Rept. No. 3278, Revision 1, Picatinny Arsenal, Dover, N.J., 1970.
24. N. E. Beach, V. C. Canfield, Compatibility of Explosives with Polymers (III), Plastec Report 40, Plastics Technical Evaluation Center, Picatinny Arsenal, Dover, N.J., 1971.
25. R. J. Graybush, F. G. May, A. C. Forsyth, *Thermochim. Acta*, *2*, 153 (1971).
26. W. L. Shimmin, J. Huntington, L. Avrami, Radiation and Shock Initiation of Lead Azide at Elevated Temperatures, Air Force Weapons Laboratory Report AFWL TR71-163 (PIFR-308), Albuquerque, N.M., 1972.
27. T. Costain, private communication.
28. H. H. Poole, *Nature*, *110*, 148 (1922).

29. H. H. Poole, *Sci. Proc. R. Soc.*, *17*, 93 (1922).
30. H. Kallmann, W. Schränkler, *Naturwissenschaften*, *21*, 379 (1933).
31. H. Muraour, *Chim. Ind.*, *30*, 39 (1933).
32. F. P. Bowden, A. D. Yoffe, *Fast Reactions in Solids*, Academic Press, New York, 1958.
33. H. Muraour, *Soc. Chim. Fr. Bull.*, *53*, 612 (1933).
34. W. E. Garner, C. H. Moon, *J. Chem. Soc.*, *1933*, 1398 (1933).
35. M. Haissinsky, R. J. Walen, *Compt. Rend.*, *208*, 2069 (1939).
36. A. C. McLaren, Ph.D. thesis, Cambridge University (1957); data in reference [32] p. 119.
37. F. P. Bowden, K. Singh, *Nature*, *172*, 378 (1953).
38. F. P. Bowden, K. Singh, *Proc. R. Soc. London*, *A227*, 22 (1954).
39. F. P. Bowden, *Proc. R. Soc. London*, *A246*, 216 (1958).
40. R. C. Long, Penetration of Explosives by Heavy Nuclear Particles, Tech. Rept. 2393, Picatinny Arsenal, Dover, N.J., 1957.
41. J. Cerny, M. S. Kirshenbaum, R. C. Nichols, *Nature*, *198*, 371 (1963).
42. E. Feenberg, *Phys. Rev.*, *55*, 980 (1939).
43. P. Fabry, C. Magnor, H. Muraour, *Compt. Rend.*, *209*, 436 (1939).
44. F. P. Bowden, H. M. Montagu-Pollock, *Nature*, *191*, 556 (1961).
45. F. P. Bowden, L. T. Chadderton, *Nature*, *192*, 31 (1961).
46. F. P. Bowden, A. D. Yoffe, *Endeavour*, *21*, 125 (1962).
47. H. M. Montagu-Pollock, *Proc. R. Soc. London*, *A269*, 219 (1962).
48. F. P. Bowden, L. T. Chadderton, *Proc. R. Soc. London*, *A269*, 143 (1962).
49. F. P. Bowden, The Initiation and Growth of Explosion in the Condensed Phase, *Proceedings of the 9th International Symposium on Combustion*, Academic Press, New York, 1963, pp. 499–516.
50. J. Cerny, J. V. R. Kaufman, *J. Chem. Phys.*, *40*, 1736 (1964).
51. J. F. Mallay, H. J. Prask, J. Cerny, *Nature*, *203*, 473 (1964).
52. H. Muraour, A. Ertaud, *Compt. Rend.*, *237*, 700 (1953).
53. Initiation of Explosives by Radiant Energy or Energetic Particles, Armour Research Foundation, Project No. A058, Final Report 29, Illinois Institute of Technology Research Institute, AD-106,621, Chicago, Ill., March 1956.
54. J. K. Raney, *Bull. Am. Phys. Soc.*, *3*, 117 (1958).
55. T. B. Flanagan, *Nature*, *181*, 42 (1958).
56. J. M. Groocock, *Proc. R. Soc.*, *A246*, 225 (1958).
57. J. E. Abel, R. W. Dreyfus, J. V. R. Kaufman, P. W. Levy, Effects of Reactor and Gamma Ray Irradiation on the Sensitivity of Lead Azide, Tech. Rept. 2500, Picatinny Arsenal, Dover, N.J., 1958.
58. J. E. Abel, R. W. Dreyfus, J. V. R. Kaufman, H. Rosenwasser, P. W. Levy, The Effect of Nuclear Radiation on Explosives and Related Materials, Tech. Rept. 2398, Picatinny Arsenal, Dover, N.J., 1957.
59. J. Jach, *Faraday Soc.*, *59*, 947 (1963).
60. J. Jach, in: *Studies in Radiation Effects*, Vol. 2, G. J. Dienes, ed., Gordon and Breach, New York, 1966.
61. M. J. Urizar, E. D. Longhran, L. C. Smith, *Explosivstoffe*, *3*, 55 (1962).
62. L. Avrami, W. E. Voreck, A Determination of Reactor-Radiation-Resistant Explosives, Propellants, and Related Materials, Tech. Rept. 3782, Picatinny Arsenal, Dover, N.J., 1969.
63. A. Mackenzie, E. W. Dalrymple, *Nucleonics*, *22*, 85 (1964).
64. A. Mackenzie, E. W. Dalrymple, F. Schwartz, Design of Pressure Vessels for Confining Explosives, Tech. Memo. 1643, Picatinny Arsenal, Dover, N.J., 1965.

65. L. Avrami, E. Dalrymple, F. Schwartz, A Large Containment Capsule for Nuclear Reactor Irradiation of Explosive Materials, Tech. Rept. 3673, Picatinny Arsenal, Dover, N.J., 1968.
66. W. E. Voreck, *Nucl. Appl.*, *6*, 582 (1969).
67. L. Avrami, H. J. Jackson, M. S. Kirshenbaum, Effects of Gamma Radiation on Selected Fluoroexplosives, Tech. Rept. 3942, Picatinny Arsenal, Dover, N.J., 1970.
68. L. Avrami, H. J. Jackson, O. Sandus, unpublished results.
69. S. R. Kurowski, E. F. Hartman, J. E. Gover, unpublished work conducted August 1973; L. Avrami, private communication.
70. J. R. Holden, Literature Survey on the Effects of Neutron and Electromagnetic Irradiation on Explosives, U.S. Naval Ordnance Laboratory NAVORD Report 4448, (AD-127340), White Oak, Md., 1957.
71. K. S. Warren, O. Sisman, Effects of Nuclear Radiation on Explosives, Oak Ridge National Laboratory Reports CF-50-4-103, April 1950; CF-50-8-54, August 1950; and CF-50-12-109, December 1950. Oak Ridge, Tenn.
72. K. S. Warren, D. E. Seeger, C. F. Dieter, O. Sisman, Effects of Gamma Radiation on Explosives, Oak Ridge National Laboratory Report ORNL-1720, Oak Ridge, Tenn., December 6, 1955.
73. J. V. R. Kaufman, *Proc. R. Soc. London*, *A246*, 219 (1958).
74. J. Berberet, The Effect of Nuclear Radiations of Explosive Solids, Air Force Systems Command, Aeronautical Systems Division Technical Report No. ASD-TDR-63-893. Eglin Air Force Base, Flda., 1963.
75. H. Rosenwasser, R. W. Dreyfus, P. W. Levy, *J. Chem Phys.*, *24*, 184 (1965).
76. R. W. Dreyfus, P. W. Levy, *Proc. R. Soc. London*, *A246*, 233 (1958).
77. P. J. Herley, P. W. Levy, Quantitative Studies on Radiation-Induced Dislocations and the Decomposition Kinetics of Ammonium Perchlorate, Proc. 7th International Symposium on Reactivity of Solids, J. S. Anderson, M. W. Roberts, F. S. Stone (eds.), Chapman and Hall, London, 1972, p. 387.
78. E. G. Prout, D. J. Moore, *Nature*, *203*, 860 (1964).
79. E. G. Prout, D. J. Moore, *Nature*, *205*, 1209 (1965).
80. L. Avrami, H. J. Jackson, M. S. Kirshenbaum, Radiation-Induced Changes in Explosive Materials, Tech. Rept. 4602, Picatinny Arsenal, Dover, N.J. (1973).
81. L. Avrami, H. J. Jackson, D. Wiegand, unpublished results.
82. P. Günther, L. Lepin, K. Andreev, *Z. Elektrochem.*, *36*, 218 (1930).
83. P. Günther, L. Lepin, K. Andreev, *Khim. Referat. Zh.*, No. 10-11, 120 (1940); *Chem. Abstr.*, *37*, 1271 (1943).
84. H. G. Heal, *Can. J. Chem.*, *31*, 1153 (1953).
85. H. G. Heal, *Trans. Faraday Soc.*, *53*, 210 (1957).
86. E. R. Johnson, *The Radiation-Induced Decomposition of Inorganic Molecular Ions*, Gorden and Breach, New York, 1970.
87. J. M. Groocock, T. R. Phillips, The Effects of High Energy X-Rays on the Thermal Detonation of Service Lead Azide, Royal Armament Research and Development Establishment Report (B) 7/56, Fort Halstead, Kent, England, 1956.
88. G. Todd, E. Parry, *Nature*, *181*, 260 (1958).
89. G. Todd, E. Parry, *Nature*, *186*, 543 (1960).
90. B. V. Erofeev, V. V. Sviridov, *Inst. Khim.* A.N.S.S.R., Nauchn. *Rabot.*, *5*, 113 (1956) (translated by U.S. Joint Publication Research Service, Picatinny Arsenal Technical Note FRL-TN-75, March 1962).
91. V. V. Sviridov, *Dokl. Akad. Nauk, Beloruss.* S.S.R., *2*, 291 (1958).
92. Y. A. Zakharov, S. M. Ryabykh, A. P. Lysykh, *Kinet. Katal.*, *9*, 679 (1968).
93. S. M. Ryabykh, A. P. Lysykh, Y. A. Zakharov, *Khim. Vys. Energ.*, *2*, 344 (1968).

94. S. M. Ryabykh, V. A. Meshkov, Y. A. Zakharov, *Izv. Vys. Khim. Tekh.*, *13*, 1558 (1970).
95. Y. A. Zakharov, S. M. Ryabykh, *Izv. Tomsk Politekh. Inst.*, *176*, 107 (1970).
96. Y. A. Zakharov, S. M. Ryabykh, V. D. Minakov, *Izv. Tomsk Politekh. Inst.*, *251*, 8 (1970).
97. J. Sawkill, *Proc. R. Soc. London*, *A229*, 135 (1955).
98. R. H. Müller, G. C. Brous, *J. Am. Chem. Soc.*, *53*, 2428 (1931).
99. R. H. Müller, G. C. Brous, *J. Chem. Phys.*, *1*, 482 (1933).
100. J. M. Groocock, F. C. Tompkins, *Proc. Roy. Soc.*, *A223*, 267 (1954).
101. J. M. Groocock, The Decomposition of Alpha-Lead Azide, Second Office of Naval Research Symposium on Detonation (AD-52144), Silver Spring, Md., 1960.
102. L. d'Ast, Initiation des Explosifs d'Amorage par Electron et par Photons, Atelier de Construction de Tarbes, Tarbes, France (Clearinghouse, U.S. Dept. of Commerce N69-13509), 1968.
103. A. Roginski, *Phys.* Z. Sowjetunion, *1*, 656 (1932).
104. J. A. Berberet, The Effects of Nuclear Radiations on Explosive Solids, U.S. Air Force Systems Command Technical Report No. ATL-TDR-64-53, Eglin Air Force Base, Fla., 1964.
105. F. A. Farber, The Effect of a High Energy Electron Beam on Explosives (AD-771527), Master's thesis, Naval Postgraduate School, Monterey, Calif., December 1972.
106. P. V. Phung, *J. Chem. Phys.*, *53*, 2906 (1970).
107. L. Avrami, F. G. Borgardt, C. F. Kooi, J. F. Riley, Initiation of Lead Azide by Pulsed Electron Beam, Proceedings of International Conference on Research in Primary Explosives, March 17–19, 1975, eds. J. M. Jenkins, J. R. White, Vol. 1–3, Proc. Exec., Ministry of Defence, Expl. Res. and Dev. Est., Waltham Abbey, Essex, England (AD B013628-30).
108. F. W. Davies, A. B. Zimmerschied, F. G. Borgardt, L. Avrami, The Hugoniot and Shock Initiation Threshold of Lead Azide, *J. Chem. Phys. 64*, 2295 (1976).
109. This mechanism of initiation previously has been recognized by J. E. Gover, E. F. Hartman and co-workers at Sandia Laboratories, Albuquerque, N.M., private communications.
110. A. N. Oblivantsev, V. M. Lykhin, V. V. Boldyrev, L. P. Eremin, V. F. Komarov, *Izv. Tomsk Politekh. Inst.*, *176*, 122 (1970).
111. V. V. Boldyrev, *Int. J. Radiat. Phys. Chem.*, *3*, 155 (1971).
112. W. E. Garner, A. S. Gomm, *J. Chem. Soc.*, *1931*, 2123 (1931).
113. W. E. Garner, J. Maggs, *Proc. R. Soc. London*, *A172*, 299 (1939).
114. N. F. Mott, *Proc. R. Soc. London*, *A172*, 325 (1939).
115. V. V. Boldyrev, A. I. Skorik, A.N.S.S.R. Dokl, *156*, 1143 (1964).
116. J. G. Dodd, Jr., Univ. of Arkansas, Final Report Contract No. DA-44-009 ENG-2439, Explosives Research Development Laboratories, Ft. Belvoir, Va., 1956.
117. J. E. Abel, P. W. Levy, Effects of Ultraviolet Irradiation on the Sensitivity of Colloidal Lead Azide, Tech. Rept. 3174, Picatinny Arsenal, Dover, N.J. (1965).
118. D. A. Wiegand, *Phys. Rev.*, *B10*, 1241 (1974).
119. J. Berchtold, J. Eggert, Naturwissenschaften, *40*, 55 (1953).
120. J. Eggert, Behavior of Endothermic Compounds on Illumination with High Intensities of Light, III, Int. Conf. Phot., London, 1953; *Chem. Abstr. 47*, 3734j (1953).
121. J. Berchtold, Behavior of Silver Nitride Towards the Effects of High Intensities of Light, and Their Use for Producing Images, Int. Conf. Phot., London, 1953.
122. J. Eggert, *Proc. R. Soc. London*, *A246*, 240 (1958).
123. J. Eggert, *J. Phys. Chem.*, *63*, 11 (1959).
124. B. Meerkämper, *Z. Elektrochem.*, *58*, 387 (1954).

125. J. Roth, *J. Chem. Phys.*, *41*, 1929 (1964).
126. J. S. Courtney-Pratt, G. T. Rogers, *Nature*, *175*, 632 (1955).
127. S. K. Deb, B. L. Evans, A. D. Yoffe, Ignition and Sensitized Reactions in the Explosive Inorganic Azide, *Eighth Symposium on Combustion.* Williams and Wilkins, Baltimore, Md., 1960, pp. 829–836.
128. R. Blanchard, *Compt. Rend.*, *256*, 2550 (1963).
129. S. S. Khlevnoi, A. P. Kalmykova, *Fiz. Goreniya Vzryva*, *4*, 122 (1968).
130. V. F. Mikheev, A. A. Koval'skii, S. S. Khlevnoi, *Fiz. Goreniyz Vzryva*, *4*, 3 (1968).
131. A. A. Brish, I. A. Galeev, B. N. Zaitsev, E. A. Sbitnev, L. V. Tatarintsev, *Fiz. Goreniya Vzryva*, *2*, 132 (1966).
132. K. Vollrath, Sur L'Amorcage des Explosife Primaires à l'Aide de Faesceaux Laser (On the Initiation of Primary Explosives with the Aid of a Laser Beam), French-German Institut de Saint-Louis Note ISL 4/65, Saint Louis, France, 1965.
133. E. G. Kessler, The Initiation of Explosives by Means of Laser Radiation, Engineering Sciences Laboratory Internal Report ESL-IR-248, Picatinny Arsenal, Dover, N.J., 1966.
134. A. A. Brish, I. A. Galeev, B. N. Zaitsev, E. A. Sbitsev, L. V. Tatarintsev, *Fiz. Goreniya Vzryva, 4*, 475 (1969).
135. M. J. Barbarisi, E. G. Kessler, Initiation of Secondary Explosives by Means of Laser Radiation, Tech. Rept. 3861, Picatinny Arsenal, Dover, N.J., 1969.
136. V. J. Menichelli, L. C. Yang, Sensitivity of Explosives to Laser Energy, Jet Propulsion Laboratory Technical Report 32-174, Pasadena, Calif., 1970.
137. V. J. Menichelli, L. C. Yang, Direct Laser Initiation of Insensitive Explosives, Proc. 7th Symposium Explosives & Pyrotechnics III-2, 1-8, Franklin Inst. Res. Labs., 1971.
138. L. C. Yang, V. J. Menichelli, *Appl. Phys. Lett.*, *19*, 473 (1971).
139. V. J. Menichelli, L. C. Yang, Initiation of Insensitive Explosives by Laser Energy, Jet Propulsion Laboratory Technical Report 32-1557 (NASA-CR-126741), Pasadena, Calif., 1972.
140. L. C. Yang, V. J. Menichelli, J. E. Earnest, Laser Initiation of Explosive Devices, *National Defense Magazine*, 344–347, Jan.–Feb. 1974.

7

The Role of Azides in Explosive Trains

W. Voreck, N. Slagg, and L. Avrami

A. INTRODUCTION

The industrial and military importance of lead and silver azide lies in their usefulness for initiating the output charge in explosive trains and in the ability to initiate them with inexpensive, low-voltage, low-current electric sources or a minimum of mechanical or thermal energy. The function of an explosive train is to accomplish the controlled augmentation of the small impulse into one of sufficient energy to cause a main charge of explosive to function [1]. Lead and silver azides are among the most powerful and stable materials for the purpose, and are used in detonators, ignition-delay units, and primers; their function in detonators is perhaps representative of the whole technology. The sensitivity of the azides increases the hazards of detonator manufacture and requires out-of-line safety and arming systems for fuzes. Since detonation is achieved after a measurable burning or reaction buildup time, the materials are not normally used where extremely precise timing (better than ± 1 μsec) is required; however, they find extensive application in the wide range of civilian and military systems for which this creates no problem.

A typical detonator has three sections, the first containing a (nondetonating) "primer"; the initiator, lead or silver azide, is in the second section; and finally, a secondary explosive forms the third section. Normally, each of the sections is filled or "loaded" by pressing the explosive into a metal sleeve or cup under pressures of several thousand pounds per square inch. The next component or element in an explosive train is an acceptor charge, containing a secondary explosive or "lead"; the output from the secondary in the detonator is required to initiate the "lead" reliably (secondary explosive acceptor).

In many detonators lead azide is found in both the priming mixture and the initiator and is followed by a secondary explosive such as PETN, RDX, or HMX. In those items that respond to mechanical stimuli, a firing pin is driven into the primer, igniting it. The flame (or blast) from the primer spreads to the azide, which builds up to a detonation and in turn initiates the secondary explosive in contact with it. A similar sequence of events occurs with explosive trains which respond to other mechanical, electrical, or thermal stimuli. The satisfactory functioning of the azide (or other energetic material) in any component or element, requires, therefore, a carefully planned sequence of response, energy release, and energy transfer within the limited dimensions of the explosive train.

In general, the design of the elements of an explosive train, in terms of a controlled deposition of energy in an element, the conversion of energy within it, and the energy transfer to the next element, is not scientifically well-developed. Modern technology continues to draw heavily on past experience as to configurations that will function adequately, and new developments continue to be based on trial-and-error experimentation. The deficiencies of this approach in terms of the optimization of designs and their reliability have recently stimulated the development of improved diagnostic techniques to identify and measure those parameters that are important in functioning and to provide data on the performance of key substances. This chapter illustrates the recent trend of the developments, specifically as they relate to the role of the explosive azides. It summarizes data on the key functional parameters and discusses some of the fundamental questions which, remaining unanswered, continue to restrict the advance of the technology to empirical approaches.

The elements of explosive trains are, in regular use, required not only to function satisfactorily in response to a given stimulus, but also to survive the consequences of varying degrees of stimulus applied accidentally, by the ambient conditions, or in deliberate attempts at destruction. In general, the principles that determine the vulnerability of elements are closely related to those that govern satisfactory functioning; they are also related to the principles that determine the sensitivity, initiation, and hazards of azides, as discussed in the preceding chapters of this volume and Chapters 4–9 of Volume 1. This chapter discusses topics that relate uniquely to explosives-train technology or lie beyond the scope of earlier discussions.

Because precise measurement techniques have not been available to determine the response and performance of detonator elements, it is not possible to state unequivocally which parameters are most important for their most effective or optimum design. Potentially important parameters can, however, be derived from well-known principles that govern the behavior of solid explosives.

The parameters may be associated with the properties of the explosive as incorporated in the detonator, such as its composition, particle size, density (both in real terms and as a function of filling pressure), and with the volume, diameter and length of the charges (again both in absolute and relative terms). Other parameters may be associated with the materials and geometry of the explosive's confinement. All these "material" parameters will be reflected in the values for the "functional" parameters of the design: the sensitivity of the explosive mixtures to the applied stimulus; the distance and time for the reaction buildup; the deflagration or detonation velocity achieved in each mixture; and the heat, shock pressures, or other forms of energy produced and transferred to the succeeding element. Other important parameters may be derived from the explosive's stability during storage or during the buildup to its intended reaction regime.

The need for data on the parameters is neither unique to the application of azides and solid explosives nor to the design of explosive-train elements, and numerous references to measurement techniques and relevant data are to be found throughout these two volumes and in standard handbooks [1]. However, the data on functional parameters, particularly within the constraints of element designs, represent a unique requirement, particularly because of the hazardous nature of the materials, the rapidity of the reactions involved, and the small quantities and dimensions available for measurement probes. In Section D of this chapter some recent techniques of measurement or observation are presented to illustrate current trends. One of the important parameters affecting the initiation and growth of reaction in the explosives is their vulnerability to shocks. Some recent techniques for quantifying these parameters are given in Section E below.

Many of the advances in diagnostics and technology which are discussed in this chapter represent a distinct approach to the better definition of phenomena that are fundamental to the understanding of azides and of explosives in general. Thus behind more macroscopic observations and the discussions reported here there lie direct links back to the microscopic or molecular approaches discussed in Volume 1.

B. CLASSES OF IGNITERS AND INITIATION ELEMENTS

As indicated above, the technology of detonators and related primary elements is generally illustrative of the many-faceted role that azides play in explosive-train technology, with respect to their satisfactory functioning, their vulnerability, and the stimuli to which they are designed to respond. In fact, elements can be classified in terms of the stimuli and, in particular, in terms of their activation by thermal, electrical, or mechanical sources of energy. Figures 1 to 5 present typical military designs; these commonly reflect greater sophistication than civilian industrial designs, and this section gives a brief discussion of their intended function and of the materials incorporated. For a more detailed discussion of civilian and military designs, including manufacturing, safety, and functioning considerations, primers, delay elements, and leads, the reader is referred to standard handbooks [1].

1. Stab- or Percussion-Sensitive Elements

In stab and percussion detonators (Figure 1), motion of the firing pin initiates a deflagration in a sensitive primary mixture, such as NOL-130 (a mixture of 40% lead styphnate, 20% lead azide, 15% antimony sulfide, 20% barium nitrate, and 5% tetrazene) [2]. The upper illustration shows a typical stab detonator. Its function is to ignite rapidly and produce enough output to detonate a lead over a small gap (~0.127 cm). Such a detonator is mounted "out-of-line" in the train, while the lead is "in-line" with a booster charge. Lead azide is initiated rapidly by the primer and detonates with sufficient impulse to detonate the output charge (RDX).

The delay detonator (Figure 1b) uses lead azide as the output charge and converts and augments the heat from a gasless delay column to a detonation. The small amount of azide used in this example would require the use of a relay or flash detonator as the next element if the sequence is to terminate in the detonation of a secondary explosive. Otherwise the output from the 20 mg of lead azide is adequate only to ignite a sensitive pyrotechnic or another lead azide charge.

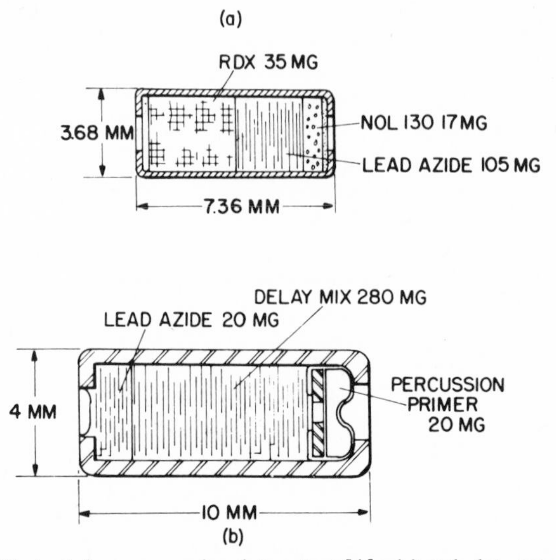

Figure 1. Lead azide in stab or percussion detonators [1]: (a) stab detonator; (b) percussion.

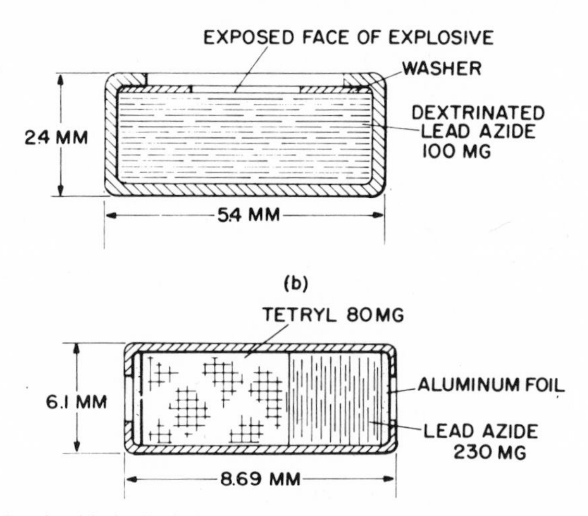

Figure 2. Lead azide in flash detonators and relays [1] : (a) relay; (b) detonator.

2. Heat- or Flash-Sensitive Elements

A typical flash detonator and relay are shown in Figure 2. These are designed to be initiated by a flame and produce a detonation output pulse. The input faces are either bare lead azide or have a thin paper covering to allow the flame to reach the azide. Again the rapid buildup in lead azide from heat to detonation is the important process involved.

3. Electrically Ignited Elements

In typical low-energy electric detonators (Figure 3) the higher sensitivity of lead styphnate to hot-wire initiation is used to minimize input energy requirements. A sensitive detonator can be made using a graphite bridge (Figure 3); however, the erratic energy requirements and the hazards associated with such sensitivity limit its use. The temperature-stable detonator (Figure 3b) takes advantage of the stability of lead azide, which passes a vacuum stability test at 200°C for 40 hr, but fails within one hour at 260°C [3,4].

Three conductive-mix detonators are shown in Figure 4. Since lead azide is an electrical insulator, a conductant is added, and flake conductants have been observed to be more effective than other particle shapes. A typical mixture contains 95% lead azide and 5% flake graphite. This type of detonator fires rapidly with low energy input; for example, the E.I. duPont de Nemours Company's product, designated X811, fires in 4 msec when initiated by a 2.2 μF

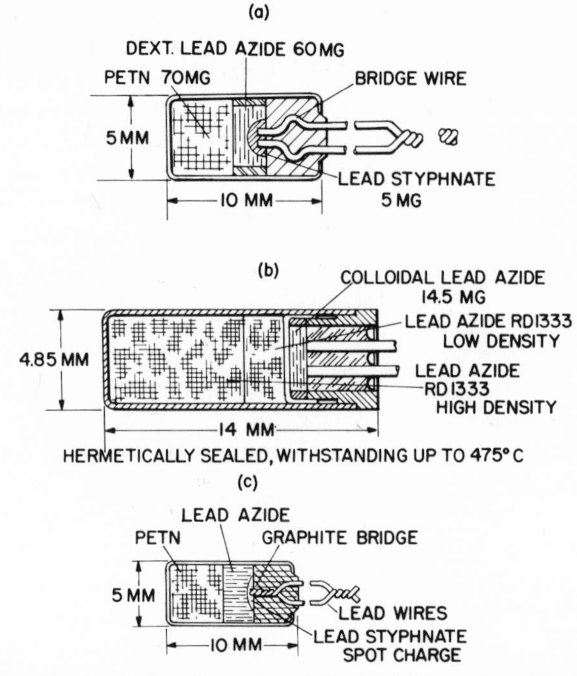

Figure 3. Lead azide in electric detonators [1]: (a) bridge wire; (b) high temperature; (c) graphite bridge.

capacitor charged to 14 V. However, conductive-mix detonators also occasionally exhibit wide variations in the energy required to initiate them, and the potential hazards associated with the phenomenon again limit their use. The electrical resistance and firing energy requirements of conductive mixtures can be easily varied over a wide range to meet initiation requirements, and the detonators are economical to manufacture because they are adaptable to mass production. The axial-gap design (Figure 4b), in particular, requires no costly electrical connections. High compaction pressures can be used without breaking "bridge wires," and for the same reason these detonators are more resistant to shock and vibration. Storage, particularly under extreme temperature variations, changes the electrical properties and may contribute to variations in the firing energy requirements.

Figure 4a shows a conventional annular-gap design, while Figure 4b shows an axial-gap design developed by Canadian Arsenal Ltd. [3]. It has two advantages over the annular-gap design in that it is less expensive to make, and the multiple paths for initiation make its resistance and firing requirements more uniform. A modification of the design developed by Picatinny Arsenal, Dover, New Jersey (Figure 4c), reduces the height of the train to 0.13 cm, exclusive of the initiation system. As will be seen in Figure 21, the minimum run distance in lead azide to initiate RDX reliably is 0.15 cm. In this detonator the output charge is, therefore, not initiated at the center, since the azide column height is only 0.038 cm. Instead, the detonation runs horizontally till it reaches the outer wall, where the pressure is doubled by reflection off the wall, and initiation of the output charge occurs in an annular ring around the outside. The radial detonation wave converges at the center of the HMX producing a high-energy jet which initiates a charge of composition A-5 over a gap of 0.31 cm. The 0.76-cm-diam, 0.25-cm-high detonator assembly is designed to fit in the top of an integrated circuit enclosure so that both the electronics and detonator can be combined in a miniature component. Initiation can be accomplished by hot wire or a heated resistive element on the integrated circuit chip, as well as by conductive mixtures.

The use of lead azide in one-ampere, no-fire detonator designs is shown in Figure 5. In these designs, the high-temperature stability of lead azide is needed if the detonators are not to fire with a 1-A current flow. To keep the azide below its initiation temperature, heat conduction paths are provided from a thin-film, resistive-bridge element (Figure 5b) through the electrical connections and the

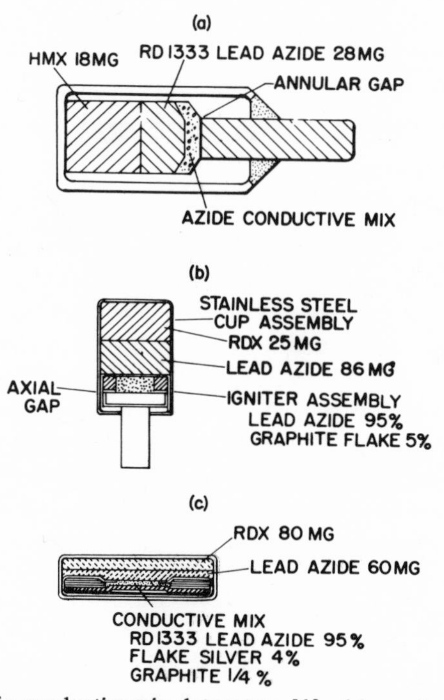

Figure 4. Lead azide in conductive-mix detonators [1]: (a) annular gap (all fire 50 V on 0.015 mF); (b) axial gap (all fire 10 V on 2.2 mF); (c) radial azide column (all fire 5 V on 33 mF).

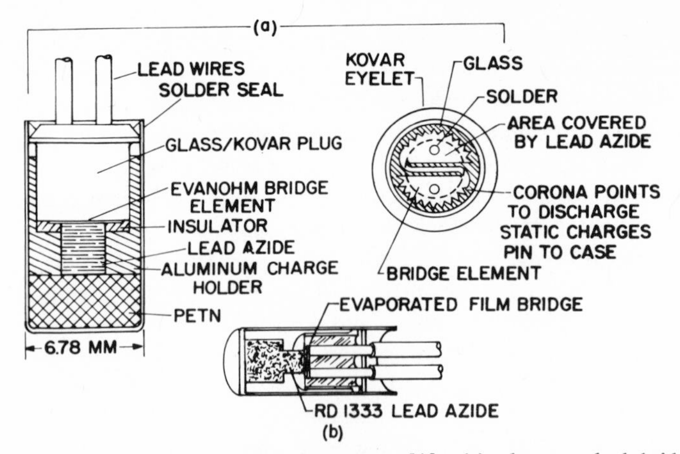

Figure 5. Lead azide in amp no-fire detonators [1]: (a) photo etched bridge (1-ohm evanohm foil); (b) thin film bridge (5000 Å chromium, 1 ohm, 0.555 × 0.08 in.)

insulator. During the firing phase, lead azide's rapid transition from burning to detonation is important in providing a sharp threshold between fire and no-fire and in minimizing the functioning time [5].

C. INITIATION OF AZIDES IN EXPLOSIVE TRAINS

1. Functioning of Primers

As discussed above, the ignition of primer mixtures in a detonator can be caused by the action of a firing pin (a stab), by heat (from a hot wire or flame), or by shock (from an exploding wire or fast-moving fragments). In most detonators containing lead azide, a stab action or electric ignition by a hot wire is used.

The output of primers includes hot gases, particles, a pressure pulse, and sometimes a shock wave. Parameters which have been measured to characterize the primer include: the volume of gas emitted, the impulse imparted to a column of mercury, the light output, the temperature rise of a calorimeter, the pressure, the conductivity between probes, and the functioning time. No general quantitative relationship between the parameters and the initiation of the next explosive in the element or train has yet emerged, although individually they may all have some importance [2,5,6].

The relative importance of a particular parameter depends on the properties of the next substance to be initiated, for example, whether it is a single-component, primary explosive such as lead azide, or a secondary explosive such as RDX. In view of the intimate contact between primer and lead azide, or between lead azide and the secondary explosive, within an element, heat conduction and shock propagation, respectively, may be expected to play significant roles in the ignition or initiation of the succeeding explosive. Between detonator and acceptor (lead) elements, fragments of the explosive and detonator case may play an additional role, but if the succeeding explosive is a secondary explosive, such as RDX, the shock characteristics and particularly the pressure–time profile are likely to be of greatest significance (see Section D.3, below). The design objective in such cases may then be high pressures for a long duration, or more specifically a maximization of the product of the square of the pressure and the duration of the pulse [7].

a. Stab Initiation

The compositions of typical sensitive stab mixtures in current use are shown in Table I. NOL-130, the most stab sensitive, also has the greatest resistance to heat and humidity. The carborundum in AN#6 (see Table I) causes a high rate of explosions during pressing operations because it grit-sensitizes the lead azide (Chapter 4). The ingredients of AN#6 also segregate more rapidly during handling than those of NOL-130. Two single-component stab-sensitive materials that have been investigated are copper chlorotetrazole and basic lead azotetrazole, but both are less sensitive than NOL-130. The least stable component in NOL-130 is tetrazene, which has a 5-sec explosion temperature of 160°C; however, if it is not incorporated, stab sensitivity decreases.

In spite of its sensitivity, flash X-ray pictures of an exploding detonator showed that the NOL-130 mixture was not fully burned by the time the lead azide and RDX layers had detonated [8]. This and the increased output due to higher loading pressure are among the first indications that current high-use detonators are not optimally designed with respect to the explosive constituents.

Table I. Typical Stab Primer Mixtures [2]

Ingredient	Common stab mixtures (%)			Special purpose mixtures (%)	
	NOL-130	PA-100	AN#6	PA-101	NOL-60
Potassium chlorate		53	33	5	
Antimony sulfide	15	17	33.3	10	10
Lead azide	20	5	28.3		
Carborundum			5		
Lead thiocyanate		25			
Basic lead styphnate	40			53	60
Barium nitrate	20			22	25
Tetracene	5			5	5
Aluminum				10	

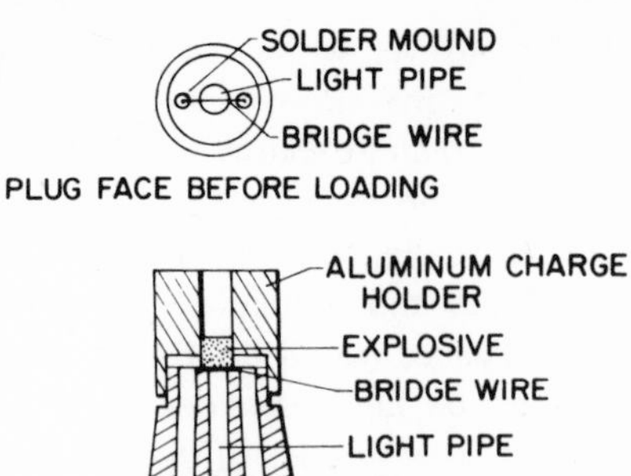

Figure 6. Assembly for hot-wire resisitivity tests [9].

b. Electrical Initiation

In the case of electrical initiation lead styphnate or an azide is in direct contact with a resistance wire and serves as a primer. The initiation of lead and silver azides by hot wires was studied in detail by Leopold [9], who used a fixture (Figure 6) to monitor both the electrical energy in the wire and the light output of the explosive. The thermal initiation times were related to wire temperature by an Arrhenius equation of the form:

$$\tau = Ae^{-E/RT}$$

where τ = time to explosion after heating begins; A = a constant; E = apparent activation energy (J); T = temperature, K; R = gas constant.

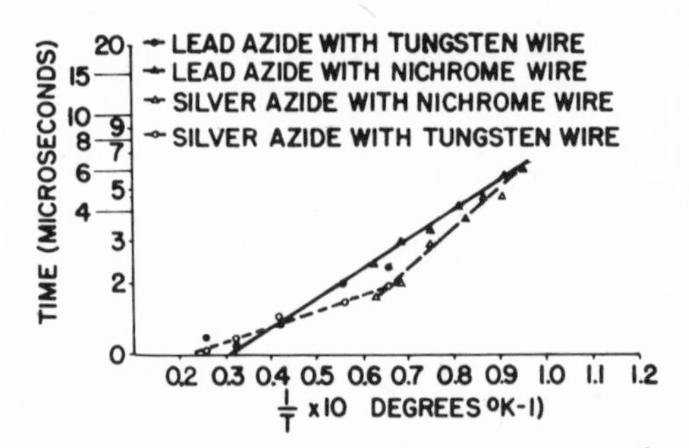

Figure 7. Explosion time vs. reciprocal of absolute temperature for primary explosives [10] using a 0.4-mil diam tungsten bridge wire and a 1-mil diam nichrome wire.

Lead and silver azides require longer times to fire at the same temperature when the wire diameter is increased from 0.00064 to 0.0025 cm, while lead styphnate initiation is not affected by wire diameter (Figures 6 and 7). In general, silver azide requires slightly less energy and fires with less time delay than lead azide. Measurements at low and high densities also showed that more energy is needed to initiate the explosives at lower densities. The activation energies (Table II) are lower than obtained with the confined explosion-temperature test at lower temperatures (Chapter 6), perhaps due to differences in heat flow, erroneous assumptions in calculating temperature, or delays in the temperature rise. However, the work of Leopold gave decreasing activation energies as the temperature was increased, which could also account for the difference. [9].

2. Growth of Reaction in the Azides

The desire to obtain high enough pressures to detonate an output charge in a particular geometry requires the achievement of detonation following the reaction started by the primer; therefore the factors that affect the growth of detonation are important in optimizing designs.

Under the conditions found in detonators, a transition to detonation is not necessarily instantaneous. As discussed in Volume 1, Chapters 8 and 9, the extent to which deflagrations or different orders of detonation occur in the azides is not clear. But there is considerable evidence that both unconfined and small confined samples of lead azide will sustain, at least transiently, propagation rates less than those of full detonation. Table III summarizes some of the data, which are dis-

Table II. Activation Energies for Initiation of Primary Explosives [9]

		Resistance heater		
Explosive	Loading pressure (psi)	Nichrome, (0.0025 cm)[a] (kJ/mole)	Tungsten, (0.01 cm)[b] (kJ/mole)	Henkin and McGill[c] (kJ/mole)
Normal lead styphnate	50,000	19.2	22.6	245
Basic lead styphnate		25.1	21.7	
Lead azide		22.9	16.3	88.6
Silver azide		31.4	12.5	100–184
Normal lead styphnate	10,000	31.7	18.4	
Basic lead styphnate		38.5	19.3	

[a]Nominal temperature range 630–1300°C.
[b]100–3400°C.
[c]Data obtained from various temperature ranges.

cussed in more detail in Chapter 8 of Volume 1. Theories exist to explain the variation of velocity in terms of sample diameter and confinement, as summarized in the Section D. Among the parameters derivable from the theories is the length of the reaction zone required to sustain the propagating reaction or shock front. Measurement techniques have not yet advanced to the stage where this quantity can be determined experimentally for azides, but Table IV summarizes values which have been computed on the basis of the current theories. Because the computations do not treat energy losses adequately, it is probable that the values for the reaction-zone lengths are higher than those calculated.

A study of the initiation of lead azide by the impact of flyer plates (Section F) showed that stress excursions behind the shock front produce pressure waves which travel through the shock-compressed azide at a velocity at least equal to the sonic velocity. The sonic velocity in the precompressed explosive is higher than in the uncompressed explosive, so the amplitude of the initial shock increases rapidly until steady-state detonation is achieved in less than 1 mm, as indicated by the data in Table IV. For an initial stress over 4 kbar, instantaneous detonation occurred; however, pressures this high are not normally present at the input to the azide in an explosive train.

D. DETONATION PROPERTIES OF THE AZIDES

Among the most important functional parameters of the azides are the detonation velocity and pressure, and these depend on the material parameters (density, azide content, particle size, etc.) and the confinement (its material, density, dimensions, etc.). Even with the most modern techniques the functional quantities are not easy to measure with reliability and precision. In this section

Table III. Propagation Rates in Lead and Silver Azide as a Function of Size

Sample	Size (mm)	Propagation rate (km/sec)	Reference
Lead Azide			
Crystal	2.4	2–5	[11]
	0.01–1.0	1–3	[12]
	0.08–0.16	2–3	[13]
Powder	0.11–1.1	1–3	[13]
	0.5–3	3–4	[14]
Pressed Sheet	0.03–0.5	2–5	[15]
Silver Azide			
Crystal	<2.0	<2.3	[10]
Powder	0.5–2.0	3.5	[16]
Pellet	0.6–0.85	2–3.5	[17]

Table IV. Reaction-Zone Length in Lead Azide

Sample	Zone length, mm	Reference
Powder	0.075[a]	[15]
	0.45	[10]
Crystals	0.2	[13]
	0.4–0.5	[11]

[a]Compressed sheets, density 3.4 g/ml.

some illustrative examples of measurement techniques are first presented; a discussion of data follows.

1. Measurement Techniques

a. Detonation Velocity Measurements

Detonation velocities in confined charges may be measured with a streak camera or by pin switches. The techniques are potentially accurate to ~0.5%. The camera has the advantage that a continuous record of buildup and detonation transfer can be obtained [18,19]. In the sample holder (Figure 8), the central hole has the diameter of a typical detonator, 0.136 in. (0.345 cm), and the initiation of the lead azide is by a hot wire. The flat window, cut and polished from a 3/8-in. (0.095-cm) -diam rod, which covers the slit provides a bright, sharp slit image through which to view the propagation of detonation with the camera.

Because of the difficulty of setting up a streak camera and of inaccuracies in reading velocities from the records, a faster, more accurate method for measuring equilibrium velocities is to use switches and electronic time-interval meters.

A pin switch apparatus for use with lead azide is shown in Figure 9. The

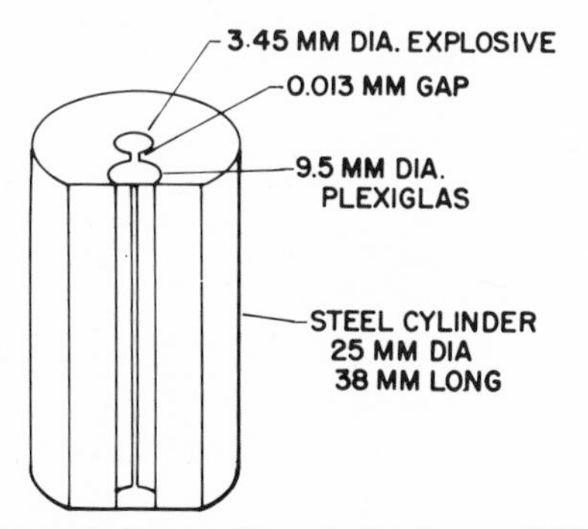

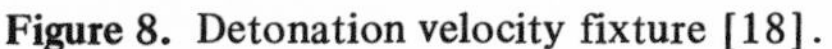
Figure 8. Detonation velocity fixture [18].

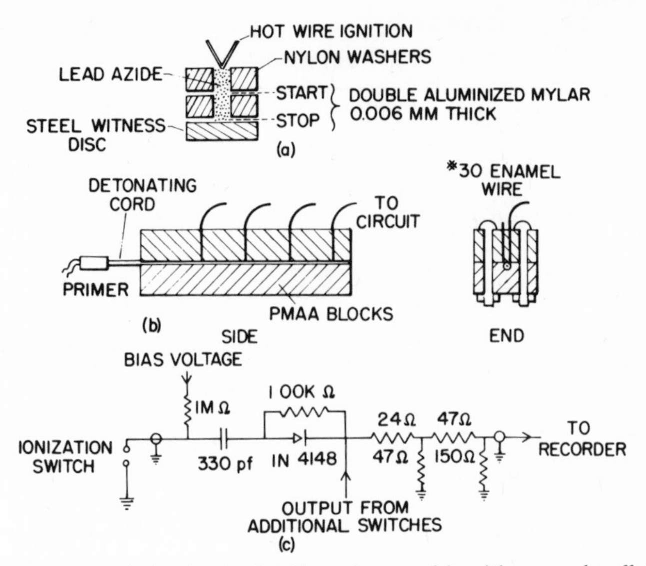

Figure 9. Ionization switch circuit details and uses: (a) with pressed pellets; (b) with detonating cord; (c) circuit details.

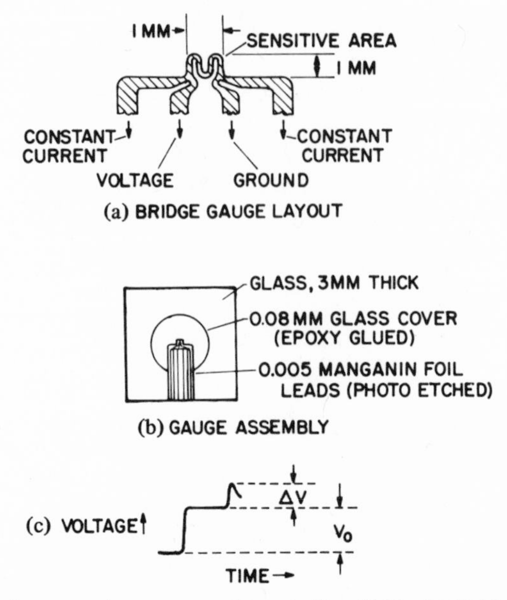

Figure 10. Manganin gauge for pressure–time profiles [10]. (a) Bridge gauge layout; (b) gauge assembly; (c) record.

azide is extruded in a lead tube which is inserted into an outer confining sleeve to prevent electrical bias on the velocity pins from initiating the lead azide. The tube is made from a metal whose sonic velocity is less than the detonation velocity. The pin signals generated by the circuit in Figure 9(c) are connected to a transient recorder accurate to 10 nsec, or to a time-interval meter accurate to 1 nsec. The long sample length provides an accurate measure of steady-state detonation velocities. For measurements of the buildup to detonation, shorter samples, pressed into nylon or steel washers with foil switches between them, are used. In such measurements a time-interval meter accurate to 1 nsec is essential [8].

b. Measurement of Output Impulses

Detonation pressure vs. time has been measured by a manganin gauge (Figure 10 and 11) made from 0.0002-in. (0.005-mm) -thick foil, insulated with glass, and with a sensitive area only 0.04 $\times$ 0.04 in. (1 mm^2). The gauge is connected to a constant current supply, and only the voltage drop across the center section is recorded to minimize the effect of lead stretching. A recording time of about 1/2 μsec is obtained from most detonators before the gauge breaks. Peak pressures are read directly, but a constantan gauge must be fired with each type of sample to provide the stretch correction for the release wave [20]. Figure 11 shows how the gauge is used in a fixture to measure the impulsive output.

The manganin gauge may be calibrated by using a gas gun with known output characteristics, and a typical gauge coefficient found by the method is 0.0024 Ω/Ω/kbar. If only peak pressure is of interest, this may be determined by using an aquarium apparatus to determine the shock velocity produced in a material whose shock Hugoniot is known (in this case water) [21]. Because of extrapola-

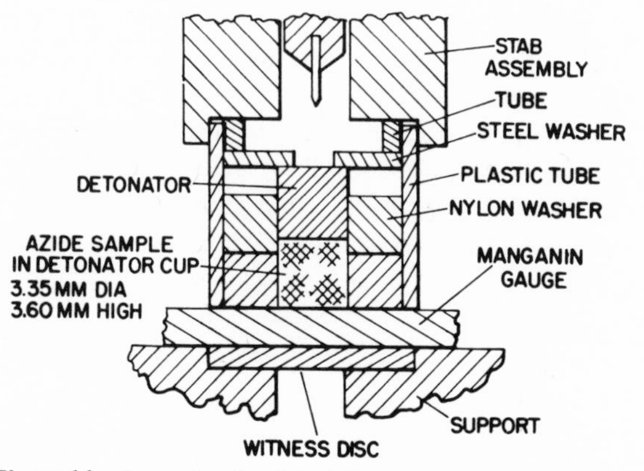

Figure 11. Apparatus for detonation pressure measurement.

tions of Hugoniot data and the curvature of the shock front, pressure measurements made by this approach are not very accurate.

2. Dependence of Detonation Velocity on Material Properties and Confinement

It has been indicated in previous discussion that the initiation of detonation in the azides is structure and confinement sensitive, and the variation in the behavior is reflected in differing detonation velocities (Table III). Notwithstanding the question whether steady detonations are achieved in some instances, available theories confirm that the detonation velocities depend on dimensions and confinement for small-diameter samples and on the density and azide content of a given mixture.

a. Dimensional Effects

Only a few measurements have been made on propagation rates in azide samples having dimensions comparable to those of detonators. In these cases the steady nature of the propagation for short lengths has not been established, and differing confinements make direct comparison of the data difficult. The single-crystal data (Table III) might be assumed to relate to azides of theoretical density; however, extrapolations of data on lead azide pellets and on the basis of available theory suggest that Chapman-Jouget detonations were not achieved in any of the cases quoted. For example, taking the theoretical density of lead azide to be 4.71 g/ml, extrapolation of data for pressed pellets leads to the following values for the ideal detonation velocity: 5.5 [14], 5.8 [22], 6.2 [13], and 6.41 km/sec [18]. Chaudhri (see Chapter 8, Volume 1) claims velocities in excess of 8 km/sec are achievable in large single crystals. Bowden and Williams [23] determined the rates of propagation in unconfined single crystals of silver azide to be 1.5–1.9 km/sec, values which are low in comparison with the data for pressed pellets (Figures 12 and 13).

The critical diameter of lead azide for unconfined powders or crystals has not been established and cannot be until one determines the pressure or absence of detonations in the small dimensions cited above. In the case of heavily confined charges, work on swaged-lead detonating cord lead to the expression [24]

$$D = 4.1 - \frac{1}{70r} \tag{2}$$

where D is the detonation velocity in km/sec, and r is the charge radius in cm.

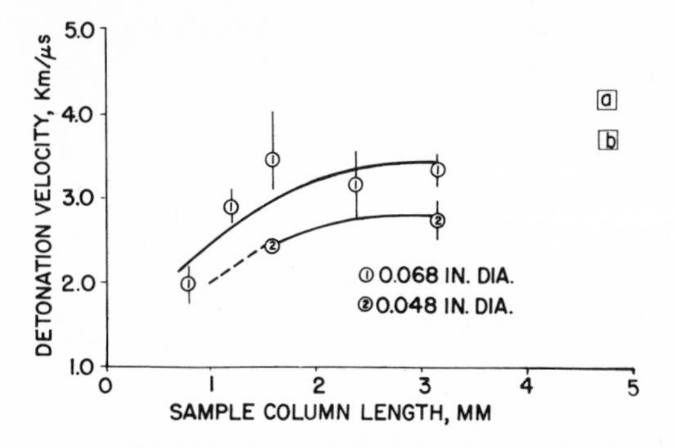

Figure 12. Detonation velocity of silver azide initiated by ~1 mg granular AgN_3. Point a represents the velocity over the last 0.15 cm of a 0.30-cm column of 0.17 cm diam. Point b similar to a except the column diameter is 0.119 cm [17].

The failure diameter for a density of 3.5 g/ml was found to be 0.06 mm with a loading of 0.05 grains/ft [24] (10^{-4} g/cm).

According to the hydrodynamic theory of reaction waves propagating in one dimension (see references to the Introduction, Volume I) the detonation velocity is expected to be less than ideal in samples of diameter d such that the observed velocity D will approach the ideal velocity, D_i, as $d \to \infty$. Eyring *et al.* [25] developed a model based on a curved shock front bounded by a burned

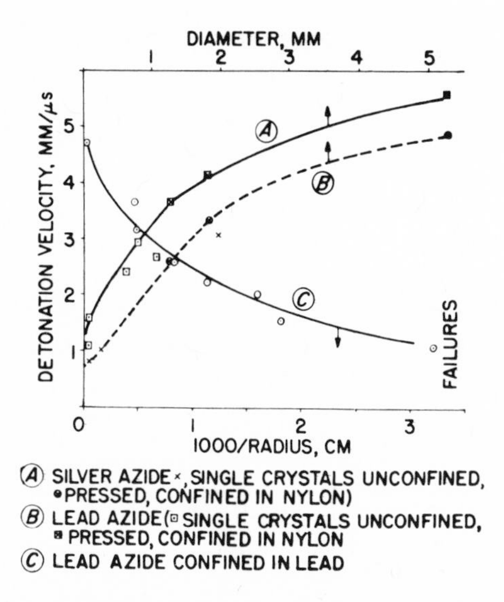

Figure 13. Detonation velocities of silver and lead azides as a function of diameter [17].

reaction zone. The front was assumed to be made up of spherical segments followed by radially divergent flow. The theory yielded the empirical equation

$$\frac{D}{D_i} = 1 - \left(\frac{a_0}{d}\right) \tag{3}$$

where a_0 is the reaction-zone length.

Jones [26] developed a model in which the shock front was planar on the cylindrical axis of a charge and curved only toward the cylinder surface. The theory yielded the relationship

$$\left(\frac{D}{D_i}\right)^2 = 1 - 0.8\left(\frac{2a_0}{d}\right)^2 \tag{4}$$

Both theories relate the observed velocity to the ideal velocity for bare, lightly cased, and heavily cased charges but treat only perturbations in the continuity of mass across the shock front and neglect transport of momentum or energy to the surface of the charge. Thus lateral losses are underestimated, and the reaction zones computed by these theories are to be viewed as probably too large.

An analysis of the consequences of divergent flow provided further insight into detonations in cylindrical charges. However, the radius of curvature, a difficult quantity to measure, is needed for application of the theory [27]. A rotating mirror camera was used to take head-on photographs of detonations and confirmed that curved fronts occur [25].

Experimental data is often plotted as D vs. $1/d$, and the ideal velocity is obtained at $1/d = 0$. Various studies have shown that the velocity depends on charge diameter more strongly for low-density charges of pressed explosives than it does for cast explosives.

Figures 26 and 27 summarize data for the dependence of the detonation velocities of lead and silver azides on the diameter and length of the samples. The data are fragmentary, and apart from indicating a general dependence on diameter (Figure 13) for both substances up to 0.14 in. diam, they show apparent inconsistencies both with respect to the comparative behavior of the substances and the dependence of detonation velocity on density. In view of the uncertainty of the measurements, it is not possible to draw firm conclusions, except to confirm that improved measurements are desirable.

For silver azide sample lengths [17] (Figure 12) up to 0.13 cm a 50–100-nsec delay occurs before a steady rate is achieved. Above 0.13 cm, lower than anticipated velocities were observed. Some unknown phenomenon may have affected the measurements, but a few samples of lead azide exhibited similar behavior.

Higher velocities were recorded when a gap was introduced into an 0.30-cm-long sample by insertion of ionization strips and the velocity was measured over the last 0.15 cm of the length. Coupled with the somewhat higher velocities for the 0.10-cm-long samples, the results suggest the existence of interactions with shock waves reflected from the side walls and the ends of the samples [17].

b. Dependence of D on Density

Using the sample holder shown in Figure 9, Schwartz *et al.* [18] obtained the dependence of the detonation velocity of lead azide on density (Figure 14). The data are in general agreement with those obtained by others [28–30]. Scott [19] found a value of 4.3 km/sec at a density of 2.86 g/ml and a semicircular cross-section of radius 0.075 in. which is in fair agreement with the Schwartz data. However, both Schwartz and Avrami *et al.* [18,29] found a considerable variation in *D* even in samples of the same (low) density.

c. Dependence of D on the Confining Material

Another factor considered by Schwartz *et al.* was the material of which the holder in Figure 9 was fabricated. On the basis of theory and experiments previously reported, only small variations are expected for differing metal containers [26,31], and no significant variations were observed for the propagation velocity of lead azide in steel, brass, or aluminum holders (Table V).

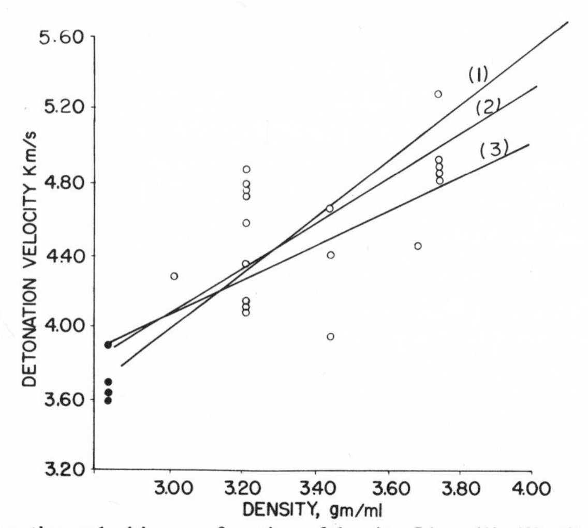

Figure 14. Detonation velocities as a function of density. Lines (1), (2), (3) from references [28], [18], and [30], respectively. •, Points from reference [29]; ○, points from reference [18].

Table V. Detonation Velocity of Lead Azide in Different Metal Holders [19][a]

Holder material	Detonation velocity (km/sec)
Steel	4.52 ± 0.25
Aluminum	4.85 ± 0.61
Brass	4.65 ± 0.31

[a]For holder shown in Figure 8. Samples pressed at 10,000 psi; density of 3.22 g/ml; with standard deviation.

3. Detonation Pressures of Lead Azide

On the basis of hydrodynamic theory, detonation velocity can be related approximately to the detonation pressure by the equation:

$$P = \frac{1}{4}\rho D^2 \tag{5}$$

where P is the detonation pressure, ρ is the density of the unreacted explosive, and D is the detonation velocity. Pressure and density are assumed to be related as follows [32]:

$$P = B\rho^3 \tag{6}$$

where B is a constant.

It is only recently that measurements have been made of the detonation pressure of lead azide, and similar data are not known for other azides. The values in Table VI were obtained using the manganin gauges described earlier, and the detonation velocities were obtained by streak photography [11].

The differences between the calculated and measured values may in part be explained by the small diameter, confinement, and the assumptions made in the derivation of the formula for pressure which applies to CHNO explosives, not necessarily azides.

Table VI. Experimental and Theoretical Detonation Pressures of Lead Azide

Density (g/ml)	Detonation velocity (km/μsec)	Peak pressure (kbar)	
		Measured	Calculated[a]
3.25	4.35	126	154
3.60	4.80		207
3.78	5.00	156	236

[a]$P = \frac{1}{4}\rho D^2$.

E. DETONATION TRANSFER FROM THE AZIDES

The efficacy of azides in initiating detonation in the succeeding elements of explosive trains may be assessed using either of two types of test. The first determines the work done or the damage induced in an inert material in contact with the terminal face of the azide sample. The work done may be related to the strain or the hardening of a metal "witness plate"; alternatively, the damage may be related to the volume of a dent or hole in the plate or to the degree of fragmentation of the witness material. A variant of the tests combines varying proportions of azide with a secondary explosive in a detonator to determine the effect on witness plates.

In the second type of test the witness material is the secondary explosive in a configuration designed to reproduce or simulate that conceived for the azide and the explosive in an actual design. The approach assesses the efficiency of the azide as an initiator by determining the quantity necessary to assure reliable initiation of the secondary explosive for a given configuration and sample quantity. The secondary explosive may, for example, be RDX [18] or PETN [19].

The use of these empirical tests is necessary because of the inadequacy of data necessary for theory to provide a computational basis for determining the output parameters necessary to achieve initiation of a secondary explosive with lead azide. The foregoing sections indicate that even empirical data on the parameters that affect the azide output are fragmentary. Nevertheless, the theory can indicate potentially important conditions necessary for an optimum detonator, and a brief discussion of the status of the theory is given at the end of this section.

1. Initiation of Secondary Explosives

Comparative measurements of the transfer from lead and silver azides have been made [8] after pressing the substances into detonator cups of 0.068 and 0.096 in. (0.173 and 0.244 cm) diam. The cups were loaded first with a 0.020-in. (0.05-cm) layer of NOL-130 primer, and the remaining 0.110 in. (0.279 cm) of the cups was filled with separate lengths of azide and RDX (Figure 15). The detonators were fired against steel witness plates, and the depths of dents so produced were used as a measure of detonator output. For both silver and lead azides the output remained constant until the azide length was less than 1.5 mm (60 mil). Thereafter, the denting ability of the lead azide RDX detonators fell off much more rapidly than those loaded with silver azide RDX, indicating that in differing degrees insufficient azide was present to produce reliable initiation of the RDX.

Silver azide was also found to be more efficient than lead azide in detonators

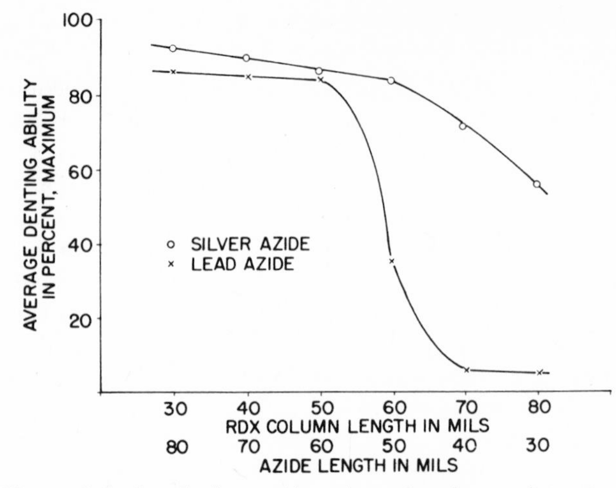

Figure 15. Effects of lead and silver azide column lengths on detonator performance: 0.172 and 0.243 cm diam; 0.28 cm total length [8].

of the type shown in Figure 1a. A specified requirement for RD1333 lead azide is that 25 mg pressed into such a detonator with an RDX charge will produce a high-order detonation and a normal plate dent.

The witness plate has also been used to determine the effect of varying the density and the quantity of lead azide in typical detonators [33]. Figure 16 illustrates the effect of lead azide quantity on a small detonator of the type shown in Figure 1a, but of dimensions 0.147 in. (0.373) diam and 0.143 in. (0.363 cm) long. The most significant consequence of reducing the weight

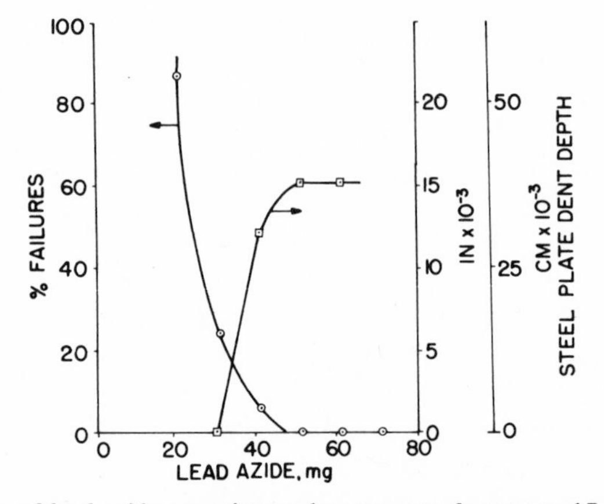

Figure 16. Effects of lead azide quantity on detonator performance: 17 mg NOL-130 used; total column length 0.35 cm; RDX filled remaining space [33].

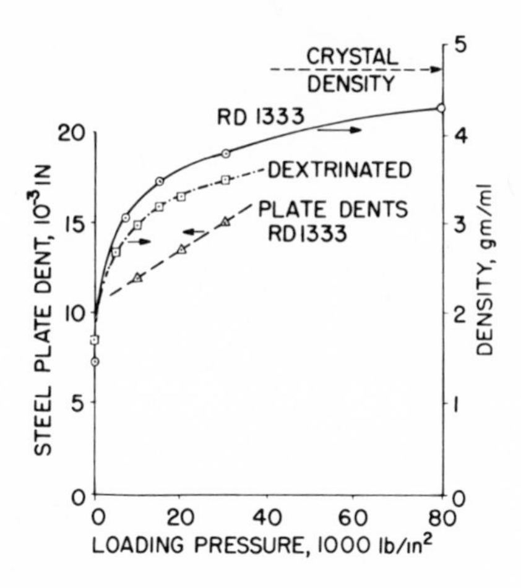

Figure 17. Effect of lead azide loading pressure on density and detonator performance [33]: 0.35 cm diam × 0.35 cm long; 17 mg NOL-130, 51 mg lead azide, 18 mg RDX.

below a critical amount (51 mg) is a rapid increase in the number of "duds," or detonations which fired but produced no measurable dent in the witness plate and presumably no detonation in the RDX.

Figure 17 shows the effect of pressure on the density and effectiveness of lead azide in the above detonators. Streak-camera and contact-pin measurements of sectioned detonators showed that high-order detonation develops in the center of the 0.060-in. (0.152-cm) -thick layer of lead azide.

Tests with 0.068- and 0.096-in. (0.173- and 0.244-cm) -diam detonators, summarized in Figure 15, showed that the weight is more important than the pellet length in initiating RDX [8,18], primarily due to increased radial losses in smaller diameters. As a result of the data in Figure 16, a minimum of 51 mg of lead azide was chosen for a 0.136-in. diam.

The increase in the density of RD1333 and dextrinated lead azide with pressure, shown in Figure 17, is also reflected in the output at higher density, as measured by the plate dent test. Detonators are usually manufactured by compacting the azide with 10,000-15,000 psi; however, 30,000 psi pressure can be used to increase output [5].

As indicated in the foregoing subsection, the dent test, when utilized with secondary explosives in a detonator, is in part a measure of the efficiency with which the azide initiates the secondary explosive.

Table VII shows how the process used to manufacture lead azide, or the consequent product, significantly affects the quantity required to initiate a standard secondary explosive, RDX, in the stab-sensitive detonator (Figure 1a). Dextrinated lead azide has a lower output because it is less compressible and has more diluent: namely, 8.5% dextrin compared to the 3.5% carboxymethyl cellulose (CMC) [34] in RD1333, 2% polyvinyl alcohol in PVA lead azide, and no binder in Service lead azide (see Chapters 1 and 2). This situation is shown quantitatively in Table VII, which shows the minimum charge weights of each

Table VII. Weights of Lead and Silver Azides Required for Initiation of Secondary Explosives

	Weight (mg)					
	RDX	HMX	Tetryl	Picric acid	PETN	TNT
Lead azide						
Dextrinated	90	300	100	260	30	270
Special Purpose	50					
PVA	25					
Service	30					
Silver azide	25		20	30	5	70

type of lead azide to produce high-order detonation of the output charge in the detonator.

The quantity of azide needed also varies significantly depending on the secondary explosive to be initiated (Table VII). These are typical examples; fuller details on the minimum amounts of dextrinated lead azide to produce high-order detonation in other explosives is given elsewhere [36].

In 1917 Wöhler and Martin [37] compared the minimum quantities of silver and lead azides required to initiate various secondary explosives of importance at that time, and their tests were conducted in blasting caps using lead witness plates. Abbreviated details of their findings are also given in Table VII.

During tests on small detonators loaded with silver azide and tetryl [38], both the consolidation pressure and the ratio of silver azide to tetryl were varied. The efficiency of silver azide increased with increasing pressure up to the maximum value used (15,000 psi). By reducing the ratio of silver azide to tetryl and allowing a larger charge of tetryl to be used, the output of the detonator was increased by 20%, as measured by the Hopkins pressure bar [1].

At the high ratios of azide to tetryl, a comparable test using lead azide gave equal output; however, at the low ratios no increase in output occurred.

2. Theory of Shock Transfer

Available evidence indicates that secondary explosives in detonators are initiated by shock waves. This evidence consists of streak traces showing an accelerating shock front passing from the azide to the secondary explosive, and flash X-ray studies show the secondary explosive detonating at the interface between the lead azide and the secondary. Since the theory of shock waves at interfaces has been adequately treated in other texts [27,32,39], only the major aspects will be presented.

Hydrodynamic theory defines the behavior of shock waves at interfaces between elements of explosive trains in terms of curves that relate the shock velocity to the particle velocity, called Hugoniot curves. Assuming that mass and momentum are conserved across the shock front, one can write

$$P = \rho_0 u_p D_s$$

where P is the shock pressure, ρ_0 the initial density, u_p the particle velocity, and D_s the shock velocity.

The empirical relationship

$$D_s = a + bu_p$$

is used to express the relation between the shock velocity and the particle velocity. However, there is no satisfactory theoretical basis for the relation [39].

At any interface between two media, shock transfer occurs such that the pressure and particle velocity across the interface are equal. The strenghs of the transmitted and reflected shock waves depend on the relative impedance which can be calculated from the Hugoniot curves (Figure 18).

A plot of P vs. u_p has the slope $\rho_0 D_s$ and passes through the origin. If the Hugoniot curve for the detonation products of lead azide at the C-J point were known (the curve for unreacted RDX is known [40]), it would be possible to determine the strength of the shock wave generated in RDX by the detonation wave from lead azide. It is to be recalled that the Chapman-Jouget model as modified calls for a reaction zone at the end of which the reactants have been completely converted to products in equilibrium and travel at the local sonic velocity. The end of the reaction zone is often called the C-J plane, and the associated pressure and temperature called the C-J pressure and temperature.

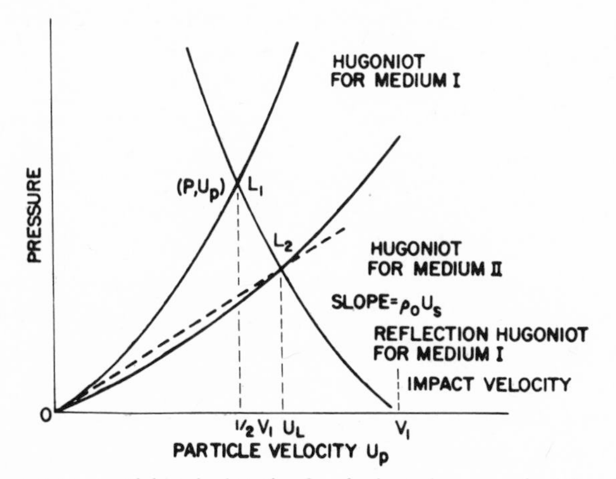

Figure 18. Pressure vs. particle velocity plot for the impedance method for shock transfer.

The detonation velocity refers to the velocity of the shock front relative to the unreacted material. Peak shock pressure needed to initiate RDX and other secondary explosives as a function of density and particle size have been measured by the small scale gap test [48]. For RDX, which is commonly used in detonators, peak pressure for initiation is found to range from 9 to 16 kbar. Since even under nonideal conditions lead azide produces over 100 kbar, as shown in Table VI, if enough azide is present, it readily initiates RDX, and other secondary explosives, as shown in Table VII.

However, it has been shown that, in addition to the peak pressure, the pulse width (pressure vs. time profile) is an important factor for shock initiation as seen from work of Walker and Wasley [7]. Shock-initiation energy requirements for secondary explosives have been determined; for RDX a minimum energy for initiation is 11 cal/cm^2 (equivalent to $P^2t = 460\rho U_s$.) Although the C-J detonation pressure of lead azide is adequate, as the column height is reduced the pulse duration decreases, thus explaining why a critical column length is needed. In actual applications where diameter and confinement are less than ideal, even more azide is needed to compensate for P^2t being less than ideal. A similar plot for lead azide is shown in Figure 19.

The response of homogeneous and heterogeneous explosives to shock waves if often viewed as being different (Figure 20). In a heterogeneous medium, a shock wave of strength above the initial value accelerates smoothly at the point of entry and develops into a stable detonation. On the other hand in a homogeneous explosive, the entering shock is initially unsupported and decays since no energy is released. After an induction period, at some point behind the initial wave a detonation starts moving through the precompressed explosive

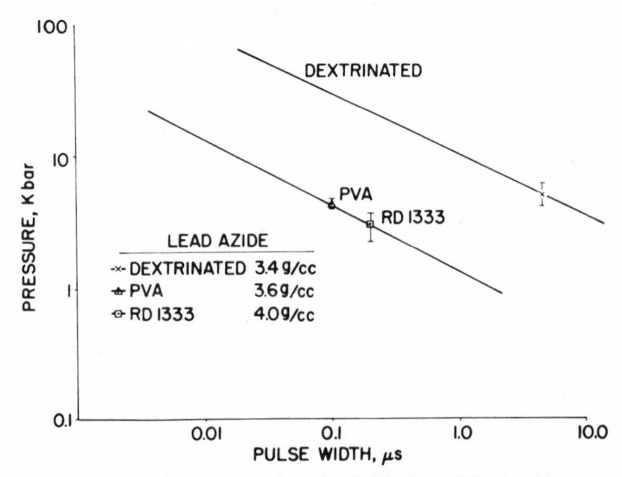

Figure 19. Pressure vs. time for initiation of lead azide.

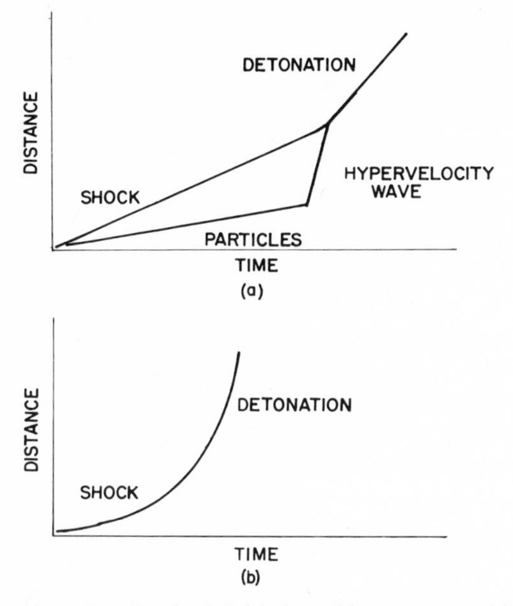

Figure 20. Distance–time plots for shock initiation of homogeneous (a) and heterogeneous (b) explosives.

with a hypervelocity (greater than the C-J velocity) [41]. This hypervelocity wave overtakes the original wave, and the newly combined waves travel as an overdriven wave, decaying to the normal detonation after a short time (~1 μsec).

According to one view the contrasting behavior of inhomogeneous and homogeneous explosives is based on the creation of hot spots in the inhomogeneous explosive as the shock passes [27]. The hot spots explode (or detonate) and transmit their energy to the shock front. The shock strength increases, creating new hot spots at higher temperatures that yield more energy. In this manner the shock strength increases until a C-J detonation is achieved.

F. EFFECTS OF STRONG SHOCKS ON LEAD AZIDE

Among the important factors determining the quality of detonator designs is their vulnerability to external stimuli, of which strong shocks generated by a variety of mechanical or nuclear sources are the most common. Data on the response of materials to strong shocks are also important in explosives technology because of their potential for providing information on the equations of state of the substances, thus quantifying the theories of detonation and of the interaction of shock waves with different media.

This section is concerned with incident shock waves with peak pressures in the 1-10-kbar range that are typical of detonation pressures required to initiate explosives. Experimental techniques to make measurements in these shock regimes are neither easily developed nor commonly available; thus much of the section emphasizes the principles and instrumental approaches utilized to obtain the data.

There are three ways to generate a shock wave in a material. A shock in one medium can be propagated directly into a second medium in contact with it. Solids can be caused to impact each other at high velocity, resulting in shock waves propagating from the point of impact. Thermal energy can be stored in a material in a spatially inhomogeneous manner; since thermal energy per unit volume is equivalent to a pressure and $\partial P/\partial x$ is equivalent to a force per unit volume, the inhomogeneous energy deposition results in a propagating shock. Each of the approaches has been applied to studies of lead azide, and they are associated with techniques utilizing a gas gun, a thin flyer plate, and a pulsed electron beam.

The investigation of the shock-initiation mechanism of lead azide is of practical and theoretical importance. In heterogeneous secondary explosives, there is considerable experimental evidence to support the hypothesis that initiation is the result of a hot-spot mechanism, i.e., surface reaction at areas of energy concentration. The measured initiating pressure profiles are far too low to effect, by shock compression, a significant temperature rise throughout the bulk of the material. In a porous explosive hot spots can be formed all along the shock path. With sufficiently strong shocks and/or a sufficiently reactive explosive, these hot spots can bring about localized chemical decompositions in time to contribute energy to the shock front and continuously increase its intensity. The shock develops into a stable detonation wave provided the explosive is of sufficient size and the reactions are not quenched by rarefaction waves.

In the case of lead azide, Andreev [42] and Bowden and Yoffe [43] suggest that lead azide detonates immediately after being ignited and that a burning regime is absent. The theory of fracture that was subsequently developed to explain the initiation of fast reaction [44,45], and the previous observations lead to the conclusion that the shock initiation mechanism of this primary explosive is not likely to exhibit the same characteristics as those exhibited by the secondary explosives. However, examination of the shock sensitivity of dextrinated and polyvinyl lead azide to pulse durations varying from 0.1 to 4.0 μsec shows that the initiation characteristics are indeed similar to those observed for heterogeneous explosives.

1. Equation of State of Unreacted Lead Azide via Gas-Gun Measurements

The Hugoniot equation of state and the sensitivity of dextrinated lead azide to shocks of long duration were determined by subjecting lead azide pellets to

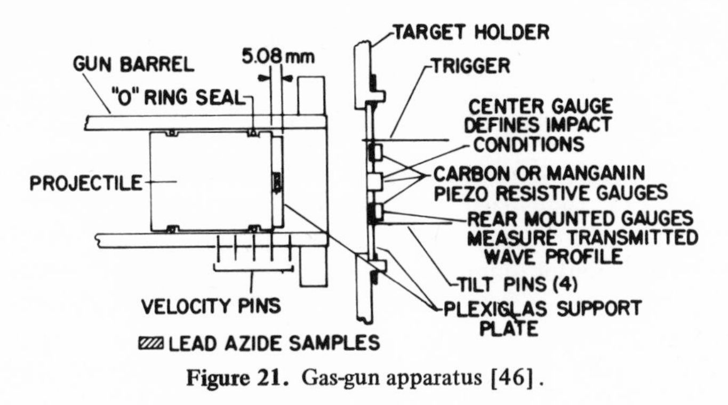

Figure 21. Gas-gun apparatus [46].

impact shocks in a 2.5-in. gas gun [46]. Using carbon and manganin gauges, the initial amplitudes of stress waves generated in the lead azide were measured as well as the wave profile after they had propagated distances of 1 and 4mm. The apparatus used a lead azide sample mounted on a projectile for impact-face measurements and an array of four azide samples and piezoresistive gauges mounted on the target for transmitted pulse measurements (Figure 21). The pulse durations were about 3.5 μsec in the four outer lead azide samples and 5 μsec in the center sample.

The samples were free standing disks of pressed dextrinated lead azide, pressed at 40,000 psi and having densities of 3.18-3.41 g/ml, depending on the thickness of the disks.

A typical signal from the center gauge in the target array is shown in Figure 22. The trace rose first to the impact stress in the acrylic and then to the stress in the explosive sample. The spike results from the impedance mismatch of the adhesive bond with the Plexiglas and lead azide.

The lead azide stress was read directly, and its particle velocity was calculated from the known acrylic Hugoniot and the impact velocity. Since continuity of stress and particle velocity at the interface between the Plexiglas and

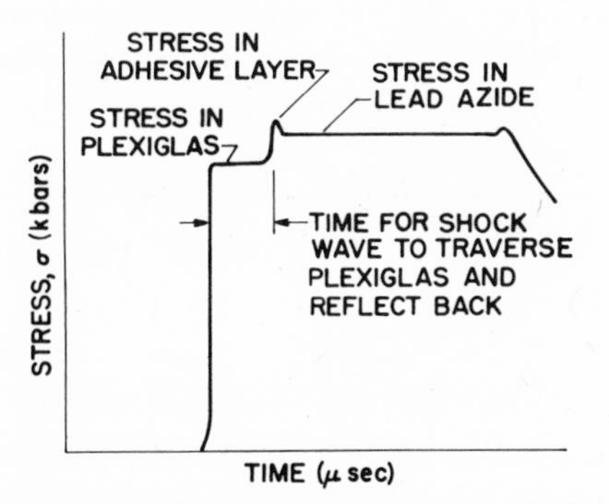

Figure 22. Typical center gauge trace [46].

the lead azide is maintained, the particle velocity in the lead azide, u_L, is given by

$$u_L = V_I - u_A$$

where V_I is the impact velocity and u_A the particle velocity in the Plexiglas when under a stress of the same magnitude as that in the explosive (Figure 18).

A summary of the data obtained in the gas-gun experiments appears in Table VIII. The Hugoniot of unreacted dextrinated lead azide of density 3.4 g/ml is plotted in the stress–particle velocity plane in Figure 23. A linear least-squares fit to this data gives the expression: $\sigma = 4.17u_p$ kbar (U_p in km/sec) with a standard deviation of 0.59. This implies a wave speed (shock velocity) of 1.23 ± 0.02 km/sec, for a stress range up to 10 kbar.

Figure 24 shows typical measured-impact and transmitted-stress profiles. The transmitted-stress profile, which is given in terms of the stress in the Plexiglas, in all cases showed marked deviations from the input wave profile, particularly at the higher stresses and after transmission through the thicker lead azide pellets. The jumps at 1–3 μsec after the first wave arrival were attributed to edge effects, i.e., disturbance propagating into the rear face-gauge through the Plexiglas holders, as their occurrence was dependent on lead azide sample thickness and not stress.

The transmitted-stress profiles showed evidence of a detonation wave follow-

Table VIII. Gas-Gun Data [46]

PbN_6 density (g/ml)	Impact velocity (km/μsec)	Acrylic stress (kbar)	PbN_6 stress (kbar)	Particle velocity (km/sec)	Calculated shock (km/sec)	Measured shock (m/sec)
		A. Dextrinated lead azide hugoniot				
3.41	0.093	1.551	1.72	0.0465	1.215	0.97
3.36	0.209	3.582	3.94	0.0946	1.240	1.10
3.41	0.306	0.306	5.4	6.00	0.1362	1.22
3.41	0.402	7.203	7.7	0.1883	1.199	1.25
3.41	0.459	8.33	8.9	0.2153	1.212	1.28

B. Transmitted stress data. Initial stress jump in lead azide (kbar) at

0 mm	1 mm	2 mm	3 mm	4 mm
1.72	1.26	1.72	1.44	1.72
3.94	3.30	3.62	3.9	3.68
6.00	5.74	5.74	14.9	>20
7.7	8.05	12.6	>20	>20
8.9	9.2	>20	>20	>20

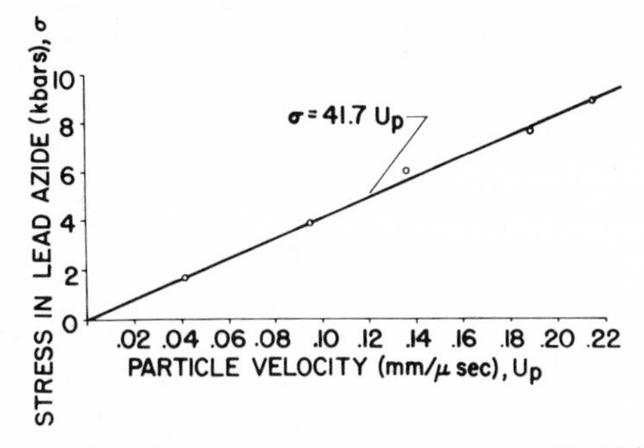

Figure 23. Hugoniot for dextrinated lead azide [46].

ing a stress wave. In Figure 24 the detonation wave is considered to arrive 0.2 μsec after the first stress wave.

The variation of the amplitude of the stress propagated through the lead azide is shown in Figure 25. No evidence of reaction in lead azide was noted below 6 kbar in the thicknesses examined, nor for 1-mm-thick specimens at any stress below 8.9 kbar.

As indicated, the experimental technique measures the impact stress directly and requires only the assumptions of continuity of mass and momentum at the impact interface to derive the Hugoniot. No steady-state assumptions were made, and the derivation of the Hugoniot was independent of the measured wave speed. Consequently, the measurements were considered to be an accurate representation of the response of the unreacted explosive.

At low stresses the measured wave velocities were less than 1.23 km/sec. At higher impact stresses both the shock velocity and the amplitude of the propagated wave increased with distance. The change in wave velocity from a value less than that for unreacted material to the detonation velocity was abrupt.

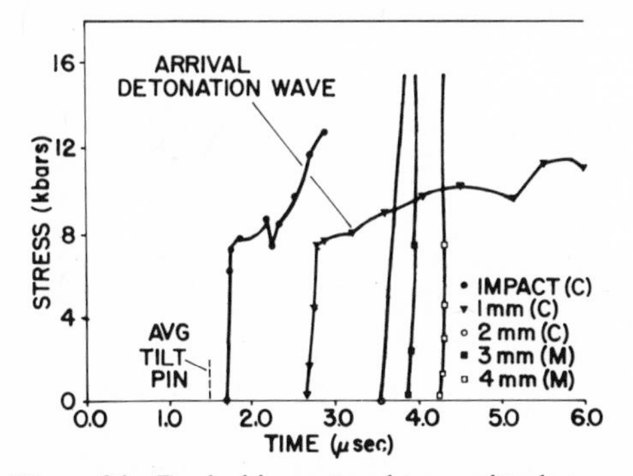

Figure 24. Typical impact and transmitted stress.

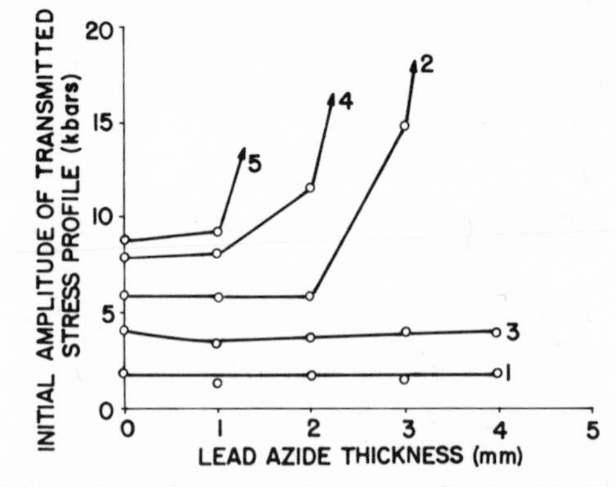

Figure 25. Variation of propagated stress amplitude with distance [46].

The data are consistent with the observations of Chaudhri and Field [12] on single crystals of lead azide (see Chapter 8, Volume 1); they concluded that the maximum propagation rate prior to detonation is the longitudinal wave speed. The data are also consistent with the fracture velocity obtained by Fox [47] (Chapter 9, Volume 1).

It appears that insofar as growth to stable detonation is concerned lead azide displays characteristics similar to those of heterogeneous secondary explosives. The delayed stress excursions evident in measured stress profiles were interpreted as reactions behind the shock front. Reactions produce pressure waves which travel through the explosive at a velocity at least equal to its velocity of sound and interact with the undecomposed explosive ahead of the reaction front, causing a nonuniform rate of growth.

The threshold for the detection of reaction in the dextrinated lead azide (3.4 g/ml) subjected to 3.5-μsec pulses was dependent upon sample thickness (Figure 25). No reaction was noted in 1-mm-thick samples even at the highest impact stresses (8.9 kbar) tested, whereas the threshold for 4-mm-thick samples was between 4 and 6 kbar.

2. Sensitivity of Lead Azide to Short Pulses via Flyer-Plate Technique

The sensitivity of dextrinated and PVA lead azides to shocks of short duration (about 0.1 μsec) was determined by impacting samples with a thin plate, observing the impact with a fast-framing camera, and measuring the transmitted stresses with a quartz gauge [46] (Figure 26).

Thin mylar or aluminum flyer plates were accelerated by a shock generated by an exploding foil. A Plexiglas moderator was placed adjacent to the foil and the flyer spalled off this plate. The technique was used so that low impact

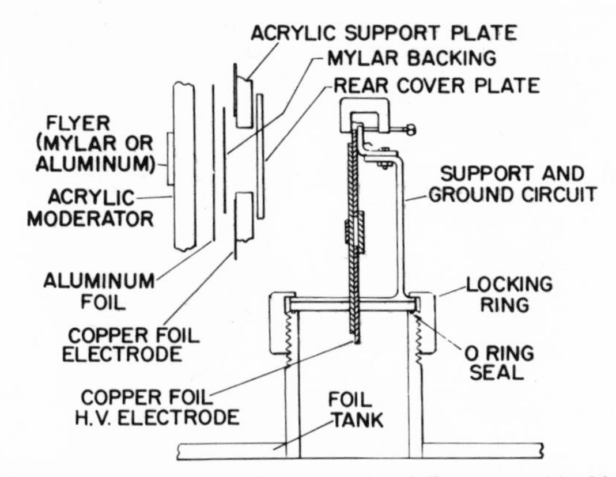

Figure 26. Exploding foil apparatus and flyer assembly [46].

velocities could be obtained. The mylar flyers were 0.127 mm thick, providing a shock duration of 0.097 μsec.

The lead azide targets (Figure 27) were pressed into lead sleeves in an aluminum cup to a nominal thickness of 2 mm with an average density of 3.6 g/ml for the PVA material and 2.95 g/ml for the dextrinated azide. The experiments were conducted at ambient temperatures and a pressure of less than 5 μm mercury.

The exploding foil apparatus permitted the calculation of the amplitude of the stress wave generated in the lead azide from the measured impact conditions which included the measured, transmitted stress amplitude and the detonation delay (Figure 28). In general, the rise time of the quartz-gauge signals exceeded the transit time through the quartz; consequently, while the records provided good time-of-arrival information, the amplitude data were questionable. The measured amplitudes of the transmitted stress waves were in all instances initially lower than the calculated impact stress.

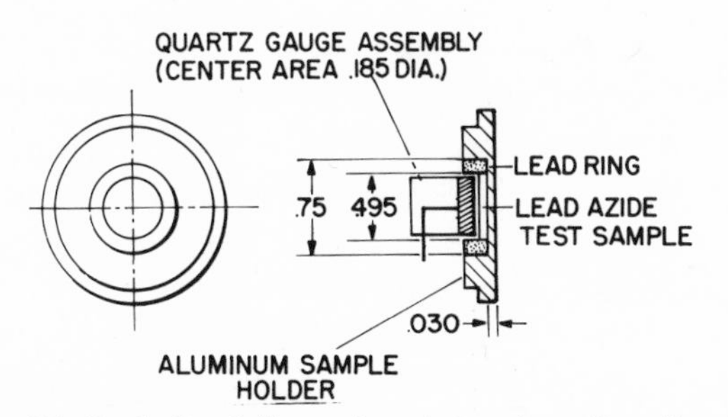

Figure 27. Exploding foil experiment–target configuration [46].

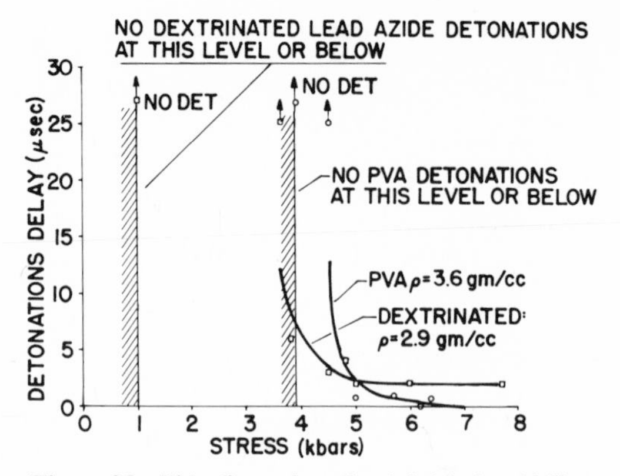

Figure 28. Thin-flyer-plate shock initiation [46].

The short-pulse initiation threshold for PVA lead azide was 4.2 kbar for a density of 3.6 g/ml, and for dextrinated lead azide was 2.1 kbar at a density of 2.7 g/ml. Almost instantaneous detonation was noted above 6.2 kbar for PVA lead azide, while even at 8 kbar dextrinated lead azide displayed a 0.2-μsec delay.

Comparison of initiation-threshold measurements suggests that there is a minimum thickness, or a run-up distance, before detonation occurs and that this is independent of pulse-width for stresses up to 10 kbars. For long pulses (3.5 μsec) in the gas gun, no evidence of detonation was detected for 1-mm-thick samples. Detonation occurred with run-up distances in the range of 1–2 mm for impact stresses of 8.9 kbar. For stresses greater than 6.0 kbar, evidence of detonation was noted after a 2-mm run. In the thin-flyer-plate experiments at stresses of 8 kbar, dextrinated lead azide displayed a 2-μsec initiation delay. Voreck and coworkers [8] determined that a similar minimum thickness of lead azide is required for complete detonation in an explosive train consisting of NOL-130, lead azide, and RDX (Figure 15).

Donor–acceptor tests, employing gaps between the donor and the acceptor, have been used to measure the shock sensitivities of explosives. The small-scale gap test is a refined version which employs a steel dent block and Lucite in the gap [1]. The gap decibang (dbg), analogous to the decibel, but relative to a reference level, is used as a measure of the stimulus:

$$X = A + 10B \log \frac{G_r}{G_t}$$

where X = stimulus, dbg; A, B = arbitrary constants; G_r = reference gap, in.; G_t = observed test gap, in. Results for lead azide are presented in Figure 29.

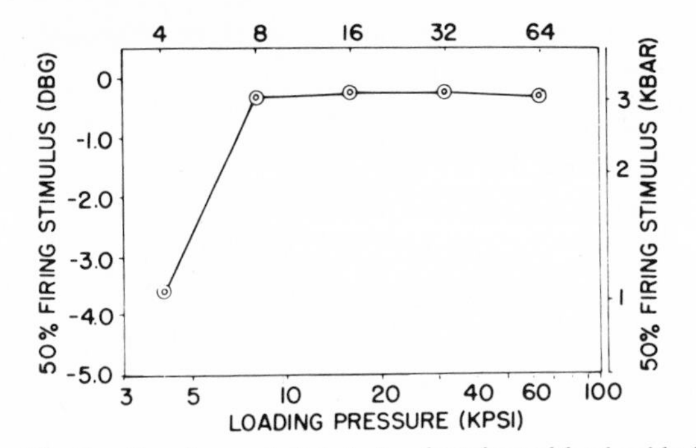

Figure 29. Small-scale gap-test data for dextrinated lead azide [48].

3. Shock Initiation of Lead Azide with an Electron Beam

Shock initiation of lead azide by an electron beam has been compared with that of potassium dinitrobenzofuroxan (KDNBF), lead styphnate, and lead mononitroresorcinate (LMNR) [49]. An aluminum slab was heated rapidly by electron deposition, generating a pressure pulse that propagated through the slab and was transmitted to a specimen bonded to its rear. The mean energy of the electrons was in the range of 900 ± keV and produced a stress pulse in the aluminum with a duration of approximately 0.2 μsec.

The RD1333 lead azide wafers were 1/4 in. (0.63 cm) diam and 0.040 in. (0.10 cm) thick, with a density of 4.0 g/ml. The wafers were bonded to aluminum overlays which were 3/4 in. (1.89 cm) diam and 1/8 in. (0.32 cm) thick.

The results obtained for lead azide are summarized in Table IX and in Figure 30. The highest temperature achieved at the aluminum–lead azide interface was less than 120°C, which is significantly below the lowest value (297°C) for the thermal initiation of lead azide. The data indicated a stress initiation threshold of 3.6 kbar for the lead azide, assuming a sound velocity of 2.5 km/sec for the explosive. The stress pulse-width was approximately 0.2 μsec. If the wave velocity (shock velocity) of 1.23 km/sec is assumed for dextrinated lead azide, then a lower bound of 2.2 kbar can be placed on the threshold for RD1333 lead azide.

The results are in reasonable agreement with some small-scale gap tests, which gave threshold initiation levels of 4–6 kbar for lead azide [50]; they are also consistent with an experiment performed by Roth [28], who observed an initiation in 95% lead azide-5% Teflon (ρ = 2.7 g/ml) subjected to an 8.5–10-kbar shock.

The shock-initiation experiments performed on KDNBF pellets are given for comparison [49,51]. The minimum thickness or run-up distance for KDNBF for

Table IX. Shock Initiation of RD1333 Lead Azide in Electron Beam Apparatus [48]

Average fluence (cal/cm^2)	Peak dose (cal/g)	Peak propagating stress in Aluminum overlap: Temperature[a] (°C)	Peak propagating stress in Aluminum overlap: Peak stress (kbar)	Azide[b] peak stress (kbar)	Response
17	65	105	6.2	4.8	Initiated (I)
13	50	88	4.7	3.6	Not initiated (NI)
16	60	103	5.7	4.4	I
15	57	98	5.4	4.2	I
19	70	118	6.6	5.1	I
17	61	108	5.8	4.5	I
10	36	73	3.5[c]	4.3[d]	C[e]
11	41	78	3.9	3.0	NI
14	50	93	4.7	3.6	I
3.2	8.8	38	0.84	0.65	NI
5.5	15	51	1.4	1.1	NI
6.2	17	55	1.6	1.2	NI
6	17	53	1.6	1.2	NI
13	36	87	3.4	2.6	NI
14	38	94	3.6	2.8	NI
4.3	12	45	1.1	0.85	NI
5.1	14	49	1.3	1.0	NI
6.4	18	55	1.7	1.3	NI

[a] Final temperature due to energy deposition; neglecting thermal losses.
[b] Assuming sound velocity of 0.25 km/μsec for RD1333 lead azide.
[c] Measured stress in quartz.
[d] Computer stress in quartz.
[e] C, calibration shot—aluminum/bond/quartz transducer.

a 0.2-μsec pulse was in the range of 0.040 in. (0.10 cm). Using the measured sound velocity of 3.0 km/sec, the initiation threshold is about 20 kbar.

Only limited data were obtained for lead styphnate and LMNR. One initiation was achieved in lead styphnate. Assuming a sound speed of 2.5 km/sec and a density of 2.75 g/ml, the stress transmitted to the lead styphnate that produced initiation was estimated at 4.3 kbar, while the highest stress that did not produce initiation was 3.1 kbar. The pulse duration was of the order of 0.4 μsec. With LMNR and the same pulse, no initiations were achieved, the highest stress transmitted to the LMNR being in the order of 8 kbar.

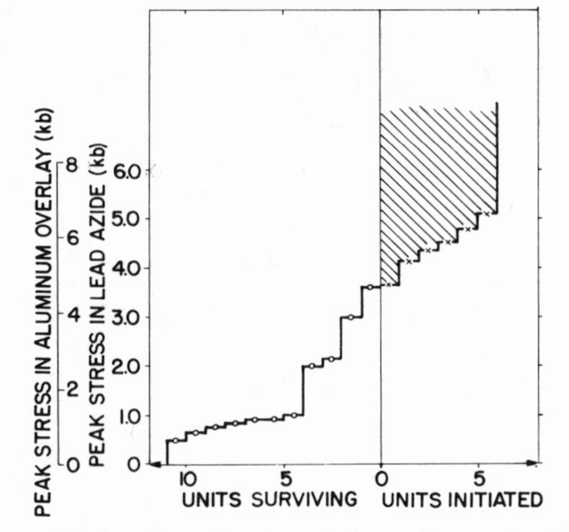

Figure 30. Initiation of lead azide pellets by a 0.2-μsec thermochemical stress pulse in an aluminum overlay [49].

Table X. Initiation Threshold for Lead Azide

Lead azide	Density (g/ml)	Pulse width (μsec)	Initiation threshold (kbar)
Dextrinated	2.9	0.1	1–3.6
	3.4	3.5	4–6
PVA	3.6	0.1	3.8–4.5
RD1333	4.0	0.2	2.2–3.6

The shock-initiation-threshold data for different types of lead azide are summarized in Table X.

4. Energy Input with No Compression

High-intensity, high-voltage, pulsed electron beams can be used to produce sudden increases in energy (and pressure) throughout a substantial volume of a solid without shock compression and with minimal departures from local thermal equilibrium. The states attained by such volume heating are well removed from the Hugoniot, thus affording the opportunity to estimate equations-of-state under conditions not associated with shock waves.

In order to characterize the state of a material following constant volume heating, the pressure–energy coupling relationship must be known. The relationship is determined through the Grüneisen coefficient, Γ, of the material, as it appears in the Mie-Gruneisen equation of state [51], and can be written in the form

$$P = f(V) + \frac{\Gamma E}{V}$$

P is the pressure, V is the specific volume, and E is the specific internal energy.

The Mie-Grüneisen equation is generally employed in finite-difference computer codes to calculate the dynamic response of homogeneous materials [52]. Porous materials are not homogeneous; however, the convenience of the codes makes it desirable to treat them as if they were. Average values for pressure, volume, and energy are then given correctly. Such formulations have been developed by Hermann [53] and Seaman and Linde [54].

An effective Grüneisen coefficient can be defined for solid materials, including substances such as granular explosives, by

$$\Gamma = V \left(\frac{\partial P}{\partial E}\right)_v$$

where the volume considered includes the voids. The coefficient thus defined is energy dependent, since internal pressure relief (still at constant volume) can occur due to the collapse of voids, especially as the shear strength of the material generally decreases with increasing temperature.

The effective Γ may be determined with the electron beam apparatus. When the sample (slab geometry) is thick enough to absorb all of the incident electrons, a compressive stress wave propagates from the irradiated region into the sample bulk. A transducer, located just beyond the deposition depth, may be used to record the stress pulse. Alternatively, the displacement or velocity of the rear surface of sample may be observed optically and used to infer the initial pressure distribution from the experimentally measured stress history. Knowledge of the energy-deposition profile then permits the determination of the Grüneisen coefficient.

Experiments with lead azide and KDNBF were conducted using the above technique and a buffer of fused quartz between the sample and the gauge [51]. Initiations occurred in most of the tests, especially with lead azide and prevented the determination of the sound speed for lead azide, but a value of 3.0 ± 0.2 km/sec was obtained for KDNBF.

There are two possible definitions of the Grüneisen parameter, one microscopic and the other macroscopic. The microscopic definition involves a model for interactions on an atomic level, while the macroscopic definition involves measurable parameters.

In the macroscopic definition, the Grüneisen constant, Γ, is defined by

$$\Gamma = \frac{1}{\rho} \left(\frac{\partial P}{\partial E}\right)_\rho$$

where P is pressure, ρ is mass density, and E is internal energy. The equation indicates that if thermal energy is deposited in a time small compared to the time necessary for significant mass motion (defined below), then the pressure in the solid changes. The quantity Γ determines the change in pressure and is a property of the structure of the material.

The time criterion mentioned above is usually taken as the time necessary for an acoustic wave to traverse an electron-scattering mean free path in the solid. Let such a time be t_1. If the pulse duration is t, then the constant-volume definition of the equation above requires

$$t < t_1$$

Physically, this means if $t < t_1$ holds, then any energy inhomogeneities introduced by the electron beam are not relieved by acoustic signals before the energy is deposited. In other words, the energy can be considered as deposited instantaneously.

The time criteria have been investigated [49–51,55], but what microscopic physics corresponds to the Γ measured in porous solid mechanics remains unanswered. While the measured "effective Γ" is a worthwhile experimental variable in that it characterizes the gross behavior of the material, it does not allow for distinguishing between porous materials as a function of their parameters.

It is possible to compute a mean Grüneisen coefficient, $\overline{\Gamma}$, for lead azide by using an effective bulk modulus, $\overline{k}$ (defined by the ratio of the applied pressure to the relative volume change), and the volume expansion coefficient, α, as

$$\overline{\Gamma} = \frac{\alpha \overline{k}}{\rho C_v} \tag{5}$$

The effective bulk modulus of orthorhombic lead azide is 4.2×10^{11} dyne/cm^2 [56]. The thermal expansion coefficients are [57]

$$\alpha^a = 5.7 \times 10^{-5}\ (^\circ\mathrm{K}^{-1})$$
$$\alpha^b = 0.3 \times 10^{-5}\ (^\circ\mathrm{K}^{-1})$$
$$\alpha^c = 1.4 \times 10^{-5}\ (^\circ\mathrm{K}^{-1})$$

with $\alpha = 7.4 \times 10^{-5}$ ($^\circ$K^{-1}). Using a specific heat of 0.11 cal/g/$^\circ$K, and a crystal density of 4.8 g/ml [27], the mean Grüneisen coefficient is computed as

$$\overline{\Gamma} = 1.4 \text{ (lead azide)}$$

G. SUMMARY AND CONCLUSIONS

It has been seen that lead azide plays both the role of donor and acceptor in explosive trains. In igniters and detonators lead azide acts as an acceptor by receiving a stimulus from a priming charge (lead azide is often one component) in the form of heat or shock. If properly designed, lead azide will detonate and initiate a detonation in the adjacent secondary explosive.

1. Primer–Lead Azide Interface Mixtures

Although the output of primer is known to consist of hot particles, a pressure pulse (in some cases a shock wave), and thermal radiation, no general quantitative requirement for initiation of lead azide is known to designers. Failures occur in igniting the primer, but once the primer burns, it always ignites the azide. Primer mixtures are complex, empirically determined compositions for which no general quantitative relationships are known. This situation makes it difficult to optimize them for a particular requirement, or to determine the cause of failures, except by trial and error.

2. Lead Azide–Secondary Explosive Interface

The requirements for shock initiation of secondary explosives are on much firmer ground scientifically and practically. However, difficulties arise in applying the knowledge to the small dimensions found in detonators and igniters. If the secondary explosive in a detonator is to properly initiate the next component in the train, a detonation-like wave must be formed by a certain point in the detonator. Otherwise the secondary will not propagate a strong enough blast wave or produce fragments of suitable velocities. Above a critical level determined by the run up to detonation distance of the azide and the sensitivity of the secondary explosive, additional azide produces little change in output from the secondary. However, below that level the output rapidly drops to zero. Generally the minimum weight of azide required to initiate the output charge remains constant as the diameter is decreased or as density is varied. The secondary explosive is needed to increase the detonation pressure and detonation velocity of the azide to a level capable of jumping a gap and initiating a booster charge. Factors such as confinement, density, and the curvature of the detonation front affect the output, but current output tests only determine if the overall detonation functioned adequately. Recent improvement in electronic, photographic, flash X-ray, gas-gun, and laser techniques are leading to greater insights. Gas-gun studies have been used to determine the unreacted Hugoniot for lead azide. Flyer-plate techniques have demonstrated the occurrence of a run-up distance for lead azide. Fast-response pressure gauges have been used to determine the detonation pressure of lead azide, and streak cameras have been employed to determine the detonation velocity of lead azide in different confinements. Further parametric studies are required in order to improve reliability, optimize the designs, and reduce the constraints on the use of the small explosive elements in particular.

3. Silver Azide

Although the data for silver azide are less extensive than for lead azide, they do indicate that silver azide is capable of slightly better performance than lead azide. This allows smaller detonators to be made; however, incompatibility with antimony sulfide makes it unsuitable for stab detonators which contain that material.

REFERENCES

1. Explosives Trains, U.S. Army Materiel Command Pamphlet, AMCP 706-179, (1965); *Blasters' Handbook*, 15th ed., E. I. duPont de Nemours Co., Wilmington, Delaware, 1969.
2. W. E. Voreck, E. W. Dalrymple, Development of an Improved Stab Sensitivity Test, Tech. Rept. 4263, Picatinny Arsenal, Dover, N.J., 1972.
3. R. M. Ferguson, Development and Design of Conductive Mix Detonators, Canadian Arsenal Ltd. Report 627-1, St. Paul L'Ermite, Quebec, Canada, 1974.
4. F. E. Walker, R. J. Wasley, Lawrence Livermore Laboratory Report UCRL-75-722, Livermore, Calif. 1974.
5. B. A. Bydal, *Ordnance*, Vol. LVI, No. 309, pp. 230–233 (1971).
6. E. R. Lake, Percussion Primers Design Requirements, McDonnell Aircraft Co., St. Louis, Mo., 1970.
7. F. E. Walker, R. J. Wasley, *Explosivstoffe 17*, 9 (1969).
8. W. Voreck, T. Costain, E. Dalrymple, Advances in Explosive Train Technology, Ninth Army Science Conference, West Point, New York, 1974.
9. H. S. Leopold, Effect of Wire Temperature Upon Initiation Times for Four Primary Explosives, Naval Ordnance Laboratory Technical Report NOL TR-72-123, White Oak, Md., 1972.
10. T. Boddington, Shock Initiation of Explosive Single Crystals, in: *Proceedings of Sixth International Congress on High Speed Photography*, J. G. A. Graff, P. Tegelar, H. D. T. Tjeek Willink, Zoon N. V., eds., Haarlem, Netherlands, 1963.
11. O. Sandus, N. Slagg, D. Wiegand, W. Garrett, Studies of the Fast and Slow Decomposition of Azides, presented at the DEA-AF-F/G-7304 Technical Meeting: Physics of Explosives, Naval Ordnance Laboratory, Silver Springs, Md., April–May, 1974.
12. M. M. Chaudhri, J. E. Field, Deflagration in Single Crystals of Lead Azide, Fifth Symposium (International) on Detonation, Office of Naval Research Report ACR-184, Arlington, Va., 1970.
13. E. Strömsöe, *Research*, *13*, 101 (1960).
14. M. A. Cook, *Science of High Explosives*, Reinhold, New York, 1958.
15. F. P. Bowden, A. C. McLaren, *Nature*, *135*, 631 (1955).
16. Y. Mizushima, *Govt. Chem. Inst. Res. Inst., Tokyo*, *59*, 204 (1964).
17. T. Costain, private communication.
18. F. Schwartz, N. Slagg, W. Voreck, E. Dalrymple, L. Millington, Detonation Transfer, Tech. Rept. 4570, Picatinny Arsenal, Dover, N.J., 1975.
19. C. Scott, Propagation of Reaction Across a Lead-Azide-PETN Interface, Proceedings of Sixth Symposium on Electroexplosive Devices, Franklin Institute, Philadelphia, Pa., 1969.

20. J. T. Rosenberg, Development of a Piezoresistant Transducer to Measure the Stress-Time Output of Small Detonators, Stanford Research Institute Report 1283, Menlo Park, Calif., 1973.
21. N. L. Colburn, T. P. Liddiard, L. A. Roslund, Dynamic Measurements of Detonator Output, Naval Ordnance Laboratory Report NOLTR 72-266, White Oak, Md., 1973.
22. B. T. Fedoroff, O. E. Sheffield, *Encyclopedia of Explosives and Related Items*, Vol. 4, Tech. Rept. 2700, Picatinny Arsenal, Dover, N.J. (1969).
23. F. P. Bowden, H. T. Williams, *Proc. R. Soc. London*, *A208*, 176 (1951).
24. R. Miller, E. I. DuPont de Nemours and Co, private communication to W. E. Voreck.
25. H. Eyring, R. E. Powell, G. H. Duff, R. B. Parlis, *Chem. Rev*., *45*, 69 (1949).
26. H. Jones, *Proc. R. Soc. London*, *189A*, 415 (1947).
27. Principles of Explosive Behavior, U.S. Army Munitions Command Pamphlet AMCP-706-180, Alexandria, Va., April 1972.
28. J. Roth, Studies on Surface Initiation of Explosives, Stanford Research Institute Technical Report AFWL-TR-65-135. Menlo Park, Calif., 1966.
29. L. Avrami, H. Jackson, O. Sandus, Pulsed Neutron Effects on Spartun Ordnance and Explosives, Tech. Rept. 4509, Picatinny Arsenal, Dover, N.J. (1973).
30. Naval Ordnance Train Designers Handbook, NOLR-1111, White Oak, Md., 1952.
31. Z. H. Akimovek, A. Ya. Aapin, *Fiz. Goreniya Vzryva*, *3*, 197 (1967).
32. Ya. B. Zeldovitch, A. S. Kampanayets, *Theory of Detonation*, Academic Press, New York, 1960.
33. D. Ellington, private communication.
34. Lead Azide RD-1333, Picatinny Arsenal Specification MIL-L-46225c, Dover, N.J., 1968.
35. R. L. Wayne, Lead Azide for Use in Detonators, Tech. Rept. 2662, Picatinny Arsenal, Dover, N.J. (1960).
36. Properties of Explosives on Military Interest, U.S. Army Material Command Pamphlet 706-177, Alexandria, Va., 1971.
37. L. Wöhler, F. Martin, *Ber. Dtsch. Chem. Ges*., *50*, 595 (1917); *J. Chem. Soc*., *112* 383 (1917).
38. R. Courant, K. O. Friedrichs, *Supersonic Flow and Shock Waves*, Interscience Publishers, New York, 1948.
39. G. E. Duvall, G. R. Fowles, in: *Shock Wave, High Pressure Physics and Chemistry*, Vol II, R. S. Bradley, ed., Academic Press, New York, 1963.
40. J. Roth, Shock Sensitivity and Shock Hugoniots of High-Density Granular Explosives, Fifth Symposium (International) on Detonation, Office of Naval Research Report ACR-184, Washington, D.C., 1970.
41. R. F. Chaiken, Comments on Hypervelocity Wave Phenomena in Condensed Explosives, Third Symposium on Detonation, Office of Naval Research, ACR-52, Vol. 1, 1955.
42. K. K. Andreev, *Combust. Flame*, 7, 175 (1963).
43. F. P. Bowden, A. D. Yoffe, *Fast Reactions in Solids*, Academic Press, New York, 1958.
44. J. L. Copp, S. E. Napier, T. Nash, W. J. Powell, H. Skelly, A. R. Ubbelohde, P. Woodward, *Phil. Trans. R. Soc. London*, *A241*, 197 (1948).
45. A. R. Ubbelohde, *Research*, *3–5*, 207 (1950).
46. F. W. Davies, A. Zimmerschied, F. Borgardt, L. Avrami, The Hugoniot and Shock Initiation Threshold of Lead Azide, *J. Chem. Phys*. *64*, 2295 (1976).
47. P. G. Fox, *J. Solid State Chem*., *2*, 491 (1970).
48. J. N. Ayres, L. J. Montesi, R. J. Bauer, Small Scale Card Gap Test, Naval Ordnance Laboratory Technical Report 73-132, White Oak, Md., 1973.
49. L. Avrami, P. Harris, J. Shea, Equation-of-State and Shock Initiation of Explosives

Using Pulsed Electron Beams, Proceedings of the Eighth Conference on the Design of Army Research Development and Testing, ARO Report 73-2, Part 2, pp. 453–474, U.S. Army Research Office, Washington, D.C., October 1973.
50. J. H. Shea, A. Mazzella, L. Avrami, Equation of State Investigation of Granular Explosives Using a Pulsed Electron Beam, Fifth Int. Symposium on Detonation, Office of Naval Research Report ACR-184, Washington, D.C., 1970.
51. V. Buck, J. Shea, L. Avrami, Pressure Energy Coupling, Sound Speed, and Shock Initiation Experiments in Explosives Using Pulsed Electron Beams, Tech. Rept. 4690, Picatinny Arsenal, Dover, N.J. (1974).
52. L. H. Bakken, P. H. Anderson, An Equation of State Handbook, Sandia Corporation Report SCL-DR-68-123, Livermore, Calif. 1969.
53. W. Hermann, *J. Appl. Phys.*, *40*, 2490 (1969).
54. L. Seaman, R. K. Linde, Distended Material Model Development, AFWL-TR-68-123, Vol. I, Air Force Weapons Laboratory, Albuquerque, N.M., 1969.
55. P. Harris, A Survey of the Physics of Shock Waves in Solids, Tech. Rept. 4345., Picatinny Arsenal, Dover, N.J. (1972).
56. C. E. Weir, S. Block, G. J. Piermarini, *J. Chem. Phys.*, *53*, 4265 (1970).
57. F. A. Mauer, C. R. Hubbard, T. A. Hahn, *J. Chem. Phys.*, *60*, 1341 (1974).

Index

α-particles, 200, 211-213
Accident, 4
Activation energy, 259
Adhesives, 144, 145, 159
Adsorbates, 203-205
Aging, 202, 223, 233
Ammonium hexanitratocerate method, 80
Antistatic agents, 106
Arc discharge
  definition, 169
  efficiency, 172
  partition of energy, 171
Arrhenius equation, 201
Assay, lead azide, 57, 65

Ball drop test, 24, 38, 51
Bismuth, 203
Black powder, 221, 226
Booster explosive, 2
Boron mixtures, 221, 222
Brisance, 1
Bruceton method, 122, 123, 147, 151
Bulk density, 24, 38, 42

Cab-O-Sil, 137
Californium-252, 216
Capacitance
  minimum, 178, 179
  optimum, 179
Charge
  relaxation time, 104
  weight, 272
Cocoon effect, 93
Color change, 212, 213, 227
Compatibility, 95, 207, 208
Concentration, 66
Conducting rubber, 174
Conductive mix detonator, 253, 254
Contact charging, 97, 98
Copper incompatibility, 79
Crystal
  mass, 206, 207, 238
  modification agent, 12
  thickness, 206, 207

Delay detonator, 252
Density, pressed, 271
Destruction method, 83, 84
Detonation, 1
  velocity, 141-144
Detonator
  conductive mix, 253, 254
  delay, 252
  miniature, 13
  one-amp no-fire, 255
  temperature stable, 253
Deuteron, 233, 234
Dielectric strength, 186
Differential thermal analysis, 208-210, 211
Discharge, 169, 171, 172
Dopants, 141-143
Drying effect, 132, 133

Effect
  cocoon, 93

Effect (*cont'd*)
  drying, 132, 133
  electric field, 125
  humidity, 102, 133
  temperature, 102, 152
Einbinder method, 123
Elcoat, 29
Electric field
  effect, 125
  induced initiation, 233 and Chapter 5
Electrolysis, 81
Electrolytic method, 84
Electron, 229-233
  donor, 204
Electrostatic
  initiation, 164, 165
  sensitivity, test, 25, 38, 42, 51 and Chapter 5
Energy deposition, 233
Equilibrium constant, hydrolysis, 92, 93
Exploding foil, 280, 281
Explosion temperature, 200, 201, 209, 210, 228, 240
  test, 24, 38
Explosive, 1, 2
  train, 2 and Chapter 7

Faraday cage, 99
Ferric chloride test, 82
Figure of insensitiveness, 122, 146, 147
Fission fragment, 216
Foil, 280-281
Frictional fuze, 144, 145
Friction test apparatus
  BAM, 150, 153
  emery paper, 147, 148, 150
  mallet, 146, 147
  pendulum, 149, 150
  sliding block, 147, 148, 154
  Yamada, 154
Functionary test, 25, 28, 42

γ-irradiation, 223-226
Gap test, 282, 283
Gas generator, 2
Glow discharge, 169
Gold particles, 238
Goodness-to-fit test, 123
Graphite bridge, 253
Griess reagent, 81
Grit, 79, 135-137, 149, 153, 158
Ground glass, 135, 136, 137, 155
Grüneisen coefficient, 285, 286, 287

Halocarbons, 138
Hazards, 75
Hot spot, 153, 231, 275, 276
  hot wire, initiation, 258, 259
Hugoniot, 273, 276, 277, 278, 279
Human resistance, 174
Humidity effect, 102, 133
Hydrazoic acid, 74, 75, 92, 95
Hydrolysis, 92, 94
Hygroscopicity test, 24, 38, 51

Ignition delay, 201, 202, 204, 206, 237, 238
Impact sensitivity test, 24, 38, 51, 231
Impact test apparatus
  ball and disk, 121, 129, 130, 136, 151
  ball drop, 120, 121, 122, 127, 128, 130, 131, 132, 134, 143
  Bureau of Mines, 114-116, 125
  drop weight, 113-121
  ERL, 116, 117, 125, 140
  falling hammer, 113-121
  Picatinny Arsenal, 117, 118, 125, 126, 128, 129, 130, 133, 137, 138, 140
  Rotter, 117, 119, 122, 124, 125, 134, 136
Impulse, 133
Impurities, 134-144, 203
Induction time, 201, 205, 219
Inductive charging, 98
Initiation
  electric field induced, 233
  electrostatic, 164, 165
  energy, 164, 165
  hot wire, 258, 259
  shock, 276, 280, 282, 283, 288
  thermal, 258
Insensitiveness, figure of, 122, 146, 147

Karber test, 123
Kinetic energy, 133
Kramer's method, 86

Lead azide
  assay, 57, 65
  critical diameter, 264

Lead azide (*cont'd*)
  phlegmatizing agent, 57
  RD1343, 44
  solubility, 57
  Special Purpose, 44, 89
Light, 233, 235, 237
  energy, 237, 283, 239, 240

Manganin gauge, 262, 263, 277
Maximum charge, 101
Miniature detonator, 13
Minimum
  capacitance, 178, 179
  charge weight, 272
Moderation, 138
Modification agent, 12
Moisture content, test, 24, 38

Neutron, 213-215
Nitrite method, 81
Nucleation, 205
  energy, 92

One-amp no-fire detonator, 255
One-shot transformed response, 123
Optimum capacitance, 179
Oxidation, 56
Oxidation-reduction method, 83

Particle
  size, 98, 101, 157
  velocity, 278
Paschen's law, 167
Percussion sensitiveness, 136
Phlegmatizer, 12, 57
Photoconductivity, 105
Pion, 216
Plastic deformation, 154, 159
Post-breakdown, 169
Potential energy, 124, 126, 132, 133
Primary explosives, 1, 2, 186
Proton, 213, 233, 234
  donor, 205
Pulse width, 274
Purity test, 24, 38, 51

Range–energy relation, 213
Reaction zone length, 260, 261, 266, 273
Reactivity, 207
Reactor radiation, 217-223
Reconversion, 85
Reduction, 56, 81
Relay, 252, 253
Resistance, human, 174
Resistivity, 106
Richter's method, 87
Rubber, conducting, 174
Rundown method, 122, 123

Secondary explosive, 1, 2
Sensitivity, 75
Shear, 145, 146, 150, 158
Shock initiation, 276, 280, 282, 283, 288
Silica, 137
Small scale gap test, 282, 283
Solubility, 134
Spark, gap, 169
Spark discharge
  definition, 169
  efficiency, 172
  partition of energy, 171
Special Purpose lead azide, 44, 89
Specifications, 56
Stability, 207
Stab mixtures, 257
Staircase method, 122, 123, 147, 151
Susan test, 145

Temperature
  effect, 102, 152
  stable detonator, 253
Tertiary explosive, 1
Test
  χ^2, 123
  ball drop, 24, 38, 51
  electrostatic sensitivity, 25, 38, 42, 51
  explosion temperature, 24, 38
  ferric chloride, 82
  friction, 146-150, 153, 154
  functionary, 25, 38, 42
  goodness-to-fit, 123
  hygroscopicity, 24, 38, 51
  impact sensitivity, 24, 38, 51, 113-140, 151
  Karber, 123
  moisture content, 24, 38
  purity, 24, 38, 51
  Susan, 145
TGA, 210, 211

Thermal
  decomposition, 215, 218
  destruction, 83
  initiation, 258
Thermogravimetric analysis, 210, 211
Time delay, 124

Ultraviolet, 233-235, 237

Up-and-down method, 122, 123, 147, 151

Vacuum stability, 201
Volhard titration, 63

Witness plate, 269

χ^2 test, 123